HVAC
Contracting

Robert & William Dries

CRAFTSMAN

Craftsman Book Company
6058 Corte del Cedro,
P. O. Box 6500
Carlsbad, California 92008

Acknowledgements

*American Society of Heating, Refrigerating and Air-Conditioning Engineers, Inc.
(ASHRAE)* — 1791 Tullie Circle NE, Atlanta, GA 30329

Amprobe Instrument Co. — P. O. Box 329, 630 Merrick Road, Lynbrook, NY 11563

Bacharach Instruments — 301 Alpha Drive, Pittsburgh, PA 15238

Burnham Corporation — 2 Main Street, Irvington, NY 10533

Carrier Corporation — SL-1, Carrier Parkway, P. O. Box 4808, Syracuse, NY 13221

Dries-Jacques Associates — 1600 Verona Street, Middleton, WI 53562

Econovent Systems Inc. — 800 Horicon Street, Mayville, WI 53050

Grundfos Pump Corp. — 2555 Clovis Avenue, P. O. Box 549, Clovis, CA 93613

Hydro Therm Inc. — Rockland Avenue, Northvale, NJ 07647

ITT Reznor — McKinley Avenue, Mercer, PA 16137

Lima Register Co. — P. O. Box 28, Lima, OH 45802

National Society of Professional Engineers — 2029 K Street NW, Washington, DC 20006

National System of Garage Ventilation, Inc. — 714 N. Church Street, P. O. Box 1146, Decatur, IL 62525

Research Products Corp. — 1015 E. Washington Avenue, Madison, WI 53703

TOPS Business Forms — 111 Marquardt Drive, Wheeling, IL 60090

The Trane Company — 3600 Pammel Creek Road, LaCrosse, WI 54601

TSI Incorporated — 500 Cardigan Road, P. O. Box 43394, St. Paul, MN 55164

United McGill Corp. — 200 E. Broadway, Westerville, OH 43081

Wisconsin Legal Blank Co. — 125 E. Wells Street, Milwaukee, WI 53202

York Environmental Systems, Borg-Warner Corp. — P. O. Box 1592, York, PA 17405

Library of Congress Cataloging-in-Publication Data

Dries, Robert.
    HVAC contracting.

    Includes index.
    1. Heating--Contracts and specifications.
2. Ventilation--Contracts and specifications.   3. Air
conditioning--Contracts and specifications.   I. Dries,
William.   II. Title.
TH7325.D75   1986   697'.0068   85-30144
ISBN 0-934041-08-3

Cover art by Bob Lee
Illustrations by Pamela A. Nesbit

Contents

Getting Started

I'm going to assume that you've picked up this book because you're anxious to learn more about the heating, ventilating and air conditioning (HVAC) business. That's good. You're ambitious. You want to improve your skills, income and professional reputation. That's exactly what this book is intended to do for you.

Maybe you already run an HVAC outfit and are looking for ways to make your business more competitive. Maybe you're a skilled installer working for an HVAC company and are considering opening a business of your own. There are lots of advantages to being your own boss. Or maybe you're just getting started in HVAC and want to sharpen your technical and business skills — planning for the day when you'll have your own company.

Whatever your experience or interest, this manual will be a good guide to the demanding and exacting work of residential and light commercial HVAC. It covers the whole trade: tools, installation, theory and design. But being a skilled craftsman isn't enough to survive in a highly competitive market. We'll also cover the contracting end of HVAC: starting and running your business, estimating costs, bidding for work and staying profitable.

How to Use this Book

There's no easy way to accumulate all the information and skills you need to be successful in the HVAC business. No single book or reference manual has it all. But there's a lot of information between the covers of this handbook — certainly more than you can absorb on a single pass. Don't expect that reading each chapter once will make you an expert on the subject covered. The recommendations and procedures in this manual will eventually become second nature. But you'll have to refer to some of the tables, drawings and charts over and over again.

First, read this book from beginning to end. Spend more time on the sections that seem more difficult or unfamiliar. When you've digested the entire book, it'll be easy to find the appropriate table or section when you need an answer to a problem on the job or in the office.

This is a practical manual. It's a working handbook for the HVAC contractor and installer. I'll keep theory to a minimum. But where theory is absolutely necessary, I'll explain it fully. The only math calculations will be very basic. You can do them all quickly and easily on any hand-held calculator that can figure squares and square roots.

If you own a personal computer or have the use of one, the design and theory chapters include sections on using computers to calculate heat loss and gain.

Surviving in the HVAC Business

According to the latest edition of the *Business/Executive Directory,* there are over 40,000 HVAC contractors in the country. That's a lot of competition. You may wonder if you can survive. Chapter 10 deals with this question in detail. For now, just understand that the HVAC business is no more competitive than construction in general. But, like any construction contracting, there's an HVAC business cycle that turns up and then down. When business is good, everyone who knows a duct from a register can make a living in the HVAC business. When construction slows, even the most profitable companies may have to cut back. But, as one veteran of the HVAC contracting trade observed, "If I can make a buck when things are slow, imagine how well I do when things are booming." And many HVAC contractors "make a buck" both in good times and bad.

Sure, it takes a lot of hard work to start and run your own company — in any type of work. But, as I said earlier, you've already shown that you're ambitious. You want to be your own boss. That puts you head and shoulders above most tradesmen. Add to that desire your experience and the know-how you'll find in this manual, and you're on your way. The rewards are good. You can make a good living. You set your own hours. Your hard work builds the asset value of your company — an asset you can cash in or pass on to a child or grandchild some day.

Codes and Standards

Every type of construction is regulated by government today. HVAC is no exception. How much regulation depends on where the job is and how big it is. For example, a large full-service restaurant in a major city may have to comply with as many as 25 different codes or sets of regulations.

Codes and standards in the HVAC industry are drafted by several professional organizations. Each group sets the standards for its area of interest. Figure 1-1 lists some of the organizations, their addresses and the subject each covers.

A more complete list of organizations and their codes and standards can be found in the American

Society of Heating, Refrigerating and Air Conditioning Engineers (ASHRAE) handbooks. These ASHRAE manuals provide the basic reference data used in HVAC design work. They're updated at regular intervals to reflect changes in the industry. The address for ASHRAE is 1791 Tullie Circle NE, Atlanta, GA 30329.

Every HVAC contractor has to comply with the building code that's been adopted by the governing city, county or state. Your local building code probably sets requirements for HVAC clearances, materials and installation practice.

The building code is like a law. It's enforced by the local building department and the inspector that visits your jobs. But unlike other laws, your building department didn't write the code. That would be a big job. Instead, the government authority in your community just adopted one of the three popular model codes: The *Uniform Building Code* (published by the International Conference of Building Officials, 5360 Workman Mill Road, Whittier, California 90601), *The Basic Building Code* (published by the Building Officials and Code Administrators International, Inc., 4051 W. Flossmer Road, Country Club Hills, Illinois 60477-5795) and the *Southern Standard Building Code* (published by the Southern Building Code Congress, 900 Montclair Road, Birmingham, Alabama 35213).

The UBC is followed in nearly all western states. The BOCA code is used in the Midwest and eastern seaboard. The SSBC is the most common code used in the Southeast.

The UBC, BOCA and SSBC codes are just models. They're only recommendations until adopted by the government agency having authority to regulate construction in your area. Then it becomes law, backed by the full authority of government.

Unfortunately, the full model code isn't adopted in every community. Many cities and counties make changes, additions or deletions to reflect local preferences when adopting the code. Some jurisdictions are still enforcing obsolete codes that were replaced years ago. The variations are endless. But no matter what code is being enforced, it's your responsibility to comply. HVAC contractors are assumed to know the code and follow it. Failure to do that can result in heavy financial penalties.

Air Conditioning Contractors of America (ACCA)
1228 17th St. NW
Washington, DC 20036

- *AC load calculation*
- *Municipal codes*
- *Heat pumps*
- *Duct selection and sizing*

Air Conditioning and Refrigeration Institute (ARI)
1815 N. Fort Myer Dr.
Arlington, VA 22209

- *Central forced-air electric heating equipment*
- *Central system humidifiers*
- *Load calculation for commercial buildings*
- *Selection, installation and servicing of humidifiers*

American Boiler Manufacturers Association (ABMA)
1500 Wilson Blvd.
Arlington, VA 22209

- *Commercial industrial boiler ratings*
- *Packaged fire tube boiler ratings*

American Gas Association (AGA)
1515 Wilson Blvd.
Arlington, VA 22209

- *Automatic gas ignition systems and components*
- *Direct vent central furnaces*
- *Draft hoods*
- *Gas appliance thermostats*
- *Gas-fired central furnaces*
- *Gas-fired duct furnaces*
- *Gas-fired gravity and fan type direct vent wall furnaces*
- *Gas-fired gravity and fan type floor furnaces*
- *Gas fired gravity and fan type vented wall furnaces*
- *National fuel gas code*
- *Separated combustion system central furnaces*

American Industrial Hygiene Association (AIHA)
475 Wolf Ledges Parkway
Akron, OH 44311

- *Industrial ventilation*
- *Fundamentals governing the design and operation of local exhaust systems*

American Society of Heating, Refrigerating and Air Conditioning Engineers, Inc. (ASHRAE)
1791 Tullie Circle, NE
Atlanta, GA 30329

- *ASHRAE handbooks*
- *Energy conservation in new building design*
- *Standards for natural and mechanical ventilation*

American Society of Mechanical Engineers (ASME)
345 East 47th St.
New York, NY 10017

- *Refrigeration piping*

Building Officials Code Administrators International (BOCA)
4051 W. Flossmer Road
Country Club Hills, IL 60477-5795

- *Basic mechanical code*
- *Basic plumbing code*
- *Basic fire prevention code*
- *One and two family dwelling code*

National Electrical Manufacturers Association (NEMA)
2101 L St. NW
Suite 300
Washington, DC 20037

- *Manual for electric comfort conditioning (estimating heat loss, heat gain and energy consumption)*

National Fire Protection Association (NFPA)
470 Atlantic Ave.
Boston, MA 02210

- *Explosion prevention of fuel oil and natural gas-fired single burner boiler furnaces*
- *Glossary of terms relating to chimneys, vents and heat-producing appliances*
- *Installation of air conditioning and ventilating systems*
- *Installation of blower and exhaust systems for dust, stock, vapor removal or conveying*
- *Installation of oil-burning equipment*
- *Installation of warm air heating and air conditioning systems*
- *National electrical code*
- *National fuel gas code*
- *Vapor removal for cooking equipment; parking structures; repair garages*

Professional organizations setting standards for HVAC
Figure 1-1

Sheet Metal and Air Conditioning Contractors National Association (SMACNA)
8224 Old Courthouse Road
Vienna, VA 22180

- *Heating and air conditioning systems installation standards for one and two family dwellings and multi-family housing*
- *Low pressure duct construction*

Underwriters Laboratories Inc. (UL)
333 Pfingsten Rd.
Northbrook, Il 60062

- *Chimneys: factory built, residential type and building heating appliance*
- *Draft equipment*
- *Electric baseboard heating equipment*
- *Electrical central air heating equipment*
- *Gas vents*
- *Oil burners*
- *Oil-fired floor furnaces*
- *Safety standards for oil-fired central furnaces*

Professional organizations setting standards for HVAC
Figure 1-1 (continued)

My advice is to make yourself thoroughly familiar with the codes that apply in your area and for the type of work you do. As your operation grows, both in area and job types, you'll need to study the codes that apply to the new areas.

Code Enforcement
Building codes are enforced in two ways. First, during approval of the plans; then by inspection during construction.

Depending on the size, use and location of the building, HVAC plans and specifications usually must be approved before work begins. Virtually all buildings open to public access fall into this category: auditoriums, restaurants, schools, and the like. Approval is also required for special-use buildings such as nursing homes and warehouses used to store flammable materials. When commercial buildings exceed a certain square footage, they'll also need this type of approval.

A state inspection office reviews and approves the plans. Sometimes city approval is also needed. Without the proper approvals, you won't get the *building permits* you need to start work.

To be sure you're complying with the approved plans, a job-site inspection is also required. This takes place when the HVAC work is almost completed, but before it's concealed by finish work. At this point you must correct any code violations. No occupancy permit will be issued until you've completed all inspections. And without an occupancy permit, the gas and electricity can't be turned on and the owners or tenants can't move in.

In some areas, small commercial jobs and residences can have HVAC systems installed without any special plan approval. The HVAC system receives no inspection other than as part of the building in general.

To determine what codes apply on a particular job, contact the building inspector for your area. He'll be able to give you the particulars. Many offices now have booklets or printed guidelines for your use.

Whatever codes are enforced in your area, here's the rule of thumb: Don't expect the inspector to bend the code in your favor. He has no authority to permit violations. But you can sometimes get an exception. If what you propose is as good or better than what the code requires, and there's a good reason for the change, your inspector may be sympathetic. But it's always better to raise the issue before work begins. If the work is already done, asking for an exception is usually asking for trouble.

If you get an exception, ask the inspector to initial the change on the plan or field sketch. Later send a letter detailing the change for a formal approval. Be sure to include the time, date and description of the change to jog the inspector's memory. He may have several others that he's working on. You wouldn't want him to forget that he said something was O.K.

A word of warning here. If you're doing your own planning and design work, you'll be responsible for any approvals needed. For some types of work, HVAC design must be done by a licensed engineer or architect. If the owner hires an engineer to do the job, you're just responsible for installation. If someone else has done the planning, that person is responsible for plan approval.

Once the plan's approved, it's very important that the installer not make even small changes without approval of the designer. Any change could be a violation of the code. This is an extremely important point. Don't be tempted to make what seems like a small change just to save a few dollars or a little time — unless you have prior approval. Besides the danger of a code violation, the change could affect system operation. Later, if the system failed to perform as expected because of your change, you may have to bear the cost of correction. If the installation is done according to plan and the design is faulty, it's the engineer or architect who's responsible.

Here's another reason not to deviate from the plan without approval: It violates the accepted principle that you get paid for changes. If a suggested change meets with the engineer's approval, he'll alter the plans and specifications to show the change. A *change order* will be issued, authorizing the work and stating who'll pay for it. We'll discuss change orders and plan modifications in detail in Chapter 11.

Sometimes code provisions are confusing and seem to contradict each other. Accept this as part of the trade. Following the code won't necessarily be the most efficient or cheapest way, but it guarantees that you've done it correctly and won't have to do it over at your own expense.

Another important factor is that adherence to the code releases you from liability if a system breakdown causes property damage or personal injury. Think of the code as your protection. If, after a loss, investigators find even one code violation in the equipment installation, they've found someone to pin blame for the loss on — you. If you've followed the code, you're clean.

If you find a contradiction or code problem on a job, always get a clarification before doing any work. And get it in writing. Interpreting codes is a tricky business. Don't assume anything. When in doubt, contact the inspector's office. Work on another part of the job until you get an explanation. Don't get burned by going ahead with the work. There's too much risk that final approval may be withheld because the installation is wrong.

When working with your inspectors, be friendly and courteous. Also be patient. That's not always easy when you're under a lot of pressure because of a job deadline. But being hasty or rude will only cause more problems. If you build a good working relationship with your inspector, he'll usually do all he can to make your job easier.

When asking for an exception or explanation, use a field sketch or the plans if possible. Think through the problem ahead of time. Plan how you can describe it quickly and clearly. It'll make the whole process easier for both of you.

Other Contractors and Unions

On smaller HVAC jobs in existing buildings, you may be the only contractor. But usually the HVAC contractor is working for the general contractor. Think of your general contractor as the coordinator for the entire project. He has been chosen by the owner, either through bidding or some other method, to carry out the complete job. As a rule, subcontractors bid to the general contractor for their part of the project.

If you've ever worked on a construction project, you know that the general contractor runs the show. Like a ringmaster at the circus, he controls the pace and sequence of performance. The general coordinates the work of the subs. He's the conduit through which information flows from the architect and owner to the job and from the job back to the architect and owner. Some general contractors have their own crews on the site. But many act only as managers and do no actual construction.

Chapter 11 details your rights and obligations as a subcontractor. But enforcing legal rights and obligations is a poor substitute for a good relationship with your general contractors.

Working alongside other subcontractors on a construction site is never easy. There'll be days when it seems everybody, including you, wants to work in the same place. Usually every sub on the project needs the only crane available exactly when you need it to unload a compressor. Most of the time you or your foreman will be able to work around these problems to everyone's satisfaction. But keep this in mind if a problem is causing real trouble: The general contractor, in the person of the job superintendent or manager, is the overall boss. He's responsible for coordination. When necessary, he'll have to decide who has priority.

Having a third party make these decisions is the only sensible way to keep the job going. It also makes your job easier when the hostility of a dissatisfied sub is directed toward the general contractor rather than you.

Of course, sometimes a general contractor will make a decision that costs you money. No one likes to see his men standing idle. And making a temporary move off one job and onto another can be expensive. No one pays for the move except you. But at times there's no alternative. Your only remedy is to avoid general contractors who are poor at coordination and insensitive to the needs of their subs.

But there's a lot you can do to avoid conflicts on the job. When you plan the job, look for isolated segments where work can be done at any time. Shift your crews to these "safety valve" areas when they have to stop working elsewhere. In time you'll be able to spot potential trouble spots before work begins — and will have "safety valves" ready when needed.

Many large jobs and most government work require union tradespeople. Union jobs sometimes require more planning to avoid problems — especially disputes over which craft union should do which type of work. For example, a new HVAC system may need piping, wiring and ductwork. Most installers can handle all three. But on a union job you may need a pipefitter, electrician and a sheet metal installer. And all three will belong to different unions. It's important to anticipate problems like that.

As a subcontractor, you have little to gain from continuous battles with either labor unions or other contractors. Tact, wit, patience and planning can help you stay on schedule and within budget. Try to anticipate problems before they become major. Find a solution that makes all involved feel like they got at least part of what they wanted. Sometimes it's best to put off a decision for a while. This is especially true if an argument develops. A calm, quiet discussion away from the job site and over a cold beer may be the best way to promote a peaceful settlement.

If you come away with less than you hoped for, consider it money invested in your education. Use what you learned on the next job or the next dispute; be willing to compromise; use your sense of fair play. Establish a reputation for being easy to work with. It's likely that you'll be working with these same contractors and employees again on another job. Make them happy to find out it's your company doing the HVAC work on that next job.

Working with Utilities and Fuel Suppliers

HVAC systems use electricity, oil or gas as an energy source. All three types of fuel aren't available in all areas. Before designing or bidding on any system in a new area, check with the local fuel supplier or utility. Don't design a natural gas system for an apartment building if the nearest gas main is miles away. If you plan to use an electric heating system in an office building, be sure there are no restrictions that make it impossible. Some utilities and codes limit the number of electric heat systems in their service area.

Utilities also have their own rules for hookups. Most offer a pamphlet that explains the procedure, and have a service representative who can answer questions. Designing and installing a system according to code doesn't necessarily guarantee that it will meet utility company standards.

Private fuel suppliers often have special requirements. In rural areas, for example, liquified petroleum (LP) gas is a common fuel for heating. Often the dealer will supply and install the tank and meter as part of his deal with the customer. You'll just install the heater and make the connection. But you must be sure that the materials you install are compatible with the dealer's fuel.

For all design and installation work, the basic rule is: *Always check with the utility or fuel supplier first.*

Planning for Safety

Safety is good business — for everyone in the HVAC business. If you've worked around construction sites for very long, you've probably heard about contractors who seem to have more than their share of accidents. Yet other contractors seldom have anyone hurt on the job. Is that just coincidence? Ask the experts — the worker's compensation insurance companies that carry the risk of loss. They know that some contractors are safety conscious. Others are accident prone and pay the price in higher insurance rates. Running a safe job doesn't just happen. It takes a commitment from everyone in your company, from the boss down to the greenest apprentice. If you're not willing to make that commitment, be prepared to suffer the consequences.

I'll go one step farther. It's my personal experience that the most professional, most profitable HVAC contractors have the best safety records. The HVAC shops that can never quite get organized, that cut corners with make-do materials, that tolerate the use of inadequate tools and poorly maintained equipment, that never have time to emphasize the importance of good safety habits, those are the outfits that have more than their share of the accidents. Decide for yourself which type of HVAC contractor you want to be. The choice seems obvious to me.

Chapter 10 explains how to protect your business with proper insurance. But even the most complete insurance coverage is no substitute for following a few good procedures and rules.

Tools and safety equipment— Good safety is based on common sense. Whether in your shop or on the job, a clean workplace prevents accidents. Power tools should be wired properly, with shields and protective devices in place. Keep the blades of shears and saws sharp. Dull blades can cause materials to jam or kick back at the user. Make sure extension cords are free from breaks or cracks in the insulation, with plugs and sockets in good shape. Always keep safety goggles, ear protectors and other safety gear readily available and in good repair.

Don't use tools or equipment that need repair. Insist on the use of hardhats and safety shoes. Make your foremen responsible for seeing that the safety equipment is used.

Hold a shop safety meeting once a month. Record the time and place of the meeting and what you discussed. If there's an accident and it turns out you never had any safety meetings, a court may conclude that you didn't consider safety very important. On some jobs, federal law requires such meetings.

Workers and passers-by on the job site have to be protected from hazards. Any excavating you do, for example, requires a sturdy fence around the hole. If some kids playing on the job on a Sunday get hurt, it's no defense that they shouldn't have been there in the first place.

Most HVAC companies need a workshop for fabricating and storing materials. Your workshop is a high risk area because it's usually in use and seems familiar to the people who work there. Keep properly-charged fire extinguishers at hand. Insist on the wearing of goggles and ear protection when necessary. Keep the floor clean and free of oil spills. The shop should be well-lighted and ventilated. Store your solvents and other flammables in a cool, fireproof location.

Ladders and scaffolding— Faulty ladders are a prime cause of accidents. Keep your ladders in good shape and have enough of each size that's needed. The top end of a rung ladder should extend well beyond the step-off point. It must be firmly anchored at the top. Be sure that your crews use ladders correctly.

On many jobs the general contractor will supply the scaffolding. Many accidents occur on and around scaffold. Inspect it carefully before using it. Be sure the scaffold you use is sturdy and is set up correctly. Scaffold that rolls on wheels presents special problems. Don't try to move the scaffold with anyone on board. Rolling a wheel into a hole in the floor could cause it to tip, unloading whoever is on top. If an accident happens, it's no excuse that the scaffold isn't yours.

Electrical hazards— Working with electricity always presents a shock hazard. It's just common sense not to stand on damp earth or concrete or in standing water when you're working with electrical wires. Some tradespeople carry rubber-backed carpet or wooden stepladders to stand on when handling wires in damp areas. Wooden pallets also work well *if* you make sure no metal bands or bolts stick through to touch the ground. Know something about the piece of equipment you're working on. Constantly be aware of where you put your hands. Never assume the power is off. Turn it off yourself. If the switch is out of sight, securely attach a clear label saying that you're working on the system and it should remain off. Better yet, throw the switch and lock it into the "off" position.

Be careful of overhead power lines when using ladders, man-lifts and scaffolding. If in doubt about any type of electrical work, call an electrician or the utility company.

Fuel hazards— Heating fuels can be dangerous if you get careless. Both LP and natural gas become highly explosive if allowed to accumulate in a closed area. You may find gas accumulations when you go on a service call to fix a defective burner. Gas has a strong odor. You won't have any doubt that gas is present. Your first step is to vent the area thoroughly. Open doors and windows and allow plenty of time for the gas to escape. If the flow of gas is constant due to a bad burner valve, you'll have to shut off the main valve. This is usually located on the gas lateral where it enters the building.

If you use a fan for venting, be sure to set it outside the room or area. Otherwise a stray electrical spark could trigger an explosion. If the gas is serviced by a utility or supplier, call them for assistance when you arrive on the scene. In cities, the fire department will also assist in dealing with gas leaks. Call them as well.

Spilled fuel oil is both a fire hazard and a prime location for a slip and fall. Clean up spills at once, especially before doing any work that requires heat or open flame.

Dealing with OSHA
On your smaller jobs, many of these safety ideas are just good sense. On larger jobs they'll be required by *OSHA* — the federal *Office of Safety and Health Administration.* Failure to comply with the lengthy OSHA requirements can mean a stiff fine and other penalties. In Chapter 10, I'll tell you where to find OSHA requirements for your area. Consult your local trade organization or businessmen's association for advice on how best to deal with an OSHA inspection.

Whether required or not, it's in your best interest to be safety-conscious. Accidents waste time, money and cause needless suffering. Every new company just getting started needs to get the most from its resources. Accidents rob you of this ability. Make safety a habit. It pays off over both the long and short haul.

This chapter has given you some of the general information you'll need to make a go of HVAC contracting. In the next chapter, we'll get down to the essentials: the tools and equipment you'll need, and the best way to set up your shop.

Tools, Equipment and the HVAC Shop

As an HVAC tradesperson, you've already invested in a good set of hand tools. But as a contractor, you'll find that you need much more. Today there's a tool or piece of equipment made for just about any job, no matter how large or small. Most of these tools aren't really necessary. They just make the work faster and easier. But since time is money in the construction business, they can be essential in increasing your profitability. You have to be careful not to "step over a dollar to save a dime." At the same time, a costly piece of equipment that sits idle after one job is a poor investment. The following sections will guide you in deciding when to buy the right tools and equipment.

Tools

When you began to learn the HVAC trade, you started with a few basic hand tools and one or two testing instruments. As your skills developed, you added more equipment as you needed it and could afford it.

The same system will work as a rule of thumb now that you're a contractor. Money spent on tools will be a big investment for your company. You'll want to spend it wisely. When you begin your first job as a contractor, it quickly becomes apparent what you really need. In the past, the company you worked for provided ladders, a cutting torch, pipe cutters, extension cords, electric power tools and the like. Now those items are your responsibility.

Before you go out on your first job, sit down and make a list of tools you think you'll need. Cross off any you already have. Then go over your list again and underline those tools you're *sure* you'll need. These are the ones you should buy. If you're not certain about a tool, wait until you get on the job and find out for sure if it's worth the investment.

When you buy tools, buy quality and durability. Then take care of them. That way they'll be on hand for many jobs in the future. Remember to mark all tools and equipment with a permanent marker as soon as you get them to discourage theft. Store them in a secure place, both on and off the job. Make sure your insurance policy covers tools and equipment in case of loss. Impress upon your crews the need to care for tools and equipment. Damage caused by careless handling means dollars lost, both in downtime and in needless repairs and replacement.

Equipment

Equipment poses special questions for the contractor. How much you need will depend on whether you're working alone or as the subcontractor. On many jobs where you're the sub, the general contractor may be willing to provide scaffolding, power lifting equipment and other items. However, depending on someone else for these items may mean delays, since other trades will also want to use them. You might be better off having your own equipment. Also, in spite of any help given by the general contractor, the final responsibility for having proper equipment for doing the HVAC work is up to you.

If you buy large equipment, such as scaffolding or a mobile crane, you'll need a secure place to store it. Also, you may only need it a few times a year, depending on the size of the jobs you get. It might be best, at least in the beginning, to rent this equipment as needed. Then if you see the need for it continuing over time, you can make the investment of buying it. Many construction equipment dealers offer rental arrangements where the rental fee can be applied to the purchase price. Savings such as these are important, especially when you're first getting started.

Ladders, scaffolding, power tools and the like should always be the best quality you can afford. They'll take a lot of abuse on the job. Keep them in good working order. Faulty or defective equipment that causes an injury will also cause a lawsuit. You can't afford that.

When thinking about the equipment you'll need, don't overlook items that aren't strictly job related. It's essential that you have a good truck or van to haul equipment and workers. And no matter what your business size, you'll need an office to handle phone calls and bookwork. You'll need a calculator, a desk and perhaps a telephone answering device if you're a one-person operation. If you can afford it, a microcomputer will be very useful. In Chapter 5, we'll talk about how you can use it to calculate heat loss and gain when designing a system. It can also keep track of business records, such as profit and loss statements, accounts payable and receivable, and payroll. As with your tools and equipment, expanding operations will soon make additional office items necessary.

Testing Instruments

The HVAC trade relies on testing instruments more than almost any other trade. These instruments help to analyze the different HVAC systems. You can then balance and adjust the system or find out where a malfunction exists. HVAC instruments are both delicate and expensive. You may invest many thousands of dollars in testing equipment over a period of time. Make that money well-spent.

Buy instruments with an eye towards quality and durability. Both the instrument and its case should be rugged. Check to see that the instrument controls are easy to grasp and dials and scales are easy to read.

Before using the instrument, read the manufacturer's operating instructions carefully. Before each use, calibrate the instrument and do the suggested tests to make sure it's operating properly. A faulty reading can send you on a wild goose chase. Many instruments use batteries for their power. Check the batteries often and carry spares.

When using the instrument, set it where it won't fall or get damaged. Check all connections to be sure they're correct. Use the instrument only for its designed purpose.

Some instruments — the anemometer, for example — measure air velocity. When using this type of instrument at a diffuser (an air outlet), it's important to position it correctly. Also, a *K-factor* must be used in the calculation. The K-factor is a mathematical constant that interprets the design variables of the diffuser itself into the air flow formula. You can obtain the K-factor from the diffuser manufacturer. Some newer instruments on the market rely less on K-factors. Always check both the instrument *and* diffuser manufacturer's instructions to insure an accurate reading.

Another important item with HVAC instruments is *calibration*. All new instruments have been calibrated at the factory for specific applications. It's very important not to change tube lengths or substitute probes or other parts from a different manufacturer's instrument. Most manufacturers offer calibration services once you own the instrument. Return the instrument to the manufacturer at the recommended time periods for calibration. That's essential for accurate readings. Always handle the instrument with care. If it's knocked around or dropped, it will often go out of calibration.

Now we'll describe some instruments basic to the HVAC trade. Chapter 9 explains how to use some of them to balance and adjust air distribution systems.

Anemometer

An anemometer is an instrument that measures wind velocity. In the HVAC trade, anemometers are commonly used to measure air flow at supply outlets and return inlets.

The need for determining air flow in a system is obvious. You can verify vague complaints by room occupants about drafts and abrupt temperature variations, for example. An air flow lower than the computed flow indicates a duct leak or other malfunction. Properly diagnosing the air flow enables you to balance and adjust the system.

The air flow in a system is measured in cubic feet per minute. This is determined by the flow area of the duct or diffuser and the velocity of the flow. This air flow relationship is expressed in a formula:

$$Q = AV$$

where
Q = air flow in cfm (cubic feet per minute)
A = flow area in square feet
V = flow velocity in fpm (feet per minute)

In reality, the K-factor used to measure air flow at the diffuser is a flow area determined by the manufacturer's testing. This area isn't expressed in any physical dimensions, but must be related to the type of instrument used. Once you have the correct K-factor, you can substitute it for the area and simplify the formula:

$$Q = KV$$

The same grille can have different K-factors, depending on the type of instrument used. However, the grille manufacturer can supply K-factors for a wide range of instruments. The manufacturer has done extensive testing to determine these factors. These tests rely on several measurements made at different locations on the grille face. Therefore it's very important to follow the manufacturer's instructions telling you where to locate the anemometer probe on the grille. Also, be sure to take the correct number of readings. Finally, check to make sure you're using the correct K-factor for your instrument when you do any calculations.

Several different types of anemometer are commonly used. A swinging vane anemometer has an airtight case which holds an aluminum vane. Air flows through the probe and the connecting tube into the case, striking the vane. The vane then deflects the pointer on the scale to show the air velocity at the probe. This type of anemometer often comes with several probes for different applications. Each probe uses a corresponding scale for calculation. The instrument has been calibrated at the factory for its probes, connecting tube and flow passage inside the instrument. It's important not to substitute another size of connector or change its length in any way.

A rotating vane anemometer uses a small, lightweight propeller. The propeller rotates with increased speed as air flows through the instrument. In the past, these instruments had to be used with a stop watch to determine velocity. Today, many of the rotating vane anemometers can be read directly in feet per minute. To get an accurate reading, you

Courtesy: TSI Incorporated
Hot-wire anemometer
Figure 2-1

must position the instrument correctly and use the proper K-factor.

A third type of anemometer is the "hot wire" anemometer (Figure 2-1). This instrument depends on the fact that the resistance of a heated wire will change with its temperature. The probe of the "hot wire" anemometer has a special wire element which gets current from batteries inside the instrument case. As air flows over the wire in the probe, the temperature of the element changes, also changing its resistance. This change in resistance is shown as a velocity on the instrument scale. "Hot wire" anemometers often have several scales that you can select, covering a wide range of velocities.

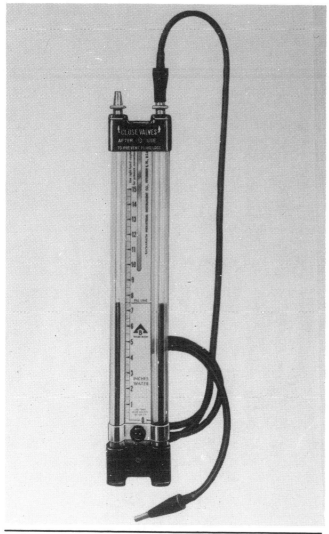

Courtesy: Bacharach Instruments

Manometer
Figure 2-2

Draft Gauge/Manometer

Anemometers measure air velocity at an outlet face. Draft gauges, also called manometers, measure the velocity of air and other gases at any point in the ductwork of the system.

A manometer, Figure 2-2, uses a transparent, U-shaped hollow tube. The tube is filled with water and calibrated in inches. A flexible rubber tube is used to connect the instrument to the duct. There are also dry-type draft gauges that use a bellows and a needle that moves across a scale. In either case, the gauge measures the difference between the pressures found inside the duct and the atmospheric pressure outside.

The HVAC contractor can find many uses for a manometer. For example, the pressure found when the tube is connected just inside the wall of the duct is *static pressure*. This is the pressure necessary to overcome the resistance caused by the friction between the air and the interior surface of the duct. By moving the tube to the center of the duct, you find the *velocity pressure*. The sum of these two readings make up the *total pressure* of the duct. This is useful in finding fan efficiency.

You can also use the manometer to determine if a filter is dirty by measuring pressures on either side of the filter. Other uses include checking flue pipes, chimneys and fire boxes, and adjusting draft regulators and basement ventilation.

Here's how to find the velocity of air or other gas in a duct using the manometer or draft gauge. First find the velocity pressure. Then find the corresponding velocity by using this formula:

$$V = 1096.4 \sqrt{\frac{VP}{d}}$$

where V = fpm (feet per minute)
VP = velocity pressure in inches of water
d = density of the test air or gas
1096.4 = mathematical constant

If the density of the test air is the standard density (0.07495 pounds per cubic foot), the formula simplifies to:

$$V = 4005 \sqrt{VP}$$

For example, if you measure the velocity pressure at 0.49 and the air is at standard density, here's the formula:

$$V = 4005 \sqrt{VP}$$
$$V = 4005 \times \sqrt{.49}$$
$$V = 2803.5 \text{ fpm}$$

If the density of the test gas is different from standard air, you must use the first formula instead of the simplified formula.

Many instrument manufacturers supply charts that give the velocity in feet per minute for a wide range of velocity pressures at standard air density. This eliminates the formula calculation. They also have correction charts for those densities that vary from standard air. However, for perfect accuracy, it's best to rely on calculating out the formulas.

As with your other instruments, follow the manufacturer's directions carefully about calibration, leveling and probe location. When finding the air flow through a large duct (over 1' diameter), it's best to take readings at several points across the duct and average the results.

Combustion Analyzer

Combustion analyzers measure flue gases for oxygen, carbon monoxide (CO), and carbon dioxide (CO_2). They also find the flue stack temperature and, for oil-fired equipment, the amount of smoke in the flue gas.

All of this information tells the skilled HVAC contractor the combustion efficiency of the heating unit. Excessive smoke, for example, has long been recognized as a sign of poor combustion in an oil unit. Low carbon dioxide in a flue gas sample means wasted fuel, since it's not burning completely and some fuel is escaping up the flue. Too high a stack temperature means an over-fired boiler.

Combustion analyzers come in a kit that includes a thermometer, a smoke tester, CO_2 indicator, draft gauge, calculation charts and carrying case. See Figure 2-3A. Newer models combine many of these functions into one digital unit, as shown in Figure 2-3B.

While the analyzing instruments will tell you a lot about the heating plant efficiency, you'll want to combine the readings with your own experience. You'll also need a set of manufacturer's recommendations for the unit you're testing. After that

it's simply a matter of fine-tuning the unit to get the best results. This adjustment, along with a good cleaning or replacement of filters and fuel jets, often means fuel savings that will please the customer. This is especially true in older, neglected units.

This is also a good time to sell your customer on a routine maintenance and inspection program, if you already haven't done so.

Thermometers

You need two types of thermometer for HVAC work — wet and dry.

A *dry bulb thermometer* primarily measures the heat contained in the air. A common type of dry bulb thermometer is the familiar glass tube filled with mercury or colored alcohol. Another type uses a flat bimetallic strip that moves an indicator needle over a scale. Still another kind is the electric thermometer that uses resistance to determine temperature, described earlier in the section on anemometers.

A *wet bulb thermometer* is also used in HVAC. Total heat content in an environment comes from both the air and the water vapor in the air. Since the dry bulb thermometer measures only the heat in the air, it's necessary to find the heat in the water vapor. Typically, a wet bulb thermometer is identical to the dry bulb, but with a cloth wick over its bulb. To use it, saturate the wick with water and swing the thermometer rapidly through the air. As water evaporates from the wick, the temperature falls below the dry bulb temperature. Take several wet bulb readings, re-soaking the cloth each time. Use the lowest reading.

After you've taken both the dry and wet readings, you can find the humidity on a psychrometric chart normally supplied with the thermometer. For example, suppose you get a dry bulb (DB) temp of 80°F and a wet bulb (WB) reading of 65°F. Using the chart, you determine that the relative humidity is 40%.

To get a reading simultaneously from both thermometers, use a *sling psychrometer*. This is simply a frame that holds both the wet and dry thermometers. Hold the handle and whirl the thermometers through the air for several minutes. Record the readings until the wet bulb temperature starts to rise. The lowest wet bulb reading taken is used as the accurate one.

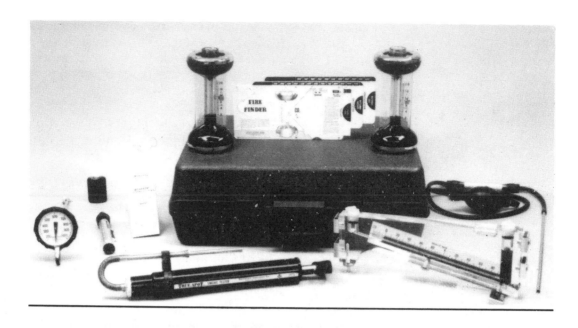

**Combustion analyzer kit
Figure 2-3A**

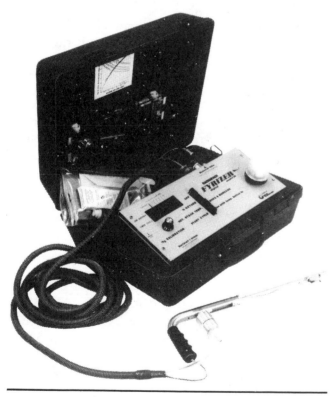

Courtesy: Bacharach Instruments

**Combustion analyzer
Figure 2-3B**

Good quality, liquid-filled thermometers are the most accurate. But you'll find plenty of uses for the handier dial type. For the best readings, buy thermometers with scales marked in no greater than one degree steps. The scale range should be as short as practical.

Treat your thermometers with care. Keep them clean and in their cases when not in use. Check them periodically against a known temperature indicator to insure their accuracy.

Electrical Instruments

HVAC systems use electric motors to power fans and pumps. The need for checking the condition of the motors is obvious. Check the amperage and voltage against the nameplate values to assure that the motors are operating safely and at the proper speed. Also, electrical measurements combined with some calculating will help you determine the brake, or useful, horsepower of the motor.

The instruments you'll use are the ammeter, voltmeter and ohmmeter. For basic motor checks, I suggest that you buy a multimeter that combines all three functions. It will save you some dollars and it's easier to carry. Whether you buy the meters separately or as one unit, the principles of using them are the same.

Voltmeters

Voltmeters use a calibrated scale to show the voltage flowing through the circuit. They have two wire leads ending in probes or clamps, which you connect in parallel, across the load, as shown in Figure 2-4. Connect the leads to the terminals as

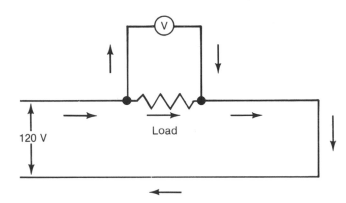

Voltmeter connected across the load
Figure 2-4

close to the motor as practical. For 115-volt circuits, connect one lead to the hot wire and the other to the neutral wire. For 230 volts, connect each lead to a hot wire. Then turn on the power to get a reading.

For three-phase circuits, the leads are connected as follows:

    Poles 1 and 2
    Poles 2 and 3
    Poles 1 and 3

Average the three readings. If the circuit is operating properly, the difference between the readings will be small. If you find a large difference, shut off the motor and call an electrician to track down the malfunction.

Here's another sign of potential motor trouble: a voltage reading delivered by any circuit which varies by more than ten percent from the nameplate rating of the motor. If this happens, shut off the motor to prevent damage and consult with an electrician to find the cause of the problem.

Ammeters

An ammeter is used to find the amperage in the motor circuit. Make sure you connect the ammeter in series *with* the load. Connecting it across the

load, like a voltmeter, will destroy the instrument.

Many ammeters have jaws which snap around the wire. This is the only connection necessary. See Figure 2-5. To check the amperage of single-phase circuits, place the jaws around the single wire. For three-phase circuits, read each wire separately and average your readings. Power must be flowing in the circuit for you to get your readings.

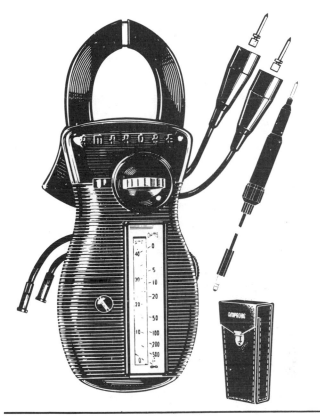

Courtesy: Amprobe Instruments

Snap-on ammeter
Figure 2-5

Ohmmeters

An ohmmeter is used to measure resistance. Unlike the voltmeter and ammeter, you never use an ohmmeter on a "hot" circuit. Instead, the ohmmeter has batteries which supply the power for the reading. Check these batteries frequently to insure the ohmmeter's accuracy.

The ohmmeter is used to track down shorts, broken wires or open motor windings, defective coils or other problems in the circuit. For example, a good conductor will show a relatively low resistance when the ohmmeter probes are applied.

If it's broken or shorted, however, an extremely high reading will appear on the meter.

Ohmmeters and other electrical testers sometimes have multiple ranges. Don't forget to adjust the instrument to the range where you'll be working. If you're not sure, adjust it to the largest range and work your way down. This will prevent damage to the instrument.

Safety Precautions

Because of the danger of shock, using electrical testers takes some extra care. Most tests have to be done on live circuits. Safety is important. Make sure your instrument is working properly by testing it on a *known* live line. You can be fooled into thinking a live line is dead if the instrument isn't working properly.

Shock is caused by current passing through you to the ground. Caution and common sense are the best way to prevent shock. When using your instrument, be careful where you stand. A wet floor or wet shoes can give you a bad jolt if you accidentally touch a terminal. Water on a roof can be hazardous when testing rooftop equipment. Drainpipes, conduit, metal ladders and other metal parts can allow electricity a nice path to ground.

You can interrupt this path by using insulating materials such as a wooden stepladder or wooden crates or pallets to stand on. Be sure they're dry and don't have metal bolts or reinforcing bands extending through them.

These precautions alone won't give you complete protection. But they'll help reduce the danger if you contact a live wire or terminal.

Not having enough room in the junction box to get the jaws around the wire is a common problem when you're using an instrument with clamp-on jaws. When this happens, your first impulse is to pull on the wire. But before doing so, throw the switch or breaker to *OFF*. Then pull on the wire carefully to avoid breaking or bending the terminal. Attach your meter and then turn on the power to get your reading.

Follow the same off-on procedure when using clamps to attach instrument leads. If you're using probes, place them carefully. Avoid too much pressure or your hands may slip off and touch a live terminal.

Your HVAC Shop

Whether you're lucky enough to have a shop when you begin HVAC contracting will depend on your finances. But as the business grows, you'll find you have to have a shop to keep up.

Other building trade contractors use their shops primarily for office space, material storage and routine maintenance of equipment and vehicles. While you need all of these functions, an HVAC shop also requires room for one other job — sheet metal fabrication.

It's possible to get by without a fabricating shop when you start your business. In fact, some HVAC contractors have never done fabrication work themselves. Instead they prefer to job it out to shops that specialize in it.

It's also possible to buy much of the ductwork you need preformed from the manufacturer. Many HVAC designs use standard sizes and shapes of ductwork that are available at local wholesalers. Figure 2-6 shows some common pieces. You can get the occasional custom-shaped piece made to order at a shop.

Doing your own fabrication, however, has several advantages. You save money over the long run. It's also a way of balancing the work load among your crews. A shop can help you avoid short layoffs between jobs and during bad weather. And it frees you from the aggravation of job delays because your supplier can't deliver your order on time.

The disadvantage of your own fabricating operation is the high start-up cost. The cost of the added floor space can be softened somewhat if you're able to find a building that's large enough but not too much over your budget. It may be a good idea to take it if you can afford it, especially if you're buying or able to get a long-term lease. That will take care of your space requirements. But you've still got equipment to consider. You'll need lights, sufficient power outlets, a drill press, bench grinder, rotary shears, squaring shears, bar folder, riveter, welder, torches, cylinder roll, grooving machine and other machine tools.

As with expensive job tools, you may have to wait until your volume of business is large and steady enough to justify the added cost. But always keep in mind your long-term goal. Acquire equipment gradually, keeping an eye out for good buys on used pieces. Often your bank will lend you funds for new equipment on the strength of the contracts you get.

Courtesy: United Sheet Metal, Div. of United McGill Corp.

Preformed ductwork
Figure 2-6

This isn't a manual of sheet metal fabrication, nor should it be. If you don't know fabrication, you're lacking one of the basics of your trade. You might want to reconsider your decision to go into this business right now. Take a crash course in sheet metal work at your local technical school. Or hire someone who knows fabrication. But you could find yourself in a lot of trouble when he quits in the middle of a big job and you can't get a replacement.

If it's been a few years since you've actually done any fabrication and you want to brush up on it, an excellent reference book is *Low Pressure Duct Construction Standards,* published by the Sheet Metal and Air Conditioning Contractors National Association. To order a copy, send $20 to:

SMACNA
P. O. Box 70
Merrifield, VA 22116

SMACNA also has manuals that detail instrument use for a wide variety of forced air applications.

Selling the HVAC System

With so many contractors available to do HVAC work, your success or failure may depend on how willing you are to go out and *sell* your services. While construction will always be a strong industry, no one expects a repeat of the nationwide building boom of past decades. There are very few companies that still have more work than they can handle. Instead, most are hiring sales and marketing people to actively go out and bring in work.

If you're just starting out, you'll have to be your own salesperson. This chapter will help you learn some of the tricks of the sales trade. Use them to make yourself effective in selling your work.

The first rule is this: Don't rely on memory alone for your sales presentation. Prepare a folder or notebook of sales material. Most people like to have something in their hands to look at. It helps you because the materials make your job of explaining easier. And it gives you a few moments to collect your thoughts while the prospects are looking it over.

If you can afford to have your company name and logo printed on the notebook, so much the better. Using this chapter as a guide, prepare a description or sketch of each of the different heating systems. Leave room for manufacturer's pamphlets and literature. Also include letters of recommendation from previous clients as soon as you can collect them. And don't forget to take along several pens, contract forms and your pocket calculator on your sales calls.

These suggestions also apply to work you bid on. When you're bidding a job, private owners reserve the right to accept what they feel is the most suitable bid. Even if your bid isn't the lowest, they may be willing to award the contract to you — *if* you've made a sales presentation that shows them you'll do the best job for the dollars spent.

Be careful how you approach the owners on a bid job, however. Explain in advance how much time you'll take and the purpose of your presentation. Then ask them if they'd be interested. Don't force the issue if they refuse, or they might feel you're trying to influence the bidding process. For government work, it may be illegal to approach the involved parties prior to the bid opening. Check this out before taking any steps.

Sell Customer Satisfaction

Customer satisfaction will depend on the system's operating costs, appearance and performance. Of course you, as the contractor, must also keep in mind your own guidelines such as profitability and the type of utility service available for your customers.

When dealing with customers, you'll need tact, patience and a sense of humor. Remember that customers will often make a good choice if you've given them correct information in a way they can understand. Since you're familiar with all the terms peculiar to the HVAC trade, you may have a tendency to slip them into your conversation. If you do, your customers probably won't understand what you're saying. Talk to them in terms they'll understand. It's often helpful to use an example to explain a point.

Whenever you're selling a job, be prepared to answer any and all questions. Be ready to give the reasons for your recommendations to the customers. Even when you're bidding a job, the buyers will want to know that the system and contractor they choose will give them what they want. Establishing good customer contact from the very beginning goes a long way towards ensuring a satisfied customer when the job is done.

Comfort Conditions

Your customers' comfort will depend on the physical conditions of the air around them — its temperature, humidity, cleanliness and velocity. These comfort conditions will vary depending on the way the building's used. The personal characteristics of the building's occupants, such as age, sex and health, will also affect how they'll react to their environment. Most people take the conditions around them for granted if they're comfortable. But when they're uncomfortable, then they'll call you.

Temperature

Temperature is the thermal condition of the air. Heat which causes the temperature of the air to increase is called *sensible heat,* since we "sense" or feel it. Another type of heat is *latent heat.* This is heat used when a substance undergoes a change. For example, the heat used to convert water from a liquid to a gas is latent heat. Aside from a few theoretical applications, HVAC work is concerned mostly with sensible heat.

The customer connection with temperature is quick and obvious. When they get too hot or cold, they'll try to adjust the system. If they can't get the results they want, they'll call you. The smart contractor sees this as more than a service call. It's also a chance to sell a whole new system or to remodel the present one. It may also be the chance to enroll the customer in a regular service and maintenance program if you're interested in that type of work.

When discussing new work with a customer, you won't usually talk about temperature unless the job involves a special situation. Most people assume that the system will maintain suitable temperatures. Be sure you understand what the customer expects in terms of temperature performance. Be alert for potential problems later on. If, for example, he expects a superior heating system with small fuel bills when his building has a lot of windows, point out that he's going to have to yield somewhere. He may have to settle for a lower operating temperature. Honesty at this stage can prevent complaints and headaches after the system's installed.

Remember that temperature is often the one item that customers take for granted, yet it's the main factor for judging system performance overall.

Humidity

Improper humidity in a system will also lead to complaints. Air at lower temperatures contains less water vapor than air at higher temperatures.

People perspire more in high humidity since the wetter air is unable to absorb moisture effectively. Besides making occupants uncomfortable, high humidity can affect work areas such as offices and drafting rooms where paperwork is handled. It can also affect the building itself, leading to mildew and rotting of the structure and its contents.

Low humidity will lead to occupant complaints of dry skin and scratchy throats. Older people may have trouble breathing. The overly dry air will also draw moisture from plaster, walls and woodwork. Furniture, paintings and musical instruments can be damaged. Any commercial building or residence where human comfort and interior furnishing are important should have the best humidity controls the owner can afford.

Use complaints about humidity problems as selling points for top-of-the-line units. Another good selling point is that the humidifier will allow better fuel economy. Air that is relatively dry will feel cool to the skin at 68°F. But the same air, properly humidified, will be comfortable at the same temperature.

Some people are under the impression that hot water or steam heating units don't need a humidifier. Explain that these systems don't add any moisture to the air. Heated air, no matter how it's heated, needs humidification for maximum comfort.

Clean Air

Clean air is vital to comfort. In many commercial buildings, you'll be asked to design and install ventilating systems for the sole purpose of removing impurities from the air. Often these systems are required by health and safety laws. For these customers, you'll need little sales work.

Homeowners, however, may need convincing. Although more people are becoming aware of the need for filtered indoor air, many still think that the panel-type filters installed in their forced air system at the factory are good enough. These filters, however, are designed primarily to protect the system equipment, removing only large particles. These large particles make up only 15% or less of all the contaminants found in the air.

Air cleaners designed specifically to clean the air, whether a part of the HVAC system or a separate unit, are much better. Depending on their rated efficiency, they can remove up to 90% of all the airborne particles passing through them. They can trap particles as small as 0.01 micron. (25,400 microns make one inch.)

There are many selling points for an air cleaner. According to the U.S. National Health Survey, respiratory illnesses are the most dominant of all illnesses reported in the country. Cleaner air in commercial buildings with large numbers of employees can reduce the number of lost workdays due to illness. Cleaner air also reduces drowsiness and makes workers feel and work better.

Around the home, an air cleaner can reduce discomfort for those afflicted with allergies and other chronic respiratory illnesses. The air cleaner will reduce household dust, tobacco smoke and odors. Walls and ceilings will stay cleaner longer.

Air cleaner installation, whether on a new or existing system, can be a good profit item for the contractor. Sell one at every opportunity. Remember that air cleaner manufacturers are quite willing to work with you in selecting the right cleaner for your job. Chapter 9 has more information on clean air systems.

Velocity

Another important comfort factor is the velocity of the air as it moves through a system. Although your customer typically won't have much input in this area, he'll certainly feel the results if they're wrong. Despite constant improvements in diffuser and duct design, many service calls are still based on complaints by room occupants of "cold drafts" and other velocity problems.

Velocity is measured in feet per minute (fpm). Calculated velocities of less than 50 fpm are acceptable. However, where human comfort is important, velocities in an occupied space should range around 15 to 20 fpm. Higher speeds may cause occupants to complain of drafts. Very low velocities will result in stagnant air and poor heating and cooling.

Air velocity will be most noticeable to room occupants when they stay in one place for any length of time. Offices, restaurants, lounges, and waiting rooms are typical. These systems need a good design, based on a careful assessment of customers' use. In such situations, you may be wise to suggest a different type of system, such as electric or hot water, instead of forced air, if it's economically feasible. Otherwise explain to the owner how he may need a more expensive forced air system to distribute the air without drafts. A single rooftop unit with a couple of large ceiling diffusers, while cheap to install, will be difficult to adjust to the occupants' comfort.

To sum up, uncomfortable operating temperatures and too high a velocity are the two conditions which lead to most customer complaints. But all of the comfort conditions provided by the system are important, both as selling points and as elements of customer satisfaction. You'll save yourself a lot of grief by paying attention to how your customers think these conditions should be maintained. If you don't agree with their ideas, try a tactful and thorough explanation. It may lead to increased sales. But make sure your customers have a clear picture of what the system will do. If they're expecting results the system can't deliver, you're bound to have problems later on.

Selling the Different Systems

In the following sections I'll review the advantages and disadvantages of the different heating systems. You'll find these points helpful during customer presentations.

Stress the positive points of the system you're hoping to sell. Keep alert during your first run-through for customer's reactions to favorable points. When you meet objections, don't duck them. Acknowledge them. Then use the clues you've picked up to repeat and reinforce the positive points. Finally, don't oversell. If you push the customers into a contract they don't really want, you may make money on the job. But it will cost you dearly in the long run.

Hot Water Systems

As a rule, hot water heating systems are controlled by one thermostat at a central location. However, they can be easily zone controlled. This allows a great deal of flexibility in commercial buildings such as apartments and offices, as well as residences.

A hot water system appeals to many homeowners because it's considered a source of "clean" heat. It has good eye appeal since the heat source is generally a small boiler and the piping eliminates large, bulky ductwork. This may be important when selling a system where the home or commercial building needs maximum square footage for usable floor space.

Perhaps the strongest appeal of hot water heat is that it eliminates any possibility of drafts, since it heats by radiation and natural convection. The owner isn't bothered with a lot of maintenance. Monthly fuel costs will depend on the market price of the type of fuel used to fire the boiler.

While we mentioned the lack of ductwork as an advantage a little earlier, it can also be a disadvantage to the buyer who wants cooling and perhaps humidity control and filtering all in one system. With a hot water system, some kind of forced air system will be needed for cooling and air filtration. Another disadvantage is that a hot water system is generally the most expensive to install.

During your sales talk, you should try to sell air conditioning, an air cleaner and a humidifier once the customer has decided on the hot water system. Point out that you can do it cheaper all at once since your crews will already be on the job. If, because of finances, the owner can only do the hot water heat, go ahead and do the work. But when the weather gets hot, be sure to follow up with a phone call suggesting the air conditioning.

One objection you may hear from time to time is that hot water heat has a lag time. That is, from the time the room cools down and the thermostat calls for heat, it may take several minutes for the hot water to circulate and radiate enough heat to make the room feel warm again. The room may feel "cool" to the occupants during this lag time. This problem was much more severe when hot water systems were first being installed many years ago. Design improvements in thermostats and piping have virtually eliminated the problem. The complaint is probably being kept alive by a competitor who's trying to sell forced air systems. Another complaint you may hear is that a hot water system is "noisy." Again, this will only be the case with a system that's poorly installed or designed.

You can counter these objections by offering the names and phone numbers of satisfied customers of yours who have hot water systems. If you're just getting started, you'll have to rely on an explanation of the improvements in hot water heat that have eliminated these problems. Point out that forced air systems may also have a certain amount of noise, and possible drafts as well.

Electric Heat

Electric heat systems use one or more unit heaters in each room. These units are separately controlled with a wall or unit thermostat and are circuited separately. Many different types are available to cover a wide range of applications. Where appearance is important, baseboard heaters are usually a good choice. Commercial, farm and industrial locations may use wall or ceiling units.

For the contractor, installing electric units is usually cheap and easy. For the owner, electric heat gives clean heat on demand, controlled on a room-by-room basis. Because of its adaptability to a wide range of comfort demands and building designs, electric heat is very popular as a supplement to less dependable heat sources such as solar and wood-burning systems. It's perfect in locations where rooms are seldom used or where only a small area needs heat. It's very useful in remodeling older buildings built on slabs or where other types of construction make piping and ductwork difficult to install. Like hot water systems, electric units don't cause drafts. Electric heat is often found in warmer climates where heat is needed only a few times a year.

One disadvantage is that the system can only heat. An all-electric system will need some sort of forced air system to cool, humidify and clean the air. But its biggest drawback is operating cost. Compared to other systems, electric heat will cost the most dollars for the Btu's delivered.

If your customer is fortunate enough to not have to worry about fuel bills, electric heat may be a good suggestion. It may be possible to reduce operating costs somewhat if the utility in your area has on-peak and off-peak service. Otherwise, you may find most of your electric installations will be supplementary units or in multi-unit housing where utility costs are shared by the government.

Be sure to check with your local utility, and check state and local codes before selling any electric heat that draws substantial wattage. In many areas of the country, availability of the service is limited.

Forced Air

Forced air systems have been one of the most popular heating systems over the years. Many old convection furnaces were modified to become forced air systems, using existing ductwork tied to a gas or oil furnace and electric fan. One advantage to you in selling forced air systems is that many customers are familiar with them and satisfied with their operation.

For the owner, forced air systems are relatively cheap to operate, since large amounts of air can be heated and moved at low cost. The system is also versatile. It can be used for air conditioning simply by putting a cooling coil in the supply duct and adding a few minor modifications. Humidifiers and filter systems are also easily added to the system.

There's another advantage in commercial buildings or boiler systems where codes require that a certain percentage of fresh air be inducted into the system. A fresh air intake and suitable damper can usually satisfy these requirements effectively.

One disadvantage is the necessary ductwork in a building where the space isn't readily available. This isn't much of a problem in new buildings since the architect can often design chases and soffits for the ductwork. Also, floor and roof trusses can be ordered with chases for ductwork built into the design, reducing space needs.

Some owners may not like the periodic maintenance of fans, filters and motors associated with a forced air system. But it's hard to escape some kind of maintenance with any system.

From your point of view, forced air systems will cost more to install than electric but less than hot water. However, because many buildings will permit the use of common duct sizes, some contractors are able to install a forced air system quickly and very economically. If you specialize in forced air, you can train your crews, purchase ductwork and other parts in volume, and incorporate other cost-cutting methods that keep installation costs down. When you run into systems that need a large amount of custom ductwork, however, price them with care.

Solar Heat

As the cost of traditional fuels has continued to climb over the last decade, solar heating has become more popular. Its chief advantage is that its energy source is free. But since sunlight isn't available on demand, solar heating has to cope with the problem of erratic supply. Solar systems compensate by storing heat with large heat sinks or storage bins made up of rocks or concrete. These take up room and many owners find them unattractive.

Passive systems typically use large south-facing windows to gather heat. Often some sort of heat sink such as a concrete wall or floor is used to store the heat. During evening hours, the windows are closed off with insulated shutters to retain the heat. Passive systems are almost always designed by a solar architect as part of the overall building architecture. Their ability to function depends on their architectural, rather than mechanical, properties.

The efficiency of passive solar systems is difficult to calculate accurately. The big disadvantage is that the large areas of glass or solar glazing that allow good heat gain when the sun shines also allow large heat loss when the sun disappears. This must be dealt with effectively for the system to work. In virtually all cases, passive systems are used to supplement more traditional types of heating. They can have a strong appeal to customers in a good solar location whose life-styles motivate them towards an energy-saving solar design.

Active systems also use the sun but have many of the mechanical parts found in forced air systems. Look at Figure 3-1. When these parts are combined with solar panels and a supplemental fossil fuel furnace, the total system cost will far exceed a forced air system alone.

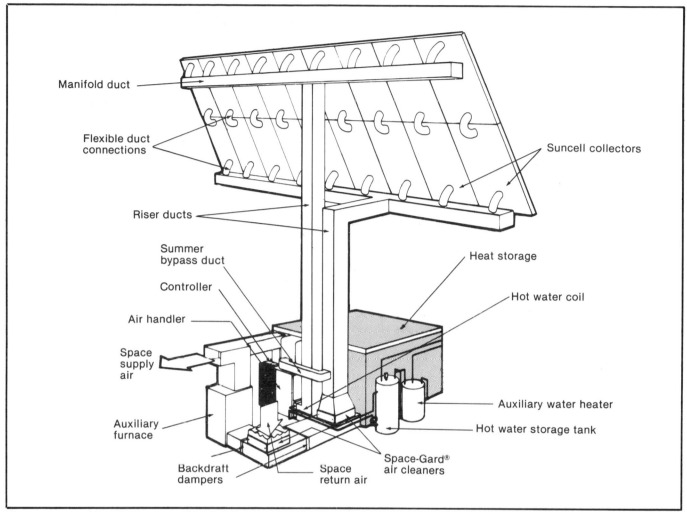

Manifold duct

Flexible duct
connections

Suncell collectors

Riser ducts

Summer
bypass duct

Controller

Air handler

Space
supply
air

Auxiliary
furnace

Backdraft
dampers

Space
return air

Space-Gard®
air cleaners

Heat storage

Hot water coil

Auxiliary water heater

Hot water storage tank

*Courtesy: **Research Products Corp.***

Active solar heat system
Figure 3-1

The question then becomes whether the system payback will justify the cost. *Payback* is the initial cost of the system compared to fuel cost savings over time. For an example, let's take the case of the owner of a shipping company who pays $3,400 annually to heat his warehouse with natural gas. You calculate that an active solar installation costing $8,000 will save him 15% of his heating costs.

15% x $3,400 = $510 annual savings
$8,000 ÷ $510 = 15.6 years payback

The rate of payback may actually be faster since it's likely the cost of natural gas will continue to in-crease over that 15-year period. Also, federal or state energy credits may apply in some areas. Tax depreciation may also result in savings. Finally, overall installation cost may be reduced if the work is done along with a major building remodeling.

Using averages based on geographical areas, you can calculate the contribution of active solar systems with a fair degree of accuracy. An excellent text that details solar systems and their efficiencies is the *ASHRAE Handbook of Experiences in the Design and Installation of Solar Heating and Cooling Systems*. It's available from the National Technical Information Service, U.S. Department of Commerce, Springfield, VA 22161.

From the contractor's point of view, the decision whether to get involved in solar work should be based on two considerations: the amount of work being done in your area, and the probability that the work can be done trouble-free and at a profit.

Availability of work will depend largely on customer acceptance in your area. If you find yourself losing a fair number of jobs to your competitor because you don't do any type of solar, you may want to reassess your position.

A bigger question for any contractor is whether or not the system can be installed at a profit. If it can, will the profit be eaten up later by constant servicing when the system doesn't perform as expected?

Here's how one contractor in Minnesota resolved these questions. When new home construction in his area slowed, he couldn't keep his crews busy. He considered branching out to solar installations in existing homes. First he spent a couple of months studying suppliers and manufacturers to find out which ones could deliver quality solar components on time. Then, working with an engineer who had experience with several solar designs, he

installed a solar system in his own home and several in customers' homes on an "experimental" basis. He installed them at cost, with the stipulation that if they didn't perform as expected, he had only limited liability. Now about 20% of his sales are in solar heating. More important, he has the flexibility to schedule his crews more efficiently.

Heat Pumps

Heat pumps are nothing new in HVAC work. Still, many customers may ask about them under the mistaken impression that they're a new type of system that will lower energy costs.

Figure 3-2 is a schematic of an air-to-air heat pump that you can use to explain the basic functions of a heat pump. Heat pumps work well in areas with moderate heating needs, where the heat load in winter is close to the cooling load in summer. They aren't very efficient in long periods of cold weather with temperatures of less than 30°F. Heat pumps may be effective in commercial buildings where large amounts of heat can be pulled from refrigerators and coolers.

Most building owners will expect a heat pump to

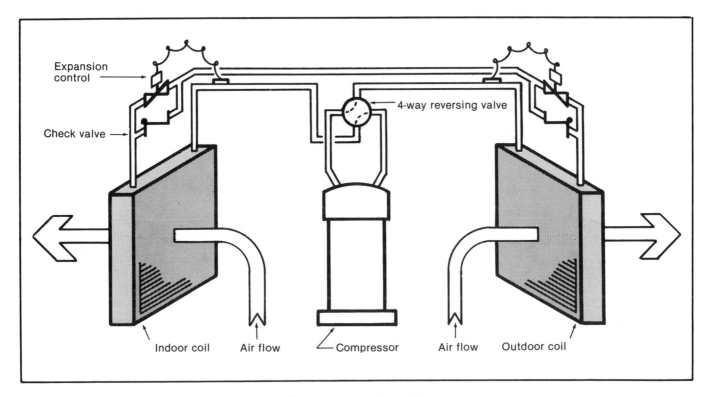

Heat pump schematic
Figure 3-2

give equal or better heating and cooling than previous systems, but at a cheaper cost. As the contractor, investigate each application thoroughly. Contact several heat pump manufacturers for advice before making any recommendations. Also check with the owners of heat pumps you've previously installed to see if they performed as expected. If a heat pump is indicated, you can calculate anticipated savings using the payback method described on page 27.

Distribution

HVAC systems will need piping or ductwork. In design work and installation, most contractors think of this distribution network as simply a part of the system. However, when dealing with customers, remember that not all of them will know about the difference in methods. Items such

as noise, cost, and more important, the appearance and location of baseboard heaters, diffusers, ductwork and piping should be "talked out" completely with your clients. Make sure that they're fully aware of what they're getting when they choose one type of system over another.

Summary

Figure 3-3 is a summary of the selling points we've discussed in this chapter. With the many types of controls and modifications available for any heating system today, endless variations can be tailored for any job. Many of the advantages and disadvantages can be reversed, depending on what you're trying to sell. This listing is not meant to imply any one system is better than another. Only you can decide what system is best for each job, and then sell it to the customer.

Type	Control zones	Air filter or humidifier?	AC or heat pump?	Owner advantages and disadvantages
Hot water	1 or more	No	No	Low maintenance, no drafts, clean heat. No AC, fresh air or humidifier. May be noisy.
Electric	1 or more	No	No	Clean heat, separately controlled. Adaptable. No AC, fresh air or humidifier. Expensive to operate.
Forced air	1 or more	Yes	Yes	Cheap to operate. Can AC, filter and humidify. Needs maintenance. May cause drafts.
Solar	1	Possible	No	Cheap to operate. May not supply total heat needed. May need daily maintenance and adjustment.

Comparing heating system features
Figure 3-3

Planning HVAC Systems

Any HVAC system, no matter how well designed, will fall short if you skimp on planning. It's very possible, as some contractors have found to their dismay, to design and install a system that works well, but isn't suited to the owners' needs. When this happens, unhappy owners may decide not to pay for the work. Then the case can end up in court where no one wins except the lawyers.

HVAC systems need both careful planning and good design. In the planning phase, you gather all the information you need: building construction and use, number of occupants, temperature ranges and system cost, among others. Then in the design phase, you calculate the system size and draw up the installation on the plan sets. We'll cover this in Chapter 5.

Because of the potential problems, some HVAC contractors prefer to use a professional design firm or architect to do their planning and design work. They don't want to put their time and energy into this part of the job. They're only interested in getting the job installed and moving on to the next one.

But planning and designing HVAC isn't that difficult. You may actually be better off doing your own planning and design, especially if you go after residential and light commercial work. Doing your own design work allows you greater control over the HVAC system. This in turn leads to opportunities to create a wider profit margin. You can save the cost of paying someone to do the design, and at the same time put yourself in position to suggest the legitimate extras that increase profitability.

When you've handled the entire HVAC design and planning package for a new customer, it's relatively easy to get a contract for regular maintenance and repair. All this means dollars in your pocket. There's nothing wrong with that. As the boss of your company, you'll have plenty of headaches and long hours. A large part of your reward will be the solid profitability of your company.

In the first half of this chapter, we'll concentrate on the basic planning steps that are common to all jobs. This consists of gathering the information you'll need to select the kind of system that'll do the job.

The second half of this chapter takes you page-by-page through typical plan sets, from the title page to the details. To plan and design HVAC systems, you must have a good working knowledge of building plans and specs. We'll talk about reading *and* drawing HVAC plans. You can skip this section if you're a whiz at reading and drawing plans, but take time to review it if you have any doubt about your ability. The best HVAC planning and design will miss the mark if you haven't drawn the plans correctly. Your installers will commit costly mistakes. Finally, even if you're only interested in installation, you'll need to know all you can about plan sets to bid competitively.

Five Basic Steps in Good Planning

A good plan pays attention to details. It covers a wide range of items, large and small. The obvious ones are building construction, location and climate. But also pay attention to owner preference as to economy, performance and appearance. Planning is the time to resolve anticipated problems and clear up any misunderstandings. If you're planning a custom job, review the selling points in Chapter 3. If it's bid work, study the chapter on Estimating and Bidding. Then begin your planning. The following five steps are essential:

● Determine the comfort zone

● Determine special needs

● Determine weather conditions

● Review building construction

● Select the system

Step 1: Determine the Comfort Zone

The occupants of the building are the end users of the HVAC system. They determine how well you've done your job. If they're unhappy, it's not a good job, no matter how well it looks or how expertly it's installed. Repeat service calls based on user complaints will eat into your job profit. Occupant satisfaction is a key goal in your business.

A large part of that satisfaction depends on the correct comfort zone. The *comfort zone* is that combination of temperature, humidity and air velocity in which the majority of room occupants feel comfortable.

We talked about temperature, humidity and air velocity in Chapter 3. These are also called *thermal climate*.

Atmospheric components are the gases and vapors typically present in the air. They also affect occupant comfort. There's the air itself, made up of oxygen, nitrogen and other gases. But there are also: *dust* — solid organic particles that settle quickly; *aerosols* — particles that remain suspended in the air; and *pollutants* — gases and vapors produced by the building, its furnishings and use, and by the occupants themselves.

Codes require that all HVAC systems handle atmospheric components to some extent, depending on the job. This insures fresh, clean air at regular intervals in the home or workplace. We'll talk about filtering and venting that are integral parts of the duct system in Chapter 6. Special applications are covered in Chapter 9. Right now we're interested in how the body relates to its thermal climate or comfort zone.

The body's relationship to the comfort zone— The average temperature of the body, for both rest and exercise, is 97°F to 100°F. This temperature is very important. Under normal temperatures and at rest, the body has no trouble maintaining its internal temperature. But under adverse conditions it has to go to great lengths to maintain this temperature, and for good reason. A temperature drop of only three degrees of the vital chest area can cause death. An increase in temperature can result in heatstroke and death.

The body creates heat through the *oxidation* of food. This process is called *metabolism.* It cools itself through a slowdown of its metabolism and through evaporation — perspiring on a hot day.

While heating and cooling itself to stay within its temperature range, the body is affected by other conditions. The type and pace of activity, the amount of clothing worn, and the surrounding external temperature all contribute to how hard the body has to work to maintain its temperature.

The comfort zone, then, is that combination of thermal conditions that requires an *acceptable amount of work from the body in order to hold its temperature.*

For example, on a 93°F day with 75% relative humidity, we complain about how "uncomfortable" we are. What we're saying is that our bodies are working harder than normal to stay at 98.6°F internal temperature. We try to help it out by slow-ing down our activities, drinking liquids or finding a cooler location. By doing these things, we reduce the amount of work our body has to do to maintain its temperature. When that effort drops back to an acceptable level, we feel "comfortable" again.

The variables of thermal climate— The parts of thermal climate — temperature, humidity and velocity — are variable. They also interact, which changes the way we're affected by them. One common case is temperature and humidity. The higher the humidity, the harder the body has to work to cool itself. The high water content already in the air slows down perspiration, the body's evaporation process. So room occupants feel "hotter" with a high humidity than they would in another room at the same temperature but with a lower humidity.

Another example is air velocity. People may feel chilled if air is moving over them even though the air temperature would be acceptable if the air were still.

Other variables are the amount of clothing worn, activity of occupants and personal metabolism rates.

With all of the above variables, calculating a comfort zone for each person, for each HVAC system, for each location, would be impossible. Fortunately, much research has been done to establish comfort zones which satisfy almost any situation.

Comfort ranges— One widely-used industry standard is a range of:

Winter: 20% rh, 68°F
Summer: 50% rh, 78°F

In other words, your goal for a typical HVAC system is to meet and maintain this comfort range, given the:

- Seasonal changes

- Building construction and use

- Number and activity of occupants

It's important to note that the above range is a guideline, dependent on several conditions. It's used for jobs where occupants are either sedentary (sitting) or doing light physical activity. It also assumes an air velocity of about 50 to 75 fpm moving over the occupants. It applies where the desired air temperature is the same as the *mean radiant temperature* (MRT): the average air temperature of the room or area.

The winter and summer comfort ranges listed above fall within those set by the Federal Energy Agency for commercial buildings. Research has shown that they'll also satisfy about 95% of all room occupants as the preferred comfort zone. The other 5% would prefer some other environment, perhaps because of physical ailments, age, or an unusual metabolic rate. But to accommodate that small minority, you'd have to install a very elaborate and expensive HVAC system. More important, meeting the requirements of the 5% would cause the 95% discomfort. So your goal remains targeted to the majority.

Not all jobs you work on will fall into the standard ranges. A warehouse for perishable foods, for example, would need cool temperatures overall. Heating units above the work stations where products are sorted or paperwork is done could provide higher comfort ranges for employees.

Guidelines for jobs that differ from the basic range can be found in the *ASHRAE Fundamentals Handbook*. ASHRAE has contracted with the Institute for Environmental Research at Kansas State University for continuous research on comfort variations. Results are graphed and updated in each new edition of the Handbook.

Also consult local codes. They'll regulate minimum and maximum ranges for a large number of buildings. They also take into account weather conditions that are unique to your area. *Always consult local codes and ordinances before doing any lengthy planning, design or installation.* Waiting until after you've done the work can mean costly "do-overs" to comply with codes.

The planning checklist— Figure 4-1 is a typical HVAC planning checklist. You can make copies and use it as it is. Or modify it to suit your preferences. The important thing is to have a checklist whenever you plan a job. It'll keep all the basic information at your fingertips. This list, along with plans, specs and your estimate sheet, will be an essential part of a well-done and profitable job.

Job # _____

Date _____

Building owner:

    Name _____

    Address _____

    Phone _____

Building location _____

    Size _____ No. stories _____

    Type of construction _____

    Manager _____ Phone _____

HVAC designer:

    Name _____

    Address _____

    Phone _____

National codes:

    Building _____

    HVAC _____

Local codes:

    Building _____

    HVAC _____

Building inspector:

    Name _____

    Address _____

    Phone _____

Sample HVAC checklist
Figure 4-1

Utility, electric:

Name _____

Address _____

Service Rep. _____ Phone_____

Service at building: Volts _____ Phase _____

Size of main switch _____

Utility, gas:

Name _____

Address _____

Service Rep. _____ Phone _____

Gas service at building: natural _____ LP _____ Pressure _____

Owner requirements:

Temperature: summer _____ winter _____

Humidity _____ Sound _____

Filter system _____

Type of fuel _____

Other needs _____

_____

_____

Outdoor design conditions:

Summer: dry bulb temp_____ wet bulb temp_____

Winter: dry bulb temp _____ wet bulb temp_____

Wind speed: winter _____ summer_____

Humidity _____

Other _____

Indoor design conditions:

Summer: dry bulb temp _____ humidity_____

Winter: dry bulb temp _____ humidity_____

Type of system _____ Manufacturer _____

Estimated cost _____ Start/finish dates _____

Checklist compiled by _____ Title_____

Sample HVAC checklist
Figure 4-1 (continued)

Record the information in the appropriate space as you get it. If an item isn't needed, place a dash or check mark in the space. This insures that you haven't overlooked any item.

Step 2: Determine Special Needs

After finding your job's comfort range and after reviewing the codes, the next step is determining any special needs. Many of these will depend on the owners' preferences and their budget. For example, an older couple building a retirement home in a northern state would probably want a high winter temperature. They'd also want a system with low maintenance requirements. Another owner may want an especially quiet system. Some owners may request a certain type or brand of equipment. Commercial buildings may need high-efficiency filtration and humidifying. These special needs can all be used as selling points. You can satisfy customer needs as well as increase job profitability. Now's the time to get them talked out and down on the planning list.

Because of the steady increase in energy costs over recent years, energy conservation has become a big concern for most home and business owners. In most cases, low-cost, efficient heating and cooling should be just one part of an overall energy management plan. Your local utility can provide details. Review the plan and then include it as part of your special needs determination. A wide variety of equipment and designs that provide comfort while saving dollars are currently on the market. They'll have good buyer appeal. You'll be doing your customers a real service by suggesting them.

Step 3: Determine Weather Conditions

Now that you know what's needed inside, you'll want to find out what's going on outside. Figure 4-2 lists design conditions for cities in all 50 states. The figures have been computed from data from the National Weather Service and other sources. Although ideally every building owner would like his heating system to handle even the coldest day of any year, such a design wouldn't be economical. Instead, the values in Figure 4-2 are intended to give the best outdoor design temperature to insure economy and comfort during weather conditions that occupy a large part of the season.

The temperatures given are for use with well-built, well-insulated buildings, with an average number of windows. For larger buildings with good thermal capacity and a small number of windows, you may be able to use higher winter and lower summer temperatures. The opposite case — a large number of windows and poor building construction — would mean going in the other direction. Also, you may want temperatures closer to your location. For these and other applications not met by Figure 4-2, a more comprehensive table of outside design conditions is included in Chapter 5. We recommend you review this chapter along with the chapter on weather data and design conditions in the ASHRAE Handbook before making a final decision. Although much of the information in ASHRAE is theoretical, it'll help you to fine-tune your choice.

Other questions also have to be considered. Is the structure exposed to high winds? Is the heating temperature absolutely critical to the occupants, such as in a nursing home? Will the building stand vacant for long periods of time? Answers to questions like these may well indicate that you should raise or lower the design temperature to get the best choice.

Local codes will help you determine local weather conditions. Keep in mind that sometimes design conditions can change drastically over a very short distance. For example, a building one block from the ocean at sea level in California may have very different design conditions than an identical building eight blocks away that's 600 feet above sea level.

If you're not familiar with an area, contact someone who's been doing weather analysis or HVAC design work in that area. This person will probably be able to fill you in on any variables you should know about. The local building inspector can also help.

After you've gathered as much weather information as possible, record the data on your checklist. You'll need this planning data to calculate heat loads and complete the other design work.

Step 4: Review Building Construction

The way a building is built determines how well it gives off and absorbs heat. The *heat loss* and *heat gain* depend on:

- Type of building material

- Air leakage (infiltration)

State	City	Winter Dry bulb	Summer Dry bulb	Summer Wet bulb	Summer mean daily range
Alabama	Mobile	25	95	80	18
Alaska	Fairbanks	-51	82	64	24
Arizona	Phoenix	31	109	76	27
Arkansas	Little Rock	15	99	80	22
California	San Francisco	35	82	65	20
Colorado	Denver	-5	93	64	28
Connecticut	Hartford	3	91	77	22
Delaware	Wilmington	11	92	79	18
District of Columbia	Washington	14	93	78	18
Florida	Fort Lauderdale	42	92	80	15
Georgia	Atlanta	17	94	77	19
Hawaii	Honolulu	62	87	76	12
Idaho	Moscow	-7	90	65	32
Illinois	Chicago	-8	91	77	20
Indiana	Indianapolis	-2	92	78	22
Iowa	Des Moines	-10	94	78	23
Kansas	Salina	0	103	78	26
Kentucky	Louisville	5	95	79	23
Louisiana	Baton Rouge	25	95	80	19
Maine	Portland	-6	87	74	22
Maryland	Baltimore	12	94	78	21
Massachusetts	Boston	6	91	75	16
Michigan	Detroit	3	91	76	20
Minnesota	Minneapolis/St. Paul	-16	92	77	22
Mississippi	Jackson	21	97	79	21
Montana	Livingston	-20	90	63	32
Nebraska	Lincoln	-5	99	78	24
Nevada	Las Vegas	25	108	71	30
New Hampshire	Manchester	-8	91	75	24
New Jersey	Trenton	11	91	78	19
New Mexico	Albuquerque	12	96	66	30
New York	Albany	-6	91	75	23
New York	New York City	11	92	76	17
North Carolina	Greensboro	14	93	77	21
North Dakota	Fargo	-22	92	76	25
Ohio	Toledo	-3	90	76	25
Oklahoma	Muskogee	10	101	79	23
Oregon	Portland	17	89	69	23
Pennsylvania	Pittsburgh	1	91	74	19
Rhode Island	Providence	5	89	75	19
South Carolina	Greenville	18	93	77	21
South Dakota	Pierre	-15	99	75	29
Tennessee	Nashville	9	97	78	21
Texas	Austin	24	100	78	22
Utah	Salt Lake City	3	97	66	32
Vermont	Burlington	-12	88	74	23
Virginia	Richmond	14	95	79	21
Washington	Olympia	16	87	67	32
West Virginia	Charleston	7	92	76	20
Wisconsin	Madison	-11	91	77	22
Wyoming	Casper	-11	92	63	31

Note: Compiled by the authors from U. S. Weather Service data.

Design conditions
Figure 4-2

- Square footage of windows and other glassed areas

- Square footage of exterior walls and roof

- Type and thickness of insulation

- Construction methods

- Interior lighting and building use

Calculating heating and cooling loads is the important first step in system design. The items listed above, along with design temperatures, are plugged into different formulas you'll find in Chapter 5. This building review step in your planning enables you to determine what numerical values will accurately represent the building.

Some values, such as temperature and square footage, are easily found since they're already numerical. Other items, like insulation and other building materials, have been given values depending on their ability to retard or transmit heat. These are called R- and U-values. They'll vary depending on the type of material and how it's put together in the building.

R-value— Technically, the R-value is defined as a unit of thermal resistance. Through extensive testing, numbers have been assigned to building materials which reflect their ability to retard the movement of heat.

U-value— The U-value is the reciprocal of thermal resistance. It's a numerical value used to calculate the amount of heat transferred through the building walls, roof and floors.

Both R- and U-values are determined experimentally and are based on the degrees of temperature lost in Btu's per hour per square foot.

In Chapter 5, you'll find tables listing these values for a wide range of construction conditions. In order to use them accurately, you'll need to know how a building is constructed. Building construction information is also important for estimating and bidding, especially for remodeling jobs. It enables you to spot potential trouble spots or profit opportunities. Most building information can be pulled together during your planning stage.

Figure 4-3 is a typical worksheet for gathering construction details. You may want to change it to suit your own purposes. When it's completed, attach it to your planning checklist.

The field check— Construction worksheets are especially helpful when you're planning a system for an existing building. Begin this type of review by contacting the owner or the architect for a set of original plans. Look them over carefully *but don't rely on them without a field check of the building.* Although plan sets are supposed to be updated to show all building changes, they are notoriously inaccurate. Remodeling and building repairs can result in a building very different from the original.

On bid jobs, this field check, sometimes called the *job-site inspection,* will be required by the bidding rules. Take along your foreman or superintendent, your worksheet, the plan set and a notebook. Don't forget a 100' tape and a good flashlight. Besides verifying the plan set, this field check is a good chance for you to scout out profit-boosting opportunities.

For new work, your construction review will depend primarily on the plans and specifications. Reading plans and specs accurately is a must for any contractor who expects to have a profitable business. Besides needing them for job planning and design, you'll find they're vital for profitable estimating and bidding. Later in this chapter, we'll give a detailed explanation of plans and specs. After reviewing the section, you'll be able to find the construction information you need without any trouble.

Besides relying on plans, specs and field checks for HVAC work, you'll need to develop a knowledge of structural design. Although chiefly the responsibility of the architect, faulty or poor building design or plan errors will affect any HVAC system you install. You can end up looking bad in your customer's eyes when it's really not your fault. On remodeling jobs, knowing structural design will help you estimate accurately how much time and effort will go into the routing of ductwork and piping, no matter how easy it appears on the plan.

The three building types— Building types can be split into three broad groups: concrete/masonry, wood frame, and steel. Some buildings are a combination of all three. It's important on each job that you know the structural integrity of the building — how it supports itself. Common methods are post and beam, steel frame and girder, and reinforced concrete.

Date_____ Reviewer_____

Building Owner _____ Phone_____
    Address _____

Job location _____
Building: new _____ existing _____
    Manager_____ Phone_____
    Plans available? _____ Where?_____
    Field check needed?_____

Architect/project manager _____ Phone_____
    Address _____
    Contact Person _____

Mechanical Designer _____ Phone_____
    Address _____
    Contact Person _____

Plumbing Contractor _____ Phone_____
    Address _____

Electrical Contractor _____ Phone_____
    Address _____

Building size, Sq. ft._____ No. of stories_____
    Type of construction_____

Foundation: Type _____ Heat required?_____
    Sq. ft. below grade _____ Sq. ft. above grade _____
    Type of insulation: _____

Construction review worksheet
Figure 4-3

Exposed walls

North: Length _____ Height _____ Sq. ft. _____

Type_____ Insulation _____

No. of windows _____ Type _____ Sq. ft. of windows _____

East: Length _____ Height _____ Sq. ft. _____

Type_____ Insulation _____

No. of windows _____ Type _____ Sq. ft. of windows _____

South: Length _____ Height _____ Sq. ft. _____

Type_____ Insulation _____

No. of windows _____ Type _____ Sq. ft. of windows _____

West: Length _____ Height _____ Sq. ft. _____

Type_____ Insulation _____

No. of windows _____ Type _____ Sq. ft. of windows _____

Roof: Length _____ Width _____ Pitch _____

Sq. ft. _____ Type _____

Insulation_____ Sq. ft. of glass_____

Bearing walls and columns:

Locations _____

Type of construction _____

No. of field sketches attached _____

Notes _____

_____

_____

_____

Construction review worksheet
Figure 4-3 (continued)

All buildings will have *bearing walls,* walls which support all or parts of the building. Architects will sometimes limit the size of openings in the bearing walls and in key beams and girders. That's an easy item to overlook. Your job as the contractor is to spot locations where maintaining the structural integrity of the building may give you installation problems. These should be talked over and worked out with the architect. Also, make sure your bid and estimate reflect the added work necessary to compensate for these problems.

Another trouble spot may be a lack of space for your ductwork or piping, or a conflict with wiring or piping to be installed by other trades. The architect is supposed to avoid these problems when drawing up the plan. But often he'll design the building first on an appearance basis. Then he'll try to fit in the shafts and crawl spaces for the mechanical systems wherever he can. You may have to fight to get your fair share.

Watch for rooms or locations that may need special equipment to deal with temperatures and humidity above or below normal operating ranges. Although usually noted by the architect on the plans and specs, he may overlook a room or assume that you'll handle it. These and other conditions can affect your job cost. If the architectural plans call for hot air ducts to run through an un-insulated area, for example, you'll have to install insulated ducts to get maximum efficiency. In the chapter on Estimating and Bidding, we'll talk about your options when a situation like this comes up on bid work. If you're competing for the work one-on-one with the potential buyer, point out these situations and explain how your solution is cost-effective. At the same time, you'll be showing him that you really care about the quality of the installation.

Finally, check the plan and specs to make sure that the building is well-built and has enough insulation. It's a touchy situation, but if the building design is substandard, you may want to call it to the owner's attention, or pass the job up. You can compensate for poor building design somewhat by installing an oversize heating plant. But it's a risky business. The heating plant will be costly to operate and probably never resolve the problem completely. You may be able to prove to the owner somewhere down the road that his high fuel bills really aren't your fault but it won't mean much to him. Especially if someone you're trying to sell calls and asks him how well he likes his HVAC system. The same thing is true for remodeling work. If a building owner expects a new furnace to keep him warm economically, but his house walls have old insulation with an R-factor of 4 (most new houses have an R of at least 13), he's asking for more than you can deliver. Instead, tell him that you'll draw up an overall energy management plan for him to review. If you don't want to do this type of work, suggest instead that both of you work with an energy expert to insure that he's satisfied.

Utility requirements— The final part of your building review will cover utility requirements. Utility companies providing gas, electricity and water have rules for service and installation based on their needs. These requirements will vary with each utility. Most of the companies have books outlining their rules. They'll also have service representatives for different areas. Note the service rep's name and phone number on your checklist, along with any special rules or prohibitions that may apply to your job.

Put the local building inspectors' names and phone numbers on the checklist, too. Be sure you understand code provisions and inspection times. Many inspection offices maintain a mailing list of contractors in their area and try to keep them posted about code changes. But the bottom line on all code compliance is that it's up to you, the contractor, to know and follow the code.

Step 5: Select the System
By now you've gathered enough information to choose the best system for the job. You know utility restrictions, owner's preferences, special needs, weather conditions, the type of building and how it will be used. If it's a custom job and you've done this kind of work before, you may feel confident enough at this point to write up an estimate and submit it to the owner along with a contract. On jobs where you're doing the design, or on bid work, you'll need to complete the design or bid before making the final proposal.

Plans and Specifications
Plans and specifications are the language of the contracting business. If you expect to do any volume of work at all, you'll have to be able to interpret and understand this language. As a journeyman or apprentice you've already followed

many sets of plans. But reading plans as a worker and reading them as a contractor are two different things. As an installer, the plans told you how the system went together. As the contractor, they'll tell you much more.

You'll get most of your work through successful estimating and bidding. For this you'll rely on plans and specifications. After you get the work, much of the job profit will be determined by how well you interpreted those plans in drawing up your estimate and bid. Bid a job too high and someone else will get the work. Bid it too low and you'll *wish* someone else had the work. Either way you won't come out ahead. Chapter 11 gives you a solid foundation for successful bidding and estimating. But it will all depend on how well you read the plans and specs.

Plan Sets

The chief purpose of any set of building plans is to relay information to the reader without confusion or misunderstanding. Since plans for a large job can easily be over 50 pages long, industry-wide standards have been set to insure clarity and order. By following these standards, even for smaller jobs, you mark yourself as a professional who knows his business.

The title page— A typical plan set will begin with a title page. This page can be simple or elaborate, depending on who's drawing the plan set. The title page contains the project name, location, and the names of the owner, architect and design firms. It can also list where the project was financed, the name of the general contractor and the subcontractors, and a directory of the plan set. Sometimes an architectural drawing of the proposed project is included to show how it will look when completed.

Each page of the plan set will have a *title block*. Like the title page, these blocks can be simple or elaborate. They should include the design firm's name and address, date, draftsperson's initials, plan approver's initials, and the page number. Page numbers are sometimes prefaced by the initial of the type of drawing. For example, the site plan would be S-1 (*S* stands for *site*).

The site plan— Next is the site plan, based on information gathered by a professional surveyor or engineer. It locates the building in relation to compass direction, streets, geography and property lines. It shows utility lines such as power lines and poles, sanitary sewers and the like. Sometimes this sheet will also show a landscape plan for the project.

The architectural drawings— After the site plan come the architectural drawings. They're the basis from which all other drawings are taken. They are numbered A-1, A-2, A-3, etc. They'll include *elevations,* drawings which show the building as viewed from the front, side and back, and *floor plans* showing the building as viewed from above.

Architectural drawings give dimensions, and the scale to which they're drawn is shown beneath the drawing's title. Sometimes an architect will change a dimension but not change the scale of the drawing. If you find this, check with the architect for the correct dimension. It'll be important later when you're scaling your HVAC design plans.

After the architectural drawings, you'll find structural plans or any special plans needed for the job.

The mechanical system plans— These follow next. The plumbing plans are first, HVAC second, and electrical third. Plans for fire protection systems and the like come last. Plumbing will be paged P-1, P-2, P-3; HVAC will be H-1, H-2, H-3 and so on. All mechanical system drawings should follow the plumbing, heating, electrical sequence — it's standard industry practice.

Plan sets can be simple or complex. The key rule is clarity. Each plan set should have a minimum of a title page, a separate sheet for architectural design, and one for each of the mechanical systems. To combine, for example, heating with plumbing will give you a plan that's hard to read, but easy to misread. There are exceptions, such as a home with all electric heat. Then the heating and electrical plans may be all on one sheet.

Floor plans and elevations give a large view of the work. Other drawings are needed for more details. *Diagrams,* sometimes called *isometrics* or *schematics,* are used for equipment, piping or ductwork. Figure 4-4 shows a typical example. These drawings aren't drawn to scale. Instead they give an approximate order or layout. They're meant to show at a glance the components needed and a basic idea of how they're connected. It's up to the contractor to lay out the installation neatly, profitably *and* according to code.

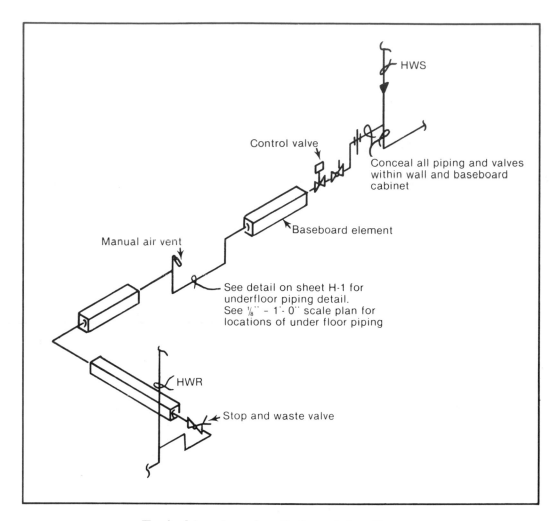

Typical baseboard radiation piping diagram
Figure 4-4

A *schedule* is a list which brings together in one convenient location information needed to understand the plan. A *symbol schedule* (Figure 4-5) identifies plan symbols. Although attempts have been made to standardize symbols for each trade, they still vary. Always check the symbol schedule of the plan you're currently reading to avoid mistakes.

An *equipment schedule* identifies a piece of equipment for a particular location. It describes the equipment and codes it by number or letter to a location on the floor plan (Figure 4-6).

Many plans will have *details*. These are enlarged drawings, drawn to scale, to explain a specific location on the plan. Details are located on the plan by number and/or letter coding, or a circle targeting the enlarged area.

A *section* is a detail that shows a building, installation or piece of equipment as if it's been cut in half from the side. See Figure 4-7.

As a rule, all details, schedules and diagrams should be on the same page as the floor plan that uses them. But with larger jobs, there may not be enough room. Then it's best to group them on a separate sheet rather than end up with a cluttered page that's too hard to read.

Drawing Plans
Drawing plans isn't too difficult if you have the right tools and methods. You might want to take a drafting course at your local technical school if you haven't done so already. If you do discover that drafting just isn't your strong point, don't despair.

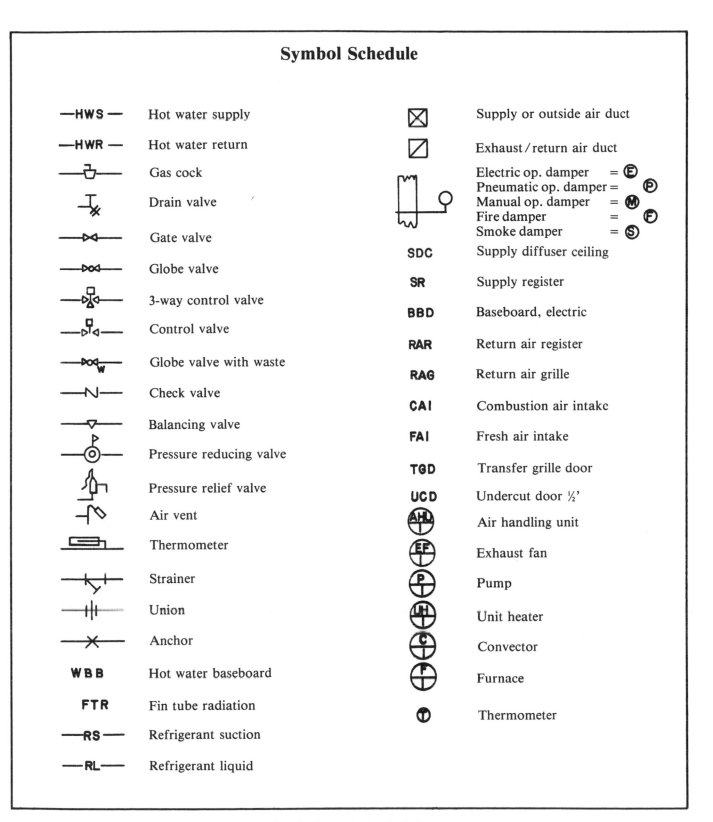

Symbol Schedule

—HWS—	Hot water supply		⊠	Supply or outside air duct
—HWR—	Hot water return		⊘	Exhaust / return air duct
	Gas cock			Electric op. damper = Ⓔ
	Drain valve			Pneumatic op. damper = Ⓟ
	Gate valve			Manual op. damper = Ⓜ
	Globe valve			Fire damper = Ⓕ
	3-way control valve			Smoke damper = Ⓢ
	Control valve		SDC	Supply diffuser ceiling
	Globe valve with waste		SR	Supply register
	Check valve		BBD	Baseboard, electric
	Balancing valve		RAR	Return air register
	Pressure reducing valve		RAG	Return air grille
	Pressure relief valve		CAI	Combustion air intake
	Air vent		FAI	Fresh air intake
	Thermometer		TGD	Transfer grille door
	Strainer		UCD	Undercut door ½'
	Union		AHU	Air handling unit
	Anchor		EF	Exhaust fan
WBB	Hot water baseboard		P	Pump
FTR	Fin tube radiation		UH	Unit heater
—RS—	Refrigerant suction		C	Convector
—RL—	Refrigerant liquid		F	Furnace
			T	Thermometer

Typical symbol schedule
Figure 4-5

Furnace Schedule

No.	CFM	External S.P. inches water	Fan size In. Dia.	In. Width	Motor HP	Drive	Heating Input MBH	Output MBH	Filters	Remarks
1	780	0.7	10	6	⅓	Direct 4 spd	48	39	Disposable	*

*Provide accessible filter rack in duct for 16 x 25 x 1 filter.

Exhaust Fan Schedule

No.	Location	CFM	S. P. in. water	HP	RPM	Max. fan tip speed FPM	Drive type	Starter	Remarks
1	Toilet	60	0.1	1/10	1550	4000	Direct	SW w/lites	
2	Range hood	200	0.1	1/10	1550	4000	Direct	SW on hood	
3	Bath	70	0.1	1/10	1550	4000	Direct	Wall SW	*

*Furnish w/250 watt R40 infrared heat lamp

Typical equipment schedules
Figure 4-6

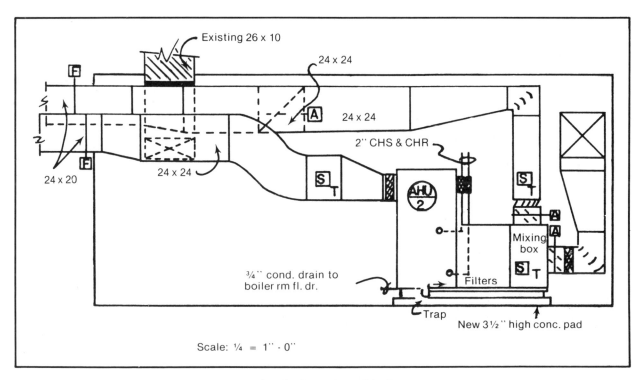

Typical section detail
Figure 4-7

It's the kind of skill that many people rely on for part-time income. You can always do the calculations and rough sketches of your design work. Then hire someone to do the drafting on an as-needed basis.

But doing your own drafting means less chasing around, and you can work according to your own schedule. It's also an excellent way of finding out just what makes up a good plan and a bad one.

Supplies and equipment— To begin your plan sets, you'll need large sheets of semi-transparent drawing paper. Sometimes the sheets are factory marked with a non-photo-reproducing blue grid, in a 1/8'' or 1/4'' scale. Draw on these sheets with special dark pencil leads or black ink. The finished drawing, called a tracing, is sent through a printing machine along with a sheet of coated paper. Light triggers a chemical reaction which causes the drawing to be reproduced on the treated sheet, which becomes the print or plan. The tracing is unharmed and stays with the architect or designer who drew it. These tracings are kept on file so additional plan sets can be run off as needed.

Some designers prefer to draw plans on a plastic sheet called mylar. Using plastic takes a special lead or ink, and requires a bit more care to avoid smearing. But it produces a fine, professional-looking print when reproduced. The mylar tracing is more durable than paper.

If you do any amount of design work, you'll want to have your own printing machine. But if you need only an occasional plan set, local blueprint shops can reproduce your tracings quickly and cheaply.

You can get the supplies you need at your local art store or blueprint shop. Tracing paper is available in different size sheets and rolls, in 1/8'' or 1/4'' scale grids. You'll need a couple of drafting pencils and leads of different hardness. A pencil sharpener, erasers, erasing shields, different size triangles and French curves will handle basic drawing. You should also get templates for different size and shape circles, triangles, squares and the like. Check with the store clerk and go through the catalogs to see what templates and supplies are available for HVAC work that the store doesn't carry.

If you'll do any drawing at all, you should have a drafting table, or at the very least, a drawing board and a T-square. A vinyl drawing board cover reduces slickness and hardness and makes drawing easier. Use masking tape to hold down the corners of the drawings.

In order to draw HVAC plans, you'll start with a set of architectural plans. You can get these from the architect or owner. You'll also need a set of floor plans of the building without the labels and dimensions found on the architectural drawings. While the architect may be willing to supply these *blanks,* he may charge you. You can trace your own set using a light table.

Making a light table— You can buy a light table, or you can easily make your own, following the plan in Figure 4-8. Take a piece of AC plywood or smooth particle board, 1/2'' x 36'' x 60''. Make sure the 36'' sides are smooth and free from burrs since your T-square will slide along them. Mount it on four legs. Two of the legs on a long side should be 36'' high. This will be the front of the table. The back legs are 44''. If you care to take the time, it's a good idea to have the legs adjustable to get the most comfortable height and slant to the work surface.

After mounting, cut an opening in the top slightly smaller than 24'' x 36'', about six inches from the front edge of the table. Pick up a piece of translucent glass, 1/4'' x 24'' x 36'', from your nearest glass shop. You can also use clear glass and spray it with a translucent paint on the underside. Rout the edges of the opening so the glass will sit flush with the table top.

Your light is a 4-lamp, 4-foot fluorescent fixture. Use hooks and hardware chain to suspend it upside down beneath the glass. You can adjust the light intensity to suit your needs by adding or removing bulbs and by moving the fixture on its chain.

Tracing a floor plan— To use the light table, take the architectural floor plan and center it *upside down* on the glass. Use your T-square and a long horizontal building line to align it. Fasten the corners with masking tape. Next place a same-size sheet of tracing paper over the architectural drawing, with the blue grid side up. This is also upside down and you'll know why in a minute. Align the tracing using a horizontal grid line so it matches the floor plan and your T-square. Tape the corners and you're all set.

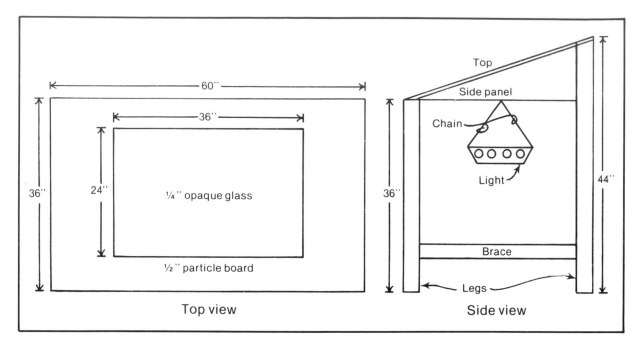

Light table plan
Figure 4-8

Using your T-square, triangles and templates, you'll be able to trace the floor plan lines. Most draftspersons prefer to use a hard lead like No. 3H, since you don't really need a heavy line. But don't push too hard on the pencil or you may cut the tracing, especially right after you've sharpened the lead.

Since the lettering on the architectural plan is upside down, you won't be able to read it. Also, some parts of the plan may not be clear since you're working through two thicknesses of paper. But don't worry. Anything you have to leave out you can add after you've retrieved the architectural drawing from beneath the tracing. The main thing you want to get is the structural outline of the floor plan.

You may discover, when you're tracing, that the architect's lines aren't quite straight. And if you leave the tracing and come back at another time, your lines may be off slightly. This is because different humidities and temperatures will cause the paper to shrink or stretch just a little. Do your best to adjust your drawing. But don't labor over it too much since small variations won't affect the overall scale of your drawing.

For large drawings, it may be necessary to do more than one layout on the table.

When you've finished tracing the floor plan, take your tracing sheet and put it on your drawing table with the grid side down. Because you did your floor plan upside down, it will now appear right side up, but on the *underside* of the sheet. This is correct. By having the floor plan on the underside of your sheet, your HVAC drawings will be on the top side. When you have to erase, you'll erase only your HVAC work. The floor plan will stay nicely intact.

Starting the drawing— Align your tracing on the drawing table, using your T-square. Fasten the corners with tape. Now you're ready to begin drawing in the HVAC system.

Here's a tip that can save you a lot of frustration. The more you handle and erase your tracings, the messier they'll get. A drafting supply store can sell you powder to sprinkle on the tracings to clean them up. But rather than starting to draw on the tracing right away, have a couple of sets of prints run off. Draw, erase and revise on these prints until you're satisfied with your design. *Then* draw it on your tracing.

You'll have to do your design work before you begin drawing the system on the floor plan.

Chapter 5 outlines the proper sequence. Basically, you'll begin with outlets and returns, the heating plant, and then the ductwork or piping. Although you'll label duct sizes and similar information on your HVAC plans, it's also good practice to draw them to the plan scale. This keeps the plan in proportion. Runs of ductwork all the same size are an exception. In this case, common practice is to use a single line to indicate where the duct runs. Typically, many dimensions that locate HVAC are left off the plans. Instead, you draw in the desired symbol using your scale ruler. This is usually sufficient for the installer to determine the location, since he'll use his ruler and scale off the plan. Ductwork, piping and equipment that's concealed is indicated by a dashed line. Mounting heights and like information can be placed in notes on the plans or in the specs. But when you begin to place items like outlets and ducts, study the architectural plans and think in three dimensions. If you're in doubt whether a certain location is suitable, check with the architect.

After finishing the basic HVAC floor plan, you'll need to do the details and diagrams. Any point on your plan that's unclear or could be misunderstood should be clarified with a detail, diagram, label or note.

At this point you should tidy up your tracings and run another set of prints. Study them carefully for mistakes and omissions. Make your corrections on the prints in red ink. When you transfer these to your tracing, cross them out in a different color ink. That way you won't miss any. Work in a clockwise motion, starting in the upper left corner.

After you're satisfied with your design, fill in the equipment and other schedules. Do the title block and any other details. Run a final set of prints and once more check these carefully. Make sure all the drawings are titled and scaled. Check that the coding between the floor plan and your details and diagrams is correct. Make sure you haven't left anything out. Check that all ductwork and equipment is sized and labeled.

The key to drawing HVAC design is to work slowly and carefully, especially on your first jobs. Run a set of prints and check your work often. It's best to start with simpler jobs before tackling a big one. On smaller jobs you may find you can divide your work up neatly into planning, designing and drawing. On more complicated work, your work order may be determined by how the information comes in. You may find yourself doing planning, designing and drawing all at the same time. Use this method only if you have to, since it's very easy to lose something in the shuffle. Instead, try to develop an orderly sequence and stick to it as much as possible for every job. This will help you avoid mistakes and omissions. Use the checklist you'll find in Chapter 6 to check your work.

Chapter 5 also has floor plans detailing typical HVAC installations. Study these carefully. You should also study all the plan sets and specifications you can get your hands on, from beginning to end. Plan sets for which you did the installation will be especially useful for studying. Then you can remember and visualize how those plans translated into the actual work. If you find a plan set that's good, ask yourself why. If you find one that doesn't do the job, analyze it to find out its shortcomings. As a contractor, you'll never know too much about plans and specifications.

Specifications

A plan set alone won't tell you everything you need to know about a job. When you do work that's been designed by a professional engineer or architect, you'll also be given a set of specifications for the job. You may also want to write your own specs for jobs where you're responsible for design work.

As the contracting business has become more complicated, more and more owners rely on specs to have the final word on job conditions, installations, and equipment. The Construction Specifications Institute (CSI) has drawn up 16 divisions for specifications covering all parts of the construction process. See Figure 4-9.

Specifications will cover both the technical and business portions of your contract with another contractor or the owner. The specs will tell you, for example, where to bid a job, bonds required, project meetings, job site conditions, payment schedules and many other business details. They'll also tell you what brand of equipment must be installed, mounting heights and locations, certain installation procedures, acceptable materials and other items which you'll accept as part of your obligations when you sign the contract.

Specifications are a binding part of any job contract and you must follow them. Chapter 11 has more on following specs. Here we're concerned with using them as a planning and design tool.

DIVISION 0 – BIDDING AND CONTRACT REQUIREMENTS

00010	PRE-BID INFORMATION
00100	INSTRUCTIONS TO BIDDERS
00200	INFORMATION AVAILABLE TO BIDDERS
00300	BID/TENDER FORMS
00400	SUPPLEMENTS TO BID/TENDER FORMS
00500	AGREEMENT FORMS
00600	BONDS AND CERTIFICATES
00700	GENERAL CONDITIONS OF THE CONTRACT
00800	SUPPLEMENTARY CONDITIONS
00950	DRAWINGS INDEX
00900	ADDENDA AND MODIFICATIONS

DIVISION 1 – GENERAL REQUIRMENTS

01010	SUMMARY OF WORK
01020	ALLOWANCES
01030	SPECIAL PROJECT PROCEDURES
01040	COORDINATION
01050	FIELD ENGINEERING
01060	REGULATORY REQUIRMENTS
01070	ABBREVIATIONS AND SYMBOLS
01080	IDENTIFICATION SYSTEMS
01100	ALTERNATES/ALTERNATIVES
01150	MEASUREMENT AND PAYMENT
01200	PROJECT MEETINGS
01300	SUBMITTALS
01400	QUALITY CONTROL
01500	CONSTRUCTION FACILITIES AND TEMPORARY CONTROLS
01600	MATERIAL AND EQUIPMENT
01650	STARTING OF SYSTEMS
01660	TESTING, ADJUSTING, AND BALANCING OF SYSTEMS
01700	CONTRACT CLOSEOUT

DIVISION 2 – SITEWORK

02010	SUBSURFACE INVESTIGATION
02050	DEMOLITION
02100	SITE PREPARATION
02150	UNDERPINNING
02200	EARTHWORK
02300	TUNNELLING
02350	PILES, CAISSONS AND COFFERDAMS
02400	DRAINAGE
02440	SITE IMPROVEMENTS
02480	LANDSCAPING
02500	PAVING AND SURFACING
02590	PONDS AND RESERVOIRS
02600	PIPED UTILITY MATERIALS AND METHODS
02700	PIPED UTILITIES
02800	POWER AND COMMUNICATION UTILITIES
02850	RAILROAD WORK
02880	MARINE WORK

DIVISION 3 – CONCRETE

03050	CONCRETING PROCEDURES
03100	CONCRETE FORMWORK
03150	FORMS
03180	FORM TIES AND ACCESSORIES
03200	CONCRETE REINFORCEMENT
03250	CONCRETE ACCESSORIES
03300	CAST-IN-PLACE CONCRETE
03350	SPECIAL CONCRETE FINISHES
03360	SPECIALLY PLACED CONCRETE
03370	CONCRETE CURING
03400	PRECAST CONCRETE
03500	CEMENTITIOUS DECKS
03600	GROUT
03700	CONCRETE RESTORATION AND CLEANING

DIVISION 4 – MASONRY

04050	MASONRY PROCEDURES
04100	MORTAR
04150	MASONRY ACCESSORIES
04200	UNIT MASONRY
04400	STONE
04500	MASONRY RESTORATION AND CLEANING
04550	REFRACTORIES
04600	CORROSION RESISTANT MASONRY

DIVISION 5 – METALS

05010	METAL MATERIALS AND METHODS
05050	METAL FASTENING
05100	STRUCTURAL METAL FRAMING
05200	METAL JOISTS
05300	METAL DECKING
05400	COLD-FORMED METAL FRAMING
05500	METAL FABRICATIONS
05700	ORNAMENTAL METAL
05800	EXPANSION CONTROL
05900	METAL FINISHES

DIVISION 6 – WOOD AND PLASTICS

06050	FASTENERS AND SUPPORTS
06100	ROUGH CARPENTRY
06130	HEAVY TIMBER CONSTRUCTION
06150	WOOD-METAL SYSTEMS
06170	PREFABRICATED STRUCTURAL WOOD
06200	FINISH CARPENTRY
06300	WOOD TREATMENT
06400	ARCHITECTURAL WOODWORK
06500	PREFABRICATED STRUCTURAL PLASTICS
06600	PLASTIC FABRICATIONS

DIVISION 7 – THERMAL AND MOISTURE PROTECTION

07100	WATERPROOFING
07150	DAMPPROOFING
07200	INSULATION
07250	FIREPROOFING
07300	SHINGLES AND ROOFING TILES
07400	PREFORMED ROOFING AND SIDING
07500	MEMBRANE ROOFING
07570	TRAFFIC TOPPING
07600	FLASHING AND SHEET METAL
07800	ROOF ACCESSORIES
07900	JOINT SEALANTS

DIVISION 8 – DOORS AND WINDOWS

08100	METAL DOORS AND FRAMES
08200	WOOD AND PLASTIC DOORS
08250	DOOR OPENING ASSEMBLIES
08300	SPECIAL DOORS
08400	ENTRANCES AND STOREFRONTS
08500	METAL WINDOWS
08600	WOOD AND PLASTIC WINDOWS
08650	SPECIAL WINDOWS
08700	HARDWARE
08800	GLAZING
08900	GLAZED CURTAIN WALLS

DIVISION 9 – FINISHES

09100	METAL SUPPORT SYSTEMS
09200	LATH AND PLASTER
09230	AGGREGATE COATINGS
09250	GYPSUM WALLBOARD
09300	TILE
09400	TERRAZZO
09500	ACOUSTICAL TREATMENT
09550	WOOD FLOORING
09600	STONE AND BRICK FLOORING
09650	RESILIENT FLOORING
09680	CARPETING
09700	SPECIAL FLOORING
09760	FLOOR TREATMENT
09800	SPECIAL COATINGS
09900	PAINTING
09950	WALL COVERING

CSI specifications index
Figure 4-9

DIVISION 10 - SPECIALITIES

10100	CHALKBOARDS AND TACKBOARDS
10150	COMPARTMENTS AND CUBICLES
10200	LOUVERS AND VENTS
10240	GRILLES AND SCREENS
10250	SERVICE WALL SYSTEMS
10260	WALL AND CORNER GUARDS
10270	ACCESS FLOORING
10280	SPECIALTY MODULES
10290	PEST CONTROL
10300	FIREPLACES AND STOVES
10340	PREFABRICATED STEEPLES, SPIRES, AND CUPOLAS
10350	FLAGPOLES
10400	IDENTIFYING DEVICES
10450	PEDESTRIAN CONTROL DEVICES
10500	LOCKERS
10520	FIRE EXTINGUISHERS, CABINETS, AND ACCESSORIES
10530	PROTECTIVE COVERS
10550	POSTAL SPECIALTIES
10600	PARTITIONS
10650	SCALES
10670	STORAGE SHELVING
10700	EXTERIOR SUN CONTROL DEVICES
10750	TELEPHONE ENCLOSURES
10800	TOILET AND BATH ACCESSORIES
10900	WARDROBE SPECIALTIES

DIVISION 11 - EQUIPMENT

11010	MAINTENANCE EQUIPMENT
11020	SECURITY AND VAULT EQUIPMENT
11030	CHECKROOM EQUIPMENT
11040	ECCLESIASTICAL EQUIPMENT
11050	LIBRARY EQUIPMENT
11060	THEATER AND STAGE EQUIPMENT
11070	MUSICAL EQUIPMENT
11080	REGISTRATION EQUIPMENT
11100	MERCANTILE EQUIPMENT
11110	COMMERCIAL LAUNDRY AND DRY CLEANING EQUIPMENT
11120	VENDING EQUIPMENT
11130	AUDIO-VISUAL EQUIPMENT
11140	SERVICE STATION EQUIPMENT
11150	PARKING EQUIPMENT
11160	LOADING DOCK EQUIPMENT
11170	WASTE HANDLING EQUIPMENT
11190	DETENTION EQUIPMENT
11200	WATER SUPPLY AND TREATMENT EQUIPMENT
11300	FLUID WASTE DISPOSAL AND TREATMENT EQUIPMENT
11400	FOOD SERVICE EQUIPMENT
11450	RESIDENTIAL EQUIPMENT
11460	UNIT KITCHENS
11470	DARKROOM EQUIPMENT
11480	ATHLETIC, RECREATIONAL, AND THERAPEUTIC EQUIPMENT
11500	INDUSTRIAL AND PROCESS EQUIPMENT
11600	LABORATORY EQUIPMENT
11650	PLANETARIUM AND OBSERVATORY EQUIPMENT
11700	MEDICAL EQUIPMENT
11780	MORTUARY EQUIPMENT
11800	TELECOMMUNICATION EQUIPMENT
11850	NAVIGATION EQUIPMENT

DIVISION 12 - FURNISHINGS

12100	ARTWORK
12300	MANUFACTURED CABINETS AND CASEWORK
12500	WINDOW TREATMENT
12550	FABRICS
12600	FURNITURE AND ACCESSORIES
12670	RUGS AND MATS
12700	MULTIPLE SEATING
12800	INTERIOR PLANTS AND PLANTINGS

DIVISION 13 - SPECIAL CONSTRUCTION

13010	AIR SUPPORTED STRUCTURES
13020	INTEGRATED ASSEMBLIES
13030	AUDIOMETRIC ROOMS
13040	CLEAN ROOMS
13050	HYPERBARIC ROOMS
13060	INSULATED ROOMS
13070	INTEGRATED CEILINGS
13080	SOUND, VIBRATION, AND SEISMIC CONTROL
13090	RADIATION PROTECTION
13100	NUCLEAR REACTORS
13110	OBSERVATORIES
13120	PRE-ENGINEERED STRUCTURES
13130	SPECIAL PURPOSE ROOMS AND BUILDINGS
13140	VAULTS
13150	POOLS
13160	ICE RINKS
13170	KENNELS AND ANIMAL SHELTERS
13200	SEISMOGRAPHIC INSTRUMENTATION
13210	STRESS RECORDING INSTRUMENTATION
13220	SOLAR AND WIND INSTRUMENTATION
13410	LIQUID AND GAS STORAGE TANKS
13510	RESTORATION OF UNDERGROUND PIPELINES
13520	FILTER UNDERDRAINS AND MEDIA
13530	DIGESTION TANK COVERS AND APPURTENANCES
13540	OXYGENATION SYSTEMS
13550	THERMAL SLUDGE CONDITIONING SYSTEMS
13560	SITE CONSTRUCTED INCINERATORS
13600	UTILITY CONTROL SYSTEMS
13700	INDUSTRIAL AND PROCESS CONTROL SYSTEMS
13800	OIL AND GAS REFINING INSTALLATIONS AND CONTROL SYSTEMS
13900	TRANSPORTATION INSTRUMENTATION
13940	BUILDING AUTOMATION SYSTEMS
13970	FIRE SUPPRESSION AND SUPERVISORY SYSTEMS
13980	SOLAR ENERGY SYSTEMS
13990	WIND ENERGY SYSTEMS

DIVISION 14 - CONVEYING SYSTEMS

14100	DUMBWAITERS
14200	ELEVATORS
14300	HOISTS AND CRANES
14400	LIFTS
14500	MATERIAL HANDLING SYSTEMS
14600	TURNTABLES
14700	MOVING STAIRS AND WALKS
14800	POWERED SCAFFOLDING
14900	TRANSPORTATION SYSTEMS

DIVISION 15 - MECHANICAL

15050	BASIC MATERIALS AND METHODS
15200	NOISE, VIBRATION, AND SEISMIC CONTROL
15250	INSULATION
15300	SPECIAL PIPING SYSTEMS
15400	PLUMBING SYSTEMS
15450	PLUMBING FIXTURES AND TRIM
15500	FIRE PROTECTION
15600	POWER OR HEAT GENERATION
15650	REFRIGERATION
15700	LIQUID HEAT TRANSFER
15800	AIR DISTRIBUTION
15900	CONTROLS AND INSTRUMENTATION

DIVISION 16 - ELECTRICAL

16050	BASIC MATERIALS AND METHODS
16200	POWER GENERATION
16300	POWER TRANSMISSION
16400	SERVICE AND DISTRIBUTION
16500	LIGHTING
16600	SPECIAL SYSTEMS
16700	COMMUNICATIONS
16850	HEATING AND COOLING
16900	CONTROLS AND INSTRUMENTATION

CSI specifications index
Figure 4-9 (continued)

Study them carefully. When you find spec requirements that are not shown on the plan, make a note on the plan. It's a good idea to sit down with your job foreman or superintendent and go over them together to avoid wrong installations.

Don't check only HVAC specs and assume you're finished. Go through the specs that deal with the general obligations and also the other trades. This will pay off. Often some HVAC work will be written up in the other trade sections if it applies to their work. But it's still your responsibility. If you don't catch it in the specs, you won't include it in your bid. It'll be a rude surprise when you first learn about it on the job, where you'll have to cover the cost out of your own pocket.

Specifications are a control to insure the quality of the equipment and work done on the job. Information in the specs should not conflict with the plan set and vice versa. If you find a conflict, you might be able to use it to your advantage. Chapter 11 describes how to deal with this situation.

Beginning the HVAC System Design

After you've finished the planning for your HVAC project, you'll use the data you've gathered for the design phase. The laws of physics and mathematics will determine the amount of heat or cooling required for each room, and the building. But you don't need a college-level math course. The math you need is the kind you use in your day-to-day work: adding, subtracting, multiplying, dividing and working simple equations.

Figures 5-1 through 5-5 are actual plans for the residence we'll use for our design work. It's a two-bedroom residence, half of a two-unit duplex. These plans have been scaled down to fit in this book. Find a similar set of architectural plans sized to actual scale. Use them to draw a set of tracings, as explained in Chapter 4. Run six sets of prints and work through each design step as we do ours for the sample residence.

After figuring heating and cooling loads in this chapter, we'll do a warm air system design in Chapter 6. In Chapter 7, we'll do a hot water system for the same residence. So you'll need two sets of tracings to draw each system and run your plan sets.

Residential Heating and Cooling Loads
For every system, the first step is to calculate the heat loss for each room. For cooling, we'll also need the heat load or gain. With this information, we can size the HVAC system and equipment. We'll follow these steps:

1) Select design conditions
2) Calculate or select U-values
3) Take off building dimensions
4) Calculate heating load factors
5) Calculate cooling load factors
6) Calculate infiltration factors
7) Calculate duct heat loss and gain
8) Summarize work sheets

This may sound like a lot of figuring, but it really isn't too bad. We'll use Figure 5-6 as our worksheet and Figure 5-7 as our calculations sheet. Figures 5-8 and 5-9 are these same sheets filled in with the data from the sample residence. Figure 5-10 lists weather conditions for the United States. Figure 5-11 is a table summarizing the formulas we'll need. They'll be explained in detail as we use them. Figures 5-12 through 5-16 are tables of values which we'll use in the calculations. Look these figures over before we start so you'll be able to find the information quickly when you need it.

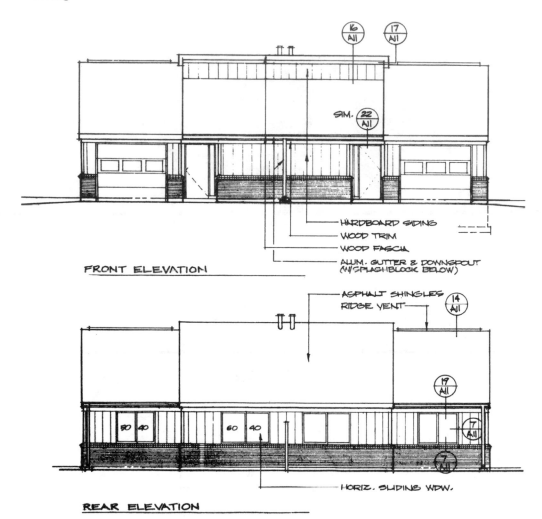

FRONT ELEVATION

HARDBOARD SIDING
WOOD TRIM
WOOD FASCIA
ALUM. GUTTER & DOWNSPOUT
(W/SPLASHBLOCK BELOW)

ASPHALT SHINGLES
RIDGE VENT

HORIZ. SLIDING WDW.

REAR ELEVATION

Front and rear elevations
Figure 5-1

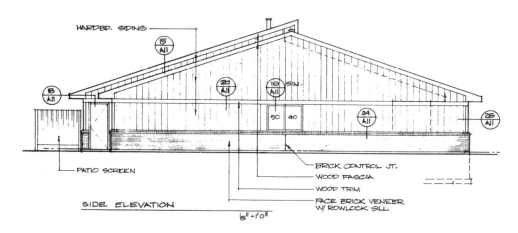

HARDBD. SIDING

PATIO SCREEN

SIDE ELEVATION

6" = 1'0"

BRICK CONTROL JT.
WOOD FASCIA
WOOD TRIM
FACE BRICK VENEER
W/ ROWLOCK SILL

Side elevation
Figure 5-2

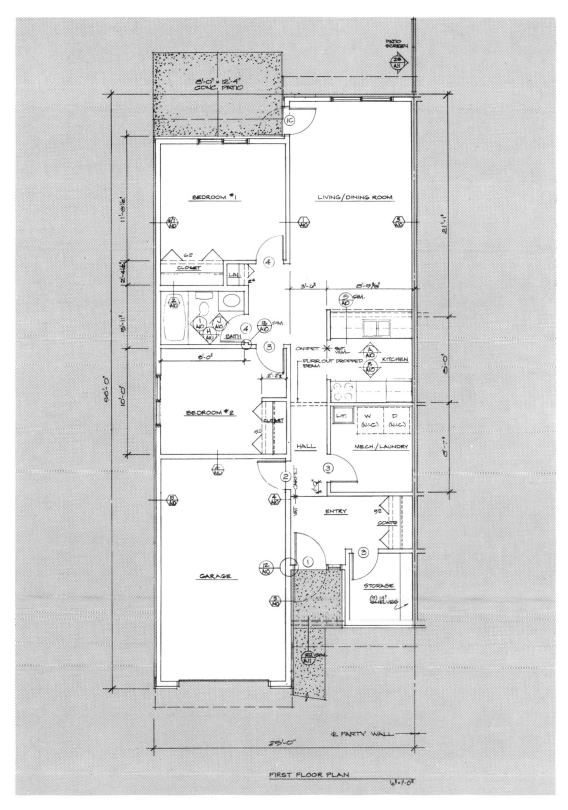

Floor plan
Figure 5-3

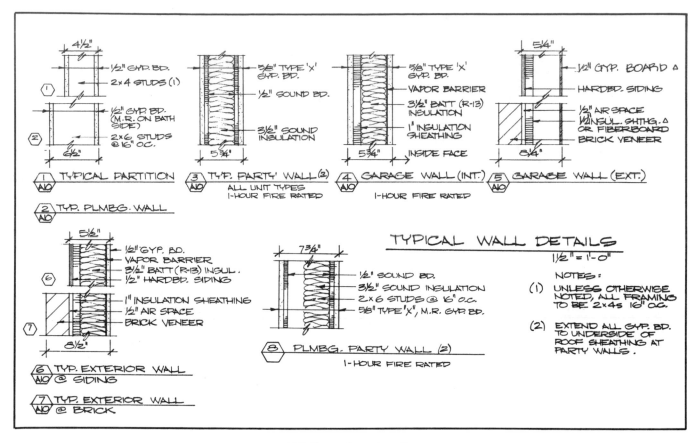

Wall section details
Figure 5-4

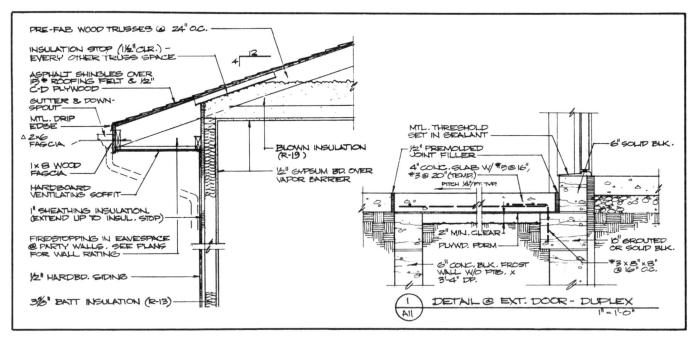

Roof and slab section details
Figure 5-5

RESIDENTIAL COOLING & HEATING
LOAD CALCULATIONS

Prepared For _____ Job No. _____ Date _____

Prepared by _____

Location _____

Latitude _____

Summary of Load Calculations for Entire House

Cooling	Heating	
		Btu/hr
_____ Btu/hr Sensible	_____ Btu/hr · F	
_____ Btu/hr Total	_____ Btu/hr · ft²	
_____ Tons	_____ ft²/ton	

Design Conditions

	Summer	Winter
Outside Temp. db, F		
Inside Temp. db, F		
Difference		

Daily Range, Deg. F _____

Temperature Swing _____ Deg. F.

Window Schedule

Type	Description	
	of Window	of Shades & Shading
A		
B		
C		
D		

Exterior Shading Calculation

Type				
Orientation				
Shade Line Factor				
Overhang, ft				
Shade Length, ft				
Distance to Top of Window ft				
Shaded ht. of Window, ft				
Width of Window, ft				
Shaded Area, ft²				
Unshaded Area, ft²				

Work Sheet

		Living Room	Dining Room	Kitchen	Bedroom No. 1	Bedroom No. 2	Bedroom No. 3	Bath No. 1	Entire House	UA Calculations U-Value	UxA	Load Factors Heating UxΔ	Cooling
WINDOW	Type ___ Facing ___ ft²												
	, ft²												
	, ft²												
	, ft²												
	, ft²												
	, ft²												
DOOR	, ft²												
	, ft²												
	, ft²												
WALL	Subtotal Area, ft²												
	Exposed, Running ft.												
	Gross Area, ft²												
	Above grade ,ft²												
	,ft²												
	,ft²												
	Below Grade ,ft²												
	,ft²												
	Subtotals												
	U_g-Value = $\frac{Sum of U \times A}{Sum of A}$												
	Roof/Ceiling ,ft²												
	,ft²												
	Floor ,ft²												
	,ft²												
	Totals												
	Overall U = $\frac{Sum of U \times A}{Ref. Area}$												

Load calculation worksheet
Figure 5-6

Load calculation datasheet
Figure 5-7

Courtesy: American Society of Heating, Refrigerating and Air-Conditioning Engineers, Inc. (ASHRAE) 1981 Fundamentals Handbook, all rights reserved.

RESIDENTIAL COOLING & HEATING
LOAD CALCULATIONS

Prepared For **Sample Residence** Job No._____ Date _____ Prepared by _____

Location **Milwaukee**

Latitude **43°N**

Design Conditions

	Summer	Winter
Outside Temp. db	90	-8
Inside Temp. db	78	67
Difference	12	15
Daily Range, Deg. F.	21	
Temperature Swing		Deg. F.

Summary of Load Calculations for Entire House

Cooling	Heating
1,905 Btu/hr Sensible	28,255 Btu/hr
10,277 Btu/hr Total	3.77 Btu/hr -F
.86 Tons 1089 ft²/ton	.30 Btu/hr -ft²

Window Schedule

Type	Description of Window	of Shades & Shading
A	Double lite	None
B		
C		
D		

Exterior Shading Calculation

Type		
Orientation		
Shade Line Factor		
Overhang, ft		
Shade Length, ft		
Distance to Top of Window, ft		
Shaded ht. of Window, ft		
Width of Window, ft		
Shaded Area, ft²		
Unshaded Area, ft²		

Work Sheet

		Living Room	Dining Room	Kitchen	Bedroom No. 1	Bedroom No. 2	Bedroom No. 3	Bath No. 1	Hall	Mech	Ent	St.	Entire House	U-Value	UxA	Heating UxΔ	Cooling
WINDOW	Type A Facing N, ft²	18			14								32	0.62	19.8	46.5	21
	A W, ft²					14							14	0.62	8.7	46.5	70
	A S, ft²										6		6	0.62	3.7	46.5	35
DOOR	A, ft²	20							20				60	0.59		44.3	0
WALL	Subtotal Area, ft²	38			14	14			20	26	1		112		32.2	40.0	
	Exposed, Running ft.	15			24	22		6	3	12	7		91	—	—		
	Gross Area, ft²	120			208	176		48	24	96	50		728	—	—		
	Above grade, ft²	82			194	162		48	4	70	56		616	.06	37	4.5	1.1
	Below Grade, ft²																
	Subtotals	120			208	176		48	24	96	56		728	.095	69.2		
	Roof/Ceiling	252		64	168	120		48	92	72	72	49	937	.055	51.5	4.1	2.2
	Floor, ft²																
	Totals												1065		120.7		
	Overall U = Sum of UxA / Ref. Area													.072			

Up-Value = Sum of UxA / Sum of A

Sample residence load calculation worksheet
Figure 5-8

RESIDENTIAL HEATING AND COOLING LOAD CALCULATIONS

COOLING LOAD CALCULATION

Description/ Orientation	Type	U-Value	Load Factor	Entire House Quant.	Entire House Btu/hr	Living Room Quant.	Living Room Btu/hr	Dining Room Quant.	Dining Room Btu/hr	Kitchen Quant.	Kitchen Btu/hr	Bedroom No. 1 Quant.	Bedroom No. 1 Btu/hr	Bedroom No. 2 Quant.	Bedroom No. 2 Btu/hr	Bedroom No. 3 Quant.	Bedroom No. 3 Btu/hr	Bath No. 1 Quant.	Bath No. 1 Btu/hr	Hall Quant.	Hall Btu/hr	Mech Quant.	Mech Btu/hr	Ent Quant.	Ent Btu/hr	Sto Quant.	Sto Btu/hr	Quant.	Btu/hr
WINDOWS: A N			21	32	672	18	378					14	294																
A W			70	14	980									14	980														
A S			35	6	210																			6	210				
DOORS																													
WALLS		.06	1.1	616	678	82	90					194	213	162	178			48	53	4	4			70	77	56	62		
ROOF		.055	2.2	937	2061	252	554			64	141	168	370	120	264			48	106	92	202	72	158	72	158	49	108		
FLOOR																													
People			225	4	900	4	900																						
Lights & Appliances					1200						1200																		
Infilt./Ventil.			.518	937	485	252	131			—	—	168	87	120	62			48	25	—	—	—	—	72	37	49	25		
Subtotal					7186		2053				1341		964		1484				184		206		158		482		195		
Duct Heat Gain @ 10%					719		205				134		96		148				18		21		16		48		20		
Room Sensible Cooling Load					7905		2258				1475		1060		1632				202		227		174		530		215		

HEATING LOAD CALCULATION

Description/ Orientation	Type	U-Value	Load Factor	Entire House Quant.	Entire House Btu/hr	Living Room Quant.	Living Room Btu/hr	Dining Room Quant.	Dining Room Btu/hr	Kitchen Quant.	Kitchen Btu/hr	Bedroom No. 1 Quant.	Bedroom No. 1 Btu/hr	Bedroom No. 2 Quant.	Bedroom No. 2 Btu/hr	Bedroom No. 3 Quant.	Bedroom No. 3 Btu/hr	Bath No. 1 Quant.	Bath No. 1 Btu/hr	Hall Quant.	Hall Btu/hr	Mech Quant.	Mech Btu/hr	Ent Quant.	Ent Btu/hr	Sto Quant.	Sto Btu/hr	Quant.	Btu/hr
WINDOW: A N			46.5	32	1488	18	837					14	651																
A W			46.5	14	651									14	651														
A S			46.5	6	279																			6	279				
DOOR			44.3	60	2658	20	886													20	886			20	886				
WALL			4.5	616	2772	82	369					194	873	162	729			48	216	4	18			70	315	56	252		
ROOF			4.1	937	3842	252	1033			64	262	168	689	120	492			48	197	92	377	72	295	72	295	49	201		
FLOOR: Edge			40	94	3760	15	600			—		26	1040	22	880			6	240	3	120	—		12	480	7	280		
Infilt./Ventil. B AC			8.64	937	8096	252	2177			—	—	168	1452	120	1037			48	415	—	—	—	—	72	622	49	423		
Subtotal					23,546		5902				262		4705		3789				1068		1401		295		2877		1156		
Duct (Piping) Heat Loss @ 20%					4,709		1180				—		941		758				214		—		—		575		231		
Room Sensible Heat Loss					28,255		7082				262		5646		4547				1282		1401		295		3452		1387		

Courtesy: *American Society of Heating, Refrigerating and Air-Conditioning Engineers, Inc. (ASHRAE)* *1981 Fundamentals Handbook, all rights reserved.*

Sample residence load calculation datasheet
Figure 5-9

Climatic Conditions for the United States[a]

Col. 1	Col. 2		Col. 3		Col. 4	Col. 5 Winter[d] Design Dry-Bulb		Col. 6 Design Dry-Bulb and Mean Coincident Wet-Bulb			Col. 7 Mean Daily	Col. 8 Design Wet-Bulb			Col. 9 Humidity Ratio	Col. 10 Ave. Winter
State and Station	Lati-tude[b]		Longi-tude[b]		Eleva-tion[c]	99%	97.5%	1%	2.5%	5%	Range	1%	2.5%	5%	2.5%	Temp.*
	°	'	°	'	Ft											
ALABAMA																
Alexander City	33	0	86	0	660	18	22	96/77	93/76	91/76	21	79	78	78	.0158	
Anniston AP	33	4	85	5	599	18	22	97/77	94/76	92/76	21	79	78	78	.0155	
Auburn	32	4	85	3	730	18	22	96/77	93/76	91/76	21	79	78	78	.0158	
Birmingham AP	33	3	86	5	610	17	21	96/74	94/75	92/74	21	78	77	76	.0146	54.2
Decatur	34	4	87	0	580	11	16	95/75	93/74	91/74	22	78	77	76	.0140	
Dothan AP	31	2	85	2	321	23	27	94/76	92/76	91/76	20	80	79	78	.0160	
Florence AP	34	5	87	4	528	17	21	97/74	94/74	92/74	22	78	77	76	.0138	
Gadsden	34	0	86	0	570	16	20	96/75	94/75	92/74	22	78	77	76	.0146	
Huntsville AP	34	4	86	4	619	11	16	95/75	93/74	91/74	23	78	77	76	.0140	51.3
Mobile AP	30	4	88	2	211	25	29	95/77	93/77	91/76	18	80	79	78	.0163	59.9
Mobile CO	30	4	88	1	119	25	29	95/77	93/77	91/76	16	80	79	78	.0163	
Montgomery AP	32	2	86	2	195	22	25	96/76	95/76	93/76	21	79	79	78	.0149	55.4
Selma-Craig AFB	32	2	87	0	207	22	26	97/78	95/77	93/77	21	81	80	79	.0158	
Talladega	33	3	86	1	565	18	22	97/77	94/76	92/76	21	79	78	78	.0155	
Tuscaloosa AP	33	1	87	4	170r	20	23	98/75	96/76	94/76	22	79	78	77	.0147	
ALASKA																
Anchorage AP	61	1	150	0	90	−23	−18	71/59	68/58	66/56	15	60	59	57	.0104	23.0
Barrow (S)	71	2	156	5	22	−45	−41	57/53	53/50	49/47	12	54	50	47	.0070	
Fairbanks AP (S)	64	5	147	5	436	−51	−47	82/62	78/60	75/59	24	64	62	60	.0071	6.7
Juneau AP	58	2	134	4	17	−4	1	74/60	70/58	67/57	15	61	59	58	.0075	32.1
Kodiak	57	3	152	3	21	10	13	69/58	65/56	62/55	10	60	58	56	.0074	
Nome AP	64	3	165	3	13	−31	−27	66/57	62/55	59/54	10	58	56	55	.0076	13.1
ARIZONA																
Douglas AP	31	3	109	3	4098	27	31	98/63	95/63	93/63	31	70	69	68	.0069	
Flagstaff AP	35	1	111	4	6973	−2	4	84/55	82/55	80/54	31	61	60	59	———	35.6
Fort Huachuca AP (S)	31	3	110	2	4664	24	28	95/62	92/62	90/62	27	69	68	67	.0071	
Kingman AP	35	2	114	0	3446	18	25	103/65	100/64	97/64	30	70	69	69	.0062	
Nogales	31	2	111	0	3800	28	32	99/64	96/64	94/64	31	71	70	69	.0074	
Phoenix AP(S)	33	3	112	0	1117	31	34	109/71	107/71	105/71	27	76	75	75	.0086	58.5
Prescott AP	34	4	112	3	5014	4	9	96/61	94/60	92/60	30	66	65	64	.0055	
Tuscon AP (S)	32	1	111	0	2584	28	32	104/66	102/66	100/66	26	72	71	71	.0067	58.1
Winslow AP	35	0	110	4	4880	5	10	97/61	95/60	93/60	32	66	65	64	.0053	43.0
Yuma AP	32	4	114	4	199	36	39	111/72	109/72	107/71	27	79	78	77	.0083	64.2
ARKANSAS																
Blytheville AFB	36	0	90	0	264	10	15	96/78	94/77	91/76	21	81	80	78	.0164	
Camden	33	4	92	5	116	18	23	98/76	96/76	94/76	21	80	79	78	.0147	
El Dorado AP	33	1	92	5	252	18	23	98/76	96/76	94/76	21	80	79	78	.0148	
Fayetteville AP	36	0	94	1	1253	7	12	97/72	94/73	92/73	23	77	76	75	.0132	
Fort Smith AP	35	2	94	2	449	12	17	101/75	98/76	95/76	24	80	79	78	.0146	50.3
Hot Springs	34	3	93	1	535	17	23	101/77	97/77	94/77	22	80	79	78	.0157	
Jonesboro	35	5	90	4	345	10	15	96/78	94/77	91/76	21	81	80	78	.0164	
Little Rock AP (S)	34	4	92	1	257	15	20	99/76	96/77	94/77	22	80	79	78	.0160	50.5
Pine Bluff AP	34	1	92	0	204	16	22	100/78	97/77	95/78	22	81	80	80	.0154	
Texarkana AP	33	3	94	0	361	18	23	98/76	96/77	93/76	21	80	79	78	.0160	54.2
CALIFORNIA																
Bakersfield AP	35	2	119	0	495	30	32	104/70	101/69	98/68	32	73	71	70	.0081	55.4
Barstow AP	34	5	116	5	2142	26	29	106/68	104/68	102/67	37	73	71	70	.0075	
Blythe AP	33	4	114	3	390	30	33	112/71	110/71	108/70	28	75	75	74	.0076	
Burbank AP	34	1	118	2	699	37	39	95/68	91/68	88/67	25	71	70	69	.0069	58.6
Chico	39	5	121	5	205	28	30	103/69	101/68	98/67	36	71	70	68	.0071	

*Oct through April, inclusive, 1976 ASHRAE Systems Handbook & Product Directory, Chapter 43

[a] Table 2.1A was prepared by ASHRAE Technical Committee 4.2, Weather Data, from data compiled from official weather stations where hourly weather observations are made by trained observers.

[b] Latitude, for use in calculating solar loads, and longitude are given to the nearest 10 minutes. For example, the latitude and longitude for Anniston, Alabama are given as 33 34 and 85 55 respectively, or 33° 40, and 85° 50.

[c] Elevations are ground elevations for each station. Temperature readings are generally made at an elevation of 5 ft above ground, except for locations marked r, indicating roof exposure of thermometer.

[d] Percentage of winter design data shows the percent of the 3-month period, December through February.

[e] Percentage of summer design data shows the percent of 4-month period, June through September.

Weather data and design conditions
Figure 5-10

Climatic Conditions for the United States[a]

State and Station	Col. 1 Latitude[b] °	'	Col. 2 Longitude[b] °	'	Col. 3 Elevation[c] Ft	Col. 4 Winter[d] Design Dry-Bulb 99%	97.5%	Col. 5 Summer[e] Design Dry-Bulb and Mean Coincident Wet-Bulb 1%	2.5%	5%	Col. 6 Mean Daily Range	Col. 7 Design Wet-Bulb 1%	2.5%	5%	Col. 8 Humidity Ratio 2.5%	Col. 9 Ave. Winter Temp*	Col. 10
Concord	38	0	122	0	195	24	27	100/69	97/68	94/67	32	71	70	68	.0080	
Covina	34	0	117	5	575	32	35	98/69	95/68	92/67	31	73	71	70	.0087	
Crescent City AP	41	5	124	0	50	31	33	68/60	65/59	63/58	18	62	60	59	.0092	
Downey	34	0	118	1	116	37	40	93/70	89/70	86/69	22	72	71	70	.0113	
El Cajon	32	4	117	0	525	42	44	83/69	80/69	78/68	30	71	70	68	.0129	
El Centro AP (S)	32	5	115	4	−30	35	38	112/74	110/74	108/74	34	81	80	78	.0097	
Escondido	33	0	117	1	660	39	41	89/68	85/68	82/68	30	71	70	69	.0110	
Eureaka/ Arcata AP	41	0	124	1	217	31	33	68/60	65/59	63/58	11	62	60	59	.0092	49.9
Fairfield- Travis AFB	38	2	122	0	72	29	32	99/68	95/67	91/66	34	70	68	67	.0077	
Fresno AP (S)	36	5	119	4	326	28	30	102/70	100/69	97/68	34	72	71	70	.0083	53.3
Hamilton AFB	38	0	122	3	3	30	32	89/68	84/66	80/65	28	72	69	67	.0096	
Laguna Beach	33	3	117	5	35	41	43	83/68	80/68	77/67	18	70	69	68	.0100	
Livermore	37	4	122	0	545	24	27	100/69	97/68	93/67	24	71	70	68	.0083	
Lompoc, Vandenburg AFB	34	4	120	3	552	35	38	75/61	70/61	67/60	20	63	61	60	.0096	
Long Beach AP	33	5	118	1	34	41	43	83/68	80/68	77/67	22	70	69	68	.0119	57.8
Los Angeles AP (S)	34	0	118	2	99	41	43	83/68	80/68	77/67	15	70	69	68	.0119	57.4
Los Angeles CO (S)	34	0	118	1	312	37	40	93/70	89/70	86/69	20	72	71	70	.0116	60.3
Merced-Castle AFB	37	2	120	3	178	29	31	102/70	99/69	96/68	36	72	71	70	.0083	
Modesto	37	4	121	0	91	28	30	101/69	98/68	95/67	36	71	70	69	.0087	
Monterey	36	4	121	5	38	35	38	75/63	71/61	68/61	20	64	62	61	.0091	
Napa	38	2	122	2	16	30	32	100/69	96/68	92/67	30	71	69	68	.0082	
Needles AP	34	5	114	4	913	30	33	112/71	110/71	108/70	27	75	75	74	.0079	
Oakland AP	37	4	122	1	3	34	36	85/64	80/63	75/62	19	66	64	63	.0088	53.5
Oceanside	33	1	117	2	30	41	43	83/68	80/68	77/67	13	70	69	68	.0119	
Ontario	34	0	117	36	995	31	33	102/70	99/69	96/67	36	74	72	71	.0088	
Oxnard	34	1	119	1	43	34	36	83/66	80/64	77/63	19	70	68	67	.0090	
Palmdale AP	34	4	118	1	2517	18	22	103/65	101/65	98/64	35	69	67	66	.0062	
Palm Springs	33	5	116	4	411	33	35	112/71	110/70	108/70	35	76	74	73	.0068	
Pasadena	34	1	118	1	864	32	35	98/69	95/68	92/67	29	73	71	70	.0090	
Petaluma	38	1	122	4	27	26	29	94/68	90/66	87/65	31	72	70	68	.0081	
Pomona CO	34	0	117	5	871	28	30	102/70	99/69	95/68	36	74	72	71	.0081	
Redding AP	40	3	122	1	495	29	31	105/68	102/67	100/66	32	71	69	68	.0064	
Redlands	34	0	117	1	1318	31	33	102/70	99/69	96/68	33	74	72	71	.0091	
Richmond	38	0	122	2	55	34	36	85/64	80/63	75/62	17	66	64	63	.0084	
Riverside- March AFB (S)	33	5	117	2	1511	29	32	100/68	98/68	95/67	37	72	71	70	.0086	
Sacramento AP	38	3	121	3	17	30	32	101/70	98/70	94/69	36	72	71	70	.0093	53.9
Salinas AP	36	4	121	4	74	30	32	74/61	70/60	67/59	24	62	61	59	.0087	
San Bernardino, Norton AFB	34	1	117	1	1125	31	33	102/70	99/69	96/68	38	74	72	71	.0088	
San Diego AP	32	4	117	1	19	42	44	83/69	80/69	78/68	12	71	70	68	.0126	59.5
San Fernando	34	1	118	3	977	37	39	95/68	91/68	88/67	38	71	70	69	.0099	
San Francisco AP	37	4	122	2	8	35	38	82/64	77/63	73/62	20	65	64	62	.0091	53.4
San Francisco CO	37	5	122	3	52	38	40	74/63	71/62	69/61	14	64	62	61	.0098	55.1
San Jose AP	37	2	122	0	70r	34	36	85/66	81/65	77/64	26	68	67	65	.0095	
San Luis Obispo	35	2	120	4	315	33	35	92/69	88/70	84/69	26	73	71	70	.0119	
Santa Ana AP	33	4	117	5	115r	37	39	89/69	85/68	82/68	28	71	70	69	.0107	
Santa Barbara MAP	34	3	119	5	10	34	36	81/67	77/66	75/65	24	68	67	66	.0111	
Santa Cruz	37	0	122	0	125	35	38	75/63	71/61	68/61	28	64	62	61	.0091	
Santa Maria AP (S)	34	5	120	3	238	31	33	81/64	76/63	73/62	23	65	64	63	.0093	54.3
Santa Monica CO	34	0	118	3	57	41	43	83/68	80/68	77/67	16	70	69	68	.0119	
Santa Paula	34	2	119	0	263	33	35	90/68	86/67	84/66	36	71	69	68	.0100	
Santa Rosa	38	3	122	5	167	27	29	99/68	95/67	91/66	34	70	68	67	.0077	
Stockton AP	37	5	121	2	28	28	30	100/69	97/68	94/67	37	71	70	68	.0080	
Ukiah	39	1	122	4	620	27	29	99/69	95/68	91/67	40	70	68	67	.0087	
Visalia	36	2	119	1	354	28	30	102/70	100/69	97/68	38	72	71	70	.0083	
Yreka	41	4	122	4	2625	13	17	95/65	92/64	89/63	38	67	65	64	.0075	
Yuba City	39	1	121	4	70	29	31	104/68	101/67	99/66	36	71	69	68	.0064	
COLORADO																
Alamosa AP	37	3	105	5	7536	−11	−6	84/57	82/57	80/57	35	62	61	60	.0074	29.7
Boulder	40	0	105	2	5385	−6	0	93/59	91/59	89/59	27	64	63	62	.0058	
Colorado Springs AP	38	5	104	4	6173	−3	2	91/58	88/57	86/57	30	63	62	61	.0053	37.3
Denver AP	39	5	104	5	5283	−5	1	93/59	91/59	89/59	28	64	63	62	.0058	37.6
Durango	37	1	107	5	6550	−6	−1	89/59	87/59	85/59	30	64	63	62	.0072	

Weather data and design conditions
Figure 5-10 (continued)

Climatic Conditions for the United States[a]

State and Station	Col. 2 Latitude °	'	Col. 3 Longitude °	'	Col. 4 Elevation Ft	Col. 5 Winter Design Dry-Bulb 99%	97.5%	Col. 6 Summer Design Dry-Bulb and Mean Coincident Wet-Bulb 1%	2.5%	5%	Col. 7 Mean Daily Range	Col. 8 Design Wet-Bulb 1%	2.5%	5%	Col. 9 Humidity Ratio 2.5%	Col. 10 Ave. Winter Temp.*
Fort Collins	40	4	105	0	5001	−5	1	93/59	91/59	89/59	28	64	63	62	.0055	
Grand Junction AP (S)	39	1	108	3	4849	2	7	96/59	94/59	92/59	29	64	63	62	.0048	
Greeley	40	3	104	4	4648	−2	4	96/60	94/60	92/60	29	65	64	63	.0053	
La Junta AP	38	0	103	3	4188	−3	3	100/68	98/68	95/67	31	72	70	69	.0101	
Leadville	39	2	106	2	10177	−18	−14	84/52	81/51	78/50	30	56	55	54		
Pueblo AP	38	2	104	2	4639	−7	0	97/61	95/61	92/61	31	67	66	65	.0057	
Sterling	40	4	103	1	3939	−7	−2	95/62	93/62	90/62	30	67	66	65	.0067	
Trinidad AP	37	2	104	2	5746	−2	3	93/61	91/61	89/61	32	66	65	64	.0069	
CONNECTICUT																
Bridgeport AP	41	1	73	1	7	6	9	86/73	84/71	81/70	18	75	74	73	.0132	39.9
Hartford, Brainard Field	41	5	72	4	15	3	7	91/74	88/73	85/72	22	77	75	74	.0140	37.3
New Haven AP	41	2	73	0	6	3	7	88/75	84/73	82/72	17	76	75	74	.0150	39.0
New London	41	2	72	1	60	5	9	88/73	85/72	83/71	16	76	75	74	.0139	
Norwalk	41	1	73	3	37	−6	9	86/73	84/71	81/70	19	75	74	73	.0132	
Norwich	41	3	72	0	20	3	7	89/75	86/73	83/72	18	76	75	74	.0144	
Waterbury	41	3	73	0	605	−4	2	88/73	85/71	82/70	21	75	74	72	.0134	
Windsor Locks, Bradley Field (S)	42	0	72	4	169	0	4	91/74	88/72	85/71	22	76	75	73	.0132	
DELAWARE																
Dover AFB	39	0	75	3	38	11	15	92/75	90/75	87/74	18	79	77	76	.0152	
Wilmington AP	39	4	75	3	78	10	14	92/74	89/74	87/73	20	77	76	75	.0146	42.5
DISTRICT OF COLUMBIA																
Andrews AFB	38	5	76	5	279	10	14	92/75	90/74	87/73	18	78	76	75	.0147	
Washington National AP	38	5	77	0	14	14	17	93/75	91/74	89/74	18	78	77	76	.0141	45.7
FLORIDA																
Belle Glade	26	4	80	4	16	41	44	92/76	91/76	89/76	16	79	78	78	.0159	
Cape Kennedy AP	28	3	80	3	16	35	38	90/78	88/78	87/78	15	80	79	79	.0184	
Daytona Beach AP	29	1	81	0	31	32	35	92/78	90/77	88/77	15	80	79	78	.0170	64.5
Fort Lauderdale	26	0	80	1	13	42	46	92/78	91/78	90/78	15	80	79	79	.0177	
Fort Myers AP	26	4	81	5	13	41	44	93/78	92/78	91/77	18	80	79	79	.0175	68.6
Fort Pierce	27	3	80	2	10	38	42	91/78	90/78	89/78	15	80	79	79	.0179	
Gainesville AP (S)	29	4	82	2	155	28	31	95/77	93/77	92/77	18	80	79	78	.0163	
Jacksonville AP	30	3	81	4	24	29	32	96/77	94/77	92/76	19	79	79	78	.0161	61.9
Key West AP	24	3	81	5	6	55	57	90/78	90/78	89/78	9	80	79	79	.0179	73.1
Lakeland CO (S)	28	0	82	0	214	39	41	93/76	91/76	89/76	17	79	78	78	.0159	66.7
Miami AP (S)	25	5	80	2	7	44	47	91/77	90/77	89/77	15	79	79	78	.0170	77.1
Miami Beach CO	25	5	80	1	9	45	48	90/77	89/77	88/77	10	79	79	78	.0172	72.5
Ocala	29	1	82	1	86	31	34	95/77	93/77	92/76	18	80	79	78	.0163	
Orlando AP	28	3	81	2	106r	35	38	94/76	93/76	91/76	17	79	78	78	.0154	65.7
Panama City, Tyndall AFB	30	0	85	4	22	29	33	92/78	90/77	89/77	14	81	80	79	.0170	
Pensacola CO	30	3	87	1	13	25	29	94/77	93/77	91/77	14	80	79	79	.0163	60.4
St. Augustine	29	5	81	2	15	31	35	92/78	89/78	87/78	16	80	79	79	.0182	
St. Petersburg	28	0	82	4	35	36	40	92/77	91/77	90/76	16	79	79	78	.0168	
Sanford	28	5	81	2	14	35	38	94/76	93/76	91/76	17	79	78	78	.0154	
Sarasota	27	2	82	3	30	39	42	93/77	92/77	90/76	17	79	79	78	.0165	
Tallahassee AP (S)	30	2	84	2	58	27	30	94/77	92/76	90/76	19	79	78	78	.0156	60.1
Tampa AP (S)	28	0	82	3	19	36	40	92/77	91/77	90/76	17	79	79	78	.0168	66.4
West Palm Beach AP	26	4	80	1	15	41	45	92/78	91/78	90/78	16	80	79	79	.0177	68.4
GEORGIA																
Albany, Turner AFB	31	3	84	1	224	25	29	97/77	95/76	93/76	20	80	79	78	.0149	
Americus	32	0	84	2	476	21	25	97/77	94/76	92/75	20	79	78	77	.0155	
Athens	34	0	83	2	700	18	22	94/74	92/74	90/74	21	78	77	76	.0138	51.8
Atlanta AP (S)	33	4	84	3	1005	17	22	94/74	92/74	90/73	19	77	76	75	.0146	51.7
Augusta AP	33	2	82	0	143	20	23	97/77	95/76	93/76	19	80	79	78	.0149	54.5
Brunswick	31	1	81	3	14	29	32	92/78	89/78	87/78	18	80	79	79	.0182	
Columbus, Lawson AFB	32	3	85	0	242	21	24	95/76	93/76	91/75	21	79	78	77	.0154	54.8
Dalton	34	5	85	0	720	17	22	94/76	93/76	91/76	22	79	78	77	.0158	
Dublin	32	3	83	0	215	21	25	96/77	93/76	91/75	20	79	78	77	.0154	
Gainesville	34	2	83	5	1254	16	21	93/74	91/74	89/73	21	77	76	75	.0152	
Griffin (S)	33	1	84	2	980	18	22	93/76	90/75	88/74	21	78	77	76	.0159	

Weather data and design conditions
Figure 5-10 (continued)

Climatic Conditions for the United States[a]

State and Station	Lat.° [b]	Lat.' [b]	Long.° [b]	Long.' [b]	Elev. Ft [c]	Winter Design Dry-Bulb 99%	Winter Design Dry-Bulb 97.5%	Summer Design Dry-Bulb and Mean Coincident Wet-Bulb 1%	2.5%	5%	Mean Daily Range	Design Wet-Bulb 1%	2.5%	5%	Humidity Ratio 2.5%	Ave. Winter Temp.*
La Grange	33	0	85	0	715	19	23	94/76	91/75	89/74	21	78	77	76	.0153	
Macon AP	32	4	83	4	356	21	25	96/77	93/76	91/75	22	79	78	77	.0158	56.2
Marietta, Dobbins AFB	34	0	84	3	1016	17	21	94/74	92/74	90/74	21	78	77	76	.0146	
Moultrie	31	1	83	4	340	27	30	97/77	95/77	92/76	20	80	79	78	.0162	
Rome AP	34	2	85	1	637	17	22	94/76	93/76	91/76	23	79	78	77	.0158	49.9
Savannah-Travis AP	32	1	81	1	52	24	27	96/77	93/77	91/77	20	80	79	78	.0163	57.8
Valdosta-Moody AFB	31	0	83	1	239	28	31	96/77	94/77	92/76	20	80	79	78	.0161	
Waycross	31	2	82	2	140	26	29	96/77	94/77	91/76	20	80	79	78	.0161	
HAWAII																
Hilo AP (S)	19	4	155	1	31	61	62	84/73	83/72	82/72	15	75	74	74	.0143	71.9
Honolulu AP	21	2	158	0	7	62	63	87/73	86/73	85/72	12	76	75	74	.0144	74.2
Kaneohe Bay MCAS	21	2	157	5	18	65	66	85/75	84/74	83/74	12	76	76	75	.0157	
Wahiawa	21	3	158	0	900	58	59	86/73	85/72	84/72	14	75	74	73	.0145	
IDAHO																
Boise AP(S)	43	3	116	1	2842	3	10	96/65	94/64	91/64	31	68	66	65	.0073	39.7
Burley	42	3	113	5	4180	−3	2	99/62	95/61	92/66	35	64	63	61	.0055	
Coeur d'Alene AP	47	5	116	5	2973	−8	−1	89/62	86/61	83/60	31	64	63	61	.0070	
Idaho Falls AP	43	3	112	0	4730r	−11	−6	89/61	87/61	84/59	38	65	53	61	.0076	
Lewiston AP	46	2	117	0	1413	−1	6	96/65	93/64	90/63	32	67	56	64	.0068	41.0
Moscow	46	4	117	0	2660	−7	0	90/63	87/62	84/61	32	65	54	62	.0073	
Mountain Home AFB	43	0	115	5	2992	6	12	99/64	97/63	94/62	36	66	65	63	.0059	
Pocatello AP	43	0	112	4	4444	−8	−1	94/61	91/60	89/59	35	64	63	61	.0060	34.8
Twin Falls (AP) (S)	42	3	114	3	4148	−3	2	99/62	95/61	92/60	34	64	63	61	.0055	
ILLINOIS																
Aurora	41	5	88	2	744	−6	−1	93/76	91/76	88/75	20	79	78	76	.0162	
Belleville, Scott AFB	38	3	89	5	447	1	6	94/76	92/76	89/75	21	79	78	76	.0160	
Bloomington	40	3	89	0	775	−6	−2	92/75	90/74	88/73	21	78	76	75	.0150	
Carbondale	37	5	89	1	380	2	7	95/77	93/77	90/76	21	80	79	77	.0167	
Champaign/Urbana	40	0	88	2	743	−3	2	95/75	92/74	90/73	21	78	77	75	.0142	
Chicago, Midway AP	41	5	87	5	610	−5	0	94/74	91/73	88/72	20	77	75	74	.0136	37.5
Chicago, O'Hare AP	42	0	87	5	658	−8	−4	91/74	89/74	86/72	20	77	76	74	.0149	35.8
Chicago CO	41	5	87	4	594	−3	2	94/75	91/74	88/73	15	79	77	75	.0145	38.9
Danville	40	1	87	4	558	−4	1	93/75	90/74	88/73	21	78	77	75	.0147	
Decatur	39	5	88	5	670	−3	2	94/75	91/74	88/73	21	78	77	75	.0145	
Dixon	41	5	89	3	696	−7	−2	93/75	90/74	88/73	23	78	77	75	.0147	
Elgin	42	0	88	2	820	−7	−2	91/75	88/74	86/73	21	78	77	75	.0155	
Freeport	42	2	89	4	780	−9	−4	91/74	89/73	87/72	24	77	76	74	.0144	
Galesburg	41	0	90	3	771	−7	−2	93/75	91/75	88/74	22	78	77	75	.0157	
Greenville	39	0	89	2	563	−1	4	94/76	92/75	89/74	21	79	78	76	.0151	
Joliet	41	3	88	1	588	−5	0	93/75	90/74	88/73	20	78	77	75	.0147	
Kankakee	41	1	87	5	625	−4	1	93/75	90/74	88/73	21	78	77	75	.0147	
La Salle/Peru	41	2	89	1	520	−7	−2	93/75	91/75	88/74	22	78	77	75	.0153	
Macomb	40	3	90	4	702	−5	0	95/76	92/76	89/75	22	79	78	76	.0160	
Moline AP	41	3	90	3	582	−9	−4	93/75	91/75	88/74	23	78	77	75	.0153	36.4
Mt Vernon	38	2	88	5	500	0	5	95/76	92/75	89/74	21	79	78	76	.0151	
Peoria AP	40	4	89	4	652	−8	−4	91/75	89/74	87/73	22	78	76	75	.0149	38.1
Quincy AP	40	0	91	1	762	−2	3	96/76	93/76	90/76	22	80	78	77	.0161	
Rantoul, Chanute AFB	40	2	88	1	740	−4	1	94/75	91/74	89/73	21	78	77	75	.0145	
Rockford	42	1	89	0	724	−9	−4	91/74	89/73	87/72	24	77	76	74	.0141	34.8
Springfield AP	39	5	89	4	587	−3	2	94/75	92/74	89/74	21	79	77	76	.0142	40.6
Waukegan	42	2	87	5	680	−6	−3	92/76	89/74	87/73	21	78	76	75	.0149	
INDIANA																
Anderson	40	0	85	4	847	0	6	95/76	92/75	89/74	22	79	78	76	.0155	
Bedford	38	5	86	3	670	0	5	95/76	92/75	89/74	22	79	78	76	.0151	
Bloomington	39	1	86	4	820	0	5	95/76	92/75	89/74	22	79	78	76	.0155	
Columbus, Bakalar AFB	39	2	85	5	661	3	7	95/76	92/75	90/74	22	79	78	76	.0151	
Crawfordsville	40	0	86	5	752	−2	3	94/75	91/74	88/73	22	79	77	76	.0148	
Evansville AP	38	0	87	3	381	4	9	95/76	93/75	91/75	22	79	78	77	.0149	45.0
Fort Wayne AP	41	0	85	1	791	−4	1	92/73	89/72	87/72	24	77	75	74	.0136	37.3
Goshen AP	41	3	85	5	823	−3	1	91/73	89/73	86/72	23	77	75	74	.0144	
Hobart	41	3	87	2	600	−4	2	91/73	88/73	85/72	21	77	75	74	.0143	
Huntington	40	4	85	3	802	−4	1	92/73	89/72	87/72	23	77	75	74	.0136	
Indianapolis AP (S)	39	4	86	2	793	−2	2	92/74	90/74	87/73	22	78	76	75	.0150	39.6
Jeffersonville	38	2	85	5	455	5	10	95/74	93/74	90/74	23	79	77	76	.0140	
Kokomo	40	3	86	1	790	−4	0	91/74	90/73	88/73	22	77	75	74	.0142	
Lafayette	40	2	86	5	600	−3	3	94/74	91/73	88/73	22	78	76	75	.0136	

Weather data and design conditions
Figure 5-10 (continued)

Climatic Conditions for the United States[a]

Col. 1	Col. 2		Col. 3		Col. 4	Col. 5		Col. 6			Col. 7	Col. 8			Col. 9	Col. 10
	Lati-tude[b]		Longi-tude[c]		Eleva-tion[c]	Winter[d] Design Dry-Bulb		Summer[e] Design Dry-Bulb and Mean Coincident Wet-Bulb			Mean Daily	Design Wet-Bulb			Humidity Ratio	Ave. Winter
State and Station	°	'	°	'	Ft	99%	97.5%	1%	2.5%	5%	Range	1%	2.5%	5%	2.5%	Temp.*
La Porte	41	3	86	4	810	−3	3	93/74	90/74	87/73	22	78	76	75	.0150	
Marion	40	3	85	4	791	−4	0	91/74	90/73	88/73	23	77	75	74	.0142	
Muncie	40	1	85	2	955	−3	2	92/74	90/73	87/73	22	76	76	75	.0142	
Peru, Bunker Hill AFB	40	4	86	1	804	−6	−1	90/74	88/73	86/73	22	77	75	74	.0146	
Richmond AP	39	5	84	5	1138	−2	2	92/74	90/74	87/73	22	78	76	75	.0150	
Shelbyville	39	3	85	5	765	−1	3	93/74	91/74	88/73	22	78	76	75	.0148	
South Bend AP	41	4	86	2	773	−3	1	91/73	89/73	86/72	22	77	75	74	.0144	36.6
Terre Haute AP	39	3	87	2	601	−2	4	95/75	92/74	89/73	22	79	77	76	.0142	
Valparaise	41	2	87	0	801	−3	3	93/74	90/74	87/73	22	78	76	75	.0150	
Vincennes	38	4	87	3	420	1	6	95/75	92/74	90/73	22	79	77	76	.0142	
IOWA																
Ames (S)	42	0	93	4	1004	−11	−6	93/74	90/74	87/73	23	78	76	75	.0150	
Burlington AP	40	5	91	1	694	−7	−3	94/74	91/75	88/73	22	78	77	75	.0153	37.6
Cedar Rapids AP	41	5	91	4	863	−10	−5	91/76	88/75	86/74	23	78	77	75	.0164	
Clinton	41	5	90	1	595	−8	−3	92/75	90/75	87/74	23	78	77	75	.0156	
Council Bluffs	41	2	95	5	1210	−8	−3	94/76	91/75	88/74	22	78	77	75	.0157	
Des Moines AP	41	3	93	4	948r	−10	−5	94/75	91/74	88/73	23	78	77	75	.0148	35.5
Dubuque	42	2	90	4	1065	−12	−7	90/74	88/73	86/72	22	77	75	74	.0146	32.7
Fort Dodge	42	3	94	1	1111	−12	−7	91/74	88/74	86/72	23	77	75	74	.0155	
Iowa City	41	4	91	3	645	−11	−6	92/76	89/76	87/74	22	80	78	76	.0167	
Keokuk	40	2	91	2	526	−5	0	95/75	92/75	89/74	22	79	77	76	.0151	
Marshalltown	42	0	92	5	898	−12	−7	92/76	90/75	88/74	23	78	77	75	.0159	
Mason City AP	43	1	93	2	1194	−15	−11	90/74	88/74	85/72	24	77	75	74	.0155	
Newton	41	4	93	0	946	−10	−5	94/75	91/74	88/73	23	78	77	75	.0148	
Ottumwa AP	41	1	92	2	842	−8	−4	94/75	91/74	88/73	22	78	77	75	.0148	
Sioux City AP	42	2	96	2	1095	−11	−7	95/74	92/74	89/73	24	78	77	75	.0146	34.0
Waterloo	42	3	92	2	868	−15	−10	91/76	89/75	86/74	23	78	77	75	.0153	32.6
KANSAS																
Atchison	39	3	95	1	945	−2	2	96/77	93/76	91/76	23	81	79	77	.0161	
Chanute AP	34	4	95	3	977	3	7	100/74	97/74	94/74	23	78	77	76	.0134	
Dodge City AP (S)	37	5	100	0	2594	0	5	100/69	97/69	95/69	25	74	73	71	.0102	42.5
El Dorado	37	5	96	5	1282	3	7	101/72	98/73	96/73	24	77	76	75	.0123	
Emporia	38	2	96	1	1209	1	5	100/74	97/74	94/73	25	78	77	76	.0134	
Garden City AP	38	0	101	0	2882	−1	4	99/69	96/69	94/69	28	74	73	71	.0107	
Goodland AP	39	2	101	4	3645	−5	0	99/66	96/65	93/66	31	71	70	68	.0079	37.8
Great Bend	38	2	98	5	1940	0	4	101/73	98/73	95/73	28	78	76	75	.0130	
Hutchinson AP	38	0	97	5	1524	4	8	102/72	99/72	97/72	28	77	75	74	.0116	
Liberal	37	0	101	0	2838	2	7	99/68	96/68	94/68	28	73	72	71	——	
Manhattan, Fort Riley (S)	39	0	96	5	1076	−1	3	99/75	95/75	92/74	24	78	77	76	.0148	
Parsons	37	2	95	3	908	5	9	100/74	97/74	94/74	23	79	77	76	.0134	
Russell AP	38	5	98	5	1864	0	4	101/73	98/73	95/73	29	78	76	75	.0130	
Salina	38	5	97	4	1271	0.	5	103/74	100/74	97/73	26	78	77	75	.0131	
Topeka AP	39	0	95	4	877	0	4	99/75	96/75	93/74	24	79	78	76	.0145	41.7
Wichita AP	37	4	97	3	1321	3	7	101/72	98/73	96/73	23	77	76	75	.0127	44.2
KENTUCKY																
Ashland	38	3	82	4	551	5	10	94/76	91/74	89/73	22	78	77	75	.0145	
Bowling Green AP	37	0	86	3	535	4	10	94/77	92/75	89/74	21	79	77	76	.0151	
Corbin AP	37	0	84	1	1175	4	9	94/73	92/73	89/72	23	77	76	75	.0137	
Covington AP	39	0	84	4	869	1	6	92/73	90/72	88/72	22	77	75	74	.0133	41.4
Hopkinsville, Campbell AFB	36	4	87	3	540	4	10	94/77	92/75	89/74	21	79	77	76	.0133	
Lexington AP (S)	38	0	84	4	979	3	8	93/73	91/74	88/72	22	77	76	75	.0139	43.8
Louisville AP	38	1	85	4	474	5	10	95/74	93/74	90/74	23	79	77	76	.0140	44.0
Madisonville	37	2	87	3	439	5	10	96/76	93/75	90/75	22	79	78	77	.0149	
Owensboro	37	5	87	1	420	5	10	97/76	94/75	91/75	23	79	78	77	.0146	
Paducah AP	37	0	88	4	398	7	12	98/76	95/75	92/75	20	79	78	77	.0144	
LOUISIANA																
Alexandria AP	31	2	92	2	92	23	27	95/77	94/77	92/77	20	80	79	78	.0161	57.5
Baton Rouge AP	30	3	91	1	64	25	29	95/77	93/77	92/77	19	80	80	79	.0163	59.8
Bogalusa	30	5	89	5	103	24	28	95/77	93/77	92/77	19	80	80	79	.0163	
Houma	29	3	90	4	13	31	35	95/78	93/78	92/77	15	81	80	79	.0172	
Lafayette AP	30	1	92	0	38	26	30	95/78	94/78	92/78	18	81	80	79	.0170	
Lake Charles AP (S)	30	1	93	1	14	27	31	95/77	93/77	92/77	17	80	79	79	.0163	60.5
Minden	32	4	93	2	250	20	25	99/77	96/76	94/76	20	79	79	78	.0151	
Monroe AP	32	3	92	0	78	20	25	99/77	96/76	94/76	20	79	79	78	.0147	
Natchitoches	31	5	93	0	120	22	26	97/77	95/77	93/77	20	80	79	78	.0158	

Weather data and design conditions
Figure 5-10 (continued)

Climatic Conditions for the United States[a]

Col. 1	Col. 2		Col. 3		Col. 4	Winter[d] Col. 5		Summer[e] Col. 6			Col. 7	Col. 8			Col. 9	Col. 10
	Lati-tude[b]		Longi-tude[c]		Eleva-tion[c]	Design Dry-Bulb		Design Dry-Bulb and Mean Coincident Wet-Bulb			Mean Daily	Design Wet-Bulb			Humidity Ratio	Ave. Winter
State and Station	°	'	°	'	Ft	99%	97.5%	1%	2.5%	5%	Range	1%	2.5%	5%	2.5%	Temp.*
New Orleans AP	30	0	90	2	3	29	33	93/78	92/78	90/77	16	81	80	79	.0175	61.0
Shreveport AP(S)	32	3	93	5	252	20	25	99/77	96/76	94/76	20	79	79	78	.0151	56.2
MAINE																
Augusta AP	44	2	69	5	350	−7	−3	88/73	85/70	82/68	22	74	72	70	.0126	
Bangor, Dow AFB	44	5	68	5	162	−11	−6	86/70	83/68	80/67	22	73	71	69	.0112	
Caribou AP (S)	46	5	68	0	624	−18	−13	84/69	81/67	78/66	21	71	69	67	.0112	24.4
Lewiston	44	0	70	1	182	−7	−2	88/73	85/70	82/68	22	74	72	70	.0123	
Millinocket AP	45	4	68	4	405	−13	−9	87/69	83/68	80/66	22	72	70	68	.0115	
Portland (S)	43	4	70	2	61	−6	−1	87/72	84/71	81/69	22	74	72	70	.0133	33.0
Waterville	44	3	69	4	89	−8	−4	87/72	84/69	81/68	22	74	72	70	.0117	
MARYLAND																
Baltimore AP	39	1	76	4	146	10	13	94/75	91/75	89/74	21	78	77	76	.0150	43.7
Baltimore CO	39	2	76	3	14	14	17	92/77	89/76	87/75	17	80	78	76	.0163	46.2
Cumberland	39	4	78	5	945	6	10	92/75	89/74	87/74	22	77	76	75	.0153	
Frederick AP	39	3	77	3	294	8	12	94/76	91/75	88/74	22	78	77	76	.0153	42.0
Hagerstown	39	4	77	4	660	8	12	94/75	91/74	89/74	22	77	76	75	.0145	
Salisbury (S)	38	2	75	3	52	12	16	93/75	91/75	88/74	18	79	77	76	.0150	
MASSACHUSETTS																
Boston AP (S)	42	2	71	0	15	6	9	91/73	88/71	85/70	16	75	74	72	.0124	40.0
Clinton	42	2	71	4	398	−2	2	90/72	87/71	84/69	17	75	73	72	.0129	
Fall River	41	4	71	1	190	5	9	87/72	84/71	81/69	18	74	73	72	.0133	
Framingham	42	2	71	3	170	3	6	89/72	86/71	83/69	17	74	73	71	.0128	
Gloucester	42	3	70	4	10	2	5	89/73	86/71	83/70	15	75	74	72	.0128	
Greenfield	42	3	72	4	205	−7	−2	88/72	85/71	82/69	23	74	73	71	.0130	
Lawrence	42	4	71	1	57	−6	0	90/73	87/72	84/70	22	76	74	73	.0134	
Lowell	42	3	71	2	90	−4	1	91/73	88/72	85/70	21	76	74	73	.0132	
New Bedford	41	4	71	0	70	5	9	85/72	82/71	80/69	19	74	73	72	.0137	
Pittsfield AP	42	3	73	2	1170	−8	−3	87/71	84/70	81/68	23	73	72	70	.0131	36.2
Springfield, Westover AFB	42	1	72	3	247	−5	0	90/72	87/71	84/69	19	75	73	72	.0126	
Taunton	41	5	71	1	20	5	9	89/73	86/72	83/70	18	75	74	73	.0136	
Worcester AP	42	2	71	5	986	0	4	87/71	84/70	81/68	18	73	72	70	.0131	34.7
MICHIGAN																
Adrian	41	5	84	0	754	−1	3	91/73	88/72	85/71	23	76	75	73	.0138	
Alpena AP	45	0	83	3	689	−11	−6	89/70	85/70	83/69	27	73	72	70	.0126	29.7
Battle Creek AP	42	2	85	2	939	1	5	92/74	88/72	85/70	23	76	74	73	.0138	
Benton Harbor AP	42	1	86	3	649	1	5	91/72	88/72	85/70	20	75	74	72	.0135	
Detroit	42	2	83	0	633	3	6	91/73	88/72	86/71	20	76	74	73	.0135	37.2
Escanaba	45	4	87	0	594	−11	−7	87/70	83/69	80/68	17	73	71	69	.0122	29.6
Flint AP	42	0	83	4	766	−4	1	90/73	87/72	85/71	25	76	74	72	.0140	33.1
Grand Rapids AP	42	5	85	3	681	1	5	91/72	88/72	85/70	24	75	74	72	.0135	34.9
Holland	42	5	86	1	612	2	6	88/72	86/71	83/70	22	75	73	72	.0131	
Jackson AP	42	2	84	2	1003	1	5	92/74	88/72	85/70	23	76	74	73	.0138	
Kalamazoo	42	1	85	3	930	1	5	92/74	88/72	85/70	23	76	74	73	.0138	
Lansing AP	42	5	84	4	852	−3	1	90/73	87/72	84/70	24	75	74	72	.0140	34.8
Marquette CO	46	3	87	3	677	−12	−8	84/70	81/69	77/66	18	72	70	68	—	30.2
Mt Pleasant	43	4	84	5	796	0	4	91/73	87/72	84/71	24	76	74	72	.0140	
Muskegon AP	43	1	86	1	627	2	6	86/72	84/70	82/70	21	75	73	72	.0128	36.0
Pontiac	42	4	83	2	974	0	4	90/73	87/72	85/71	21	76	74	73	.0140	
Port Huron	43	0	82	3	586	0	4	90/73	87/72	83/71	21	76	74	73	.0145	
Saginaw AP	43	3	84	1	662	0	4	91/73	87/72	84/71	23	76	74	72	.0137	
Sault Ste. Marie AP (S)	46	3	84	2	721	−12	−8	84/70	81/69	77/66	23	72	70	68	.0127	27.7
Traverse City AP	44	4	85	4	618	−3	1	89/72	86/71	83/69	22	75	73	71	.0131	
Yipsilanti	42	1	83	3	777	1	5	92/72	89/71	86/70	22	75	74	72	.0127	
MINNESOTA																
Albert Lea	43	4	93	2	1235	−17	−12	90/74	87/72	84/71	24	77	75	73	.0140	
Alexandria AP	45	5	95	2	1421	−22	−16	91/72	88/72	85/70	24	76	74	72	.0141	
Bemidji AP	47	3	95	0	1392	−31	−26	88/69	85/69	81/67	24	73	71	69	.0124	
Brainerd	46	2	94	2	1214	−20	−16	90/73	87/71	84/69	24	75	73	71	.0132	
Duluth AP	46	5	92	1	1426	−21	−16	85/70	82/68	79/66	22	72	70	68	.0123	23.4
Fairbault	44	2	93	2	1190	−17	−12	91/74	88/72	85/71	24	77	75	73	.0138	
Fergus Falls	46	1	96	0	1210	−21	−17	91/72	88/72	85/70	24	76	74	72	.0138	
International Falls AP	48	3	93	2	1179	−29	−25	85/68	83/68	80/66	26	71	70	68	.0118	

Weather data and design conditions
Figure 5-10 (continued)

Climatic Conditions for the United States[a]

Col. 1	Col. 2		Col. 3		Col. 4	Col. 5 Winter[d]		Col. 6 Summer[e]			Col. 7	Col. 8			Col. 9	Col. 10
State and Station	Lati-tude[b]		Longi-tude[c]		Eleva-tion[c]	Design Dry-Bulb		Design Dry-Bulb and Mean Coincident Wet-Bulb			Mean Daily Range	Design Wet-Bulb			Humidity Ratio	Ave. Winter Temp.*
	°	'	°	'	Ft	99%	97.5%	1%	2.5%	5%		1%	2.5%	5%	2.5%	
Mankato	44	1	94	0	785	−17	−12	91/72	88/72	85/70	24	77	75	73	.0138	
Minneapolis/ St Paul AP	44	5	93	1	822	−16	−12	92/75	89/73	86/71	22	77	75	73	.0144	28.3
Rochester AP	44	0	92	3	1297	−17	−12	90/74	87/72	84/71	24	77	75	73	.0143	28.8
St Cloud AP (S)	45	4	94	1	1034	−15	−11	91/74	88/72	85/70	24	76	74	72	.0138	
Virginia	47	3	92	3	1435	−25	−21	85/69	83/68	80/66	23	71	70	68	.0120	
Willmar	45	1	95	0	1133	−15	−11	91/74	88/72	85/71	24	76	74	72	.0138	
Winona	44	1	91	4	652	−14	−10	91/75	88/73	85/72	24	77	75	74	.0143	
MISSISSIPPI																
Biloxi, Keesler AFB	30	2	89	0	25	28	31	94/79	92/79	90/78	16	82	81	80	.0184	
Clarksdale	34	1	90	3	178	14	19	96/77	94/77	92/76	21	80	79	78	.0161	
Columbus AFB	33	4	88	3	224	15	20	95/77	93/77	91/76	22	80	79	78	.0163	
Greenville AFB	33	3	91	1	139	15	20	95/77	93/77	91/76	21	80	79	78	.0163	
Greenwood	33	3	90	1	128	15	20	95/77	93/77	91/76	21	80	79	78	.0163	
Hattiesburg	31	2	89	2	200	24	27	96/78	94/77	92/77	21	81	80	79	.0165	
Jackson AP	32	2	90	1	330	21	25	97/76	95/76	93/76	21	79	78	78	.0153	55.7
Laurel	31	4	89	1	264	24	27	96/78	94/77	92/77	21	81	80	79	.0164	
McComb AP	31	2	90	3	458	21	26	96/77	94/76	92/76	18	80	79	78	.0160	
Meridian AP	32	2	88	5	294	19	23	97/77	95/76	93/76	22	80	79	78	.0153	55.4
Natchez	31	4	91	3	168	23	27	96/78	94/78	92/77	21	81	80	79	.0170	
Tupelo	34	2	88	4	289	14	19	96/77	94/77	92/76	22	80	79	78	.0164	
Vicksburg CO	32	2	91	0	234	22	26	97/78	95/78	93/77	21	81	80	79	.0167	56.9
MISSOURI																
Cape Girardeau	37	1	89	3	330	8	13	98/76	95/75	92/75	21	79	78	77	.0144	42.3
Columbia AP (S)	39	0	92	2	778	−1	4	97/74	94/74	91/73	22	78	77	76	.0141	
Farmington AP	37	5	90	3	928	3	8	96/76	93/75	90/74	22	78	77	75	.0152	
Hannibal	39	4	91	2	489	−2	3	96/76	93/76	90/76	22	80	78	77	.0158	
Jefferson City	38	4	92	1	640	2	7	98/75	95/74	92/74	23	78	77	76	.0135	
Joplin AP	37	1	94	3	982	6	10	100/73	97/73	94/73	24	78	77	76	.0125	
Kansas City AP	39	1	94	4	742	2	6	99/75	96/74	93/74	20	78	77	76	.0136	43.9
Kirksville AP	40	1	92	4	966	−5	0	96/74	93/74	90/73	24	78	77	76	.0143	
Mexico	39	1	92	0	775	−1	4	97/74	94/74	91/73	22	78	77	76	.0141	
Moberly	39	3	92	3	850	−2	3	97/74	94/74	91/73	23	78	77	76	.0141	
Poplar Bluff	36	5	90	3	322	11	16	98/78	95/76	92/76	22	81	79	78	.0153	
Rolla	38	0	91	5	1202	3	9	94/77	91/75	89/74	22	78	77	76	.0157	
St Joseph AP	39	5	95	0	809	−3	2	96/77	93/76	91/76	23	81	79	77	.0161	40.3
St Louis AP	38	5	90	2	535	2	6	97/75	94/75	91/74	21	78	77	76	.0146	43.1
St Louis CO	38	4	90	2	465	3	8	98/75	94/75	91/74	18	78	77	76	.0146	
Sedalia, Whiteman AFB	38	4	93	3	838	−1	4	95/76	92/76	90/75	22	79	78	76	.0164	
Sikeston	36	5	89	3	318	9	15	98/77	95/76	92/75	21	80	78	77	.0153	
Springfield AP	37	1	93	2	1265	3	9	96/73	93/74	91/74	23	78	77	75	——	44.5
MONTANA																
Billings AP	45	5	108	3	3567	−15	−10	94/64	91/64	88/63	31	67	66	64	.0083	34.5
Bozeman	45	5	111	0	4856	−20	−14	90/61	87/60	84/59	32	63	62	60	.0071	
Butte AP	46	0	112	3	5526r	−24	−17	86/58	83/56	80/56	35	60	58	57	——	
Cut Bank AP	48	4	112	2	3838r	−25	−20	88/61	85/61	82/60	33	64	62	61	.0061	
Glasgow AP (S)	48	1	106	4	2277	−22	−18	92/64	89/63	85/62	29	68	66	64	.0075	26.4
Glendive	47	1	104	4	2076	−18	−13	95/66	92/64	89/62	29	69	67	65	.0073	
Great Falls AP (S)	47	3	111	2	3664r	−21	−15	91/60	88/60	85/59	28	64	62	60	.0062	32.8
Havre	48	3	109	4	2488	−18	−11	94/65	90/64	87/63	33	68	66	65	.0080	28.1
Helena AP	46	4	112	0	3893	−21	−16	91/60	88/60	85/59	32	64	62	61	.0064	31.1
Kalispell AP	48	2	114	2	2965	−14	−7	91/62	87/61	84/60	34	65	63	62	.0068	31.4
Lewiston AP	47	0	109	3	4132	−22	−16	90/62	87/61	83/60	30	65	63	62	.0073	
Livingston AP	45	4	110	3	4653	−20	−14	90/61	87/60	84/59	32	63	62	60	.0069	
Miles City AP	46	3	105	5	2629	−20	−15	98/66	95/66	92/65	30	70	68	67	.0083	31.2
Missoula AP	46	5	114	1	3200	−13	−6	92/62	88/61	85/60	36	65	63	62	.0066	31.5
NEBRASKA																
Beatrice	40	2	96	5	1235	−5	−2	99/75	95/74	92/74	24	78	77	76	.0139	
Chadron AP	42	5	103	0	3300	−8	−3	97/66	94/65	91/65	30	71	69	68	.0084	
Columbus	41	3	97	2	1442	−6	−2	98/74	95/73	92/73	25	77	76	75	.0133	
Fremont	41	3	96	3	1203	−6	−2	98/75	95/74	92/74	22	78	77	76	.0139	

Weather data and design conditions
Figure 5-10 (continued)

Climatic Conditions for the United States[a]

State and Station	Lati-tude[b] °	'	Longi-tude[c] °	'	Eleva-tion[c] Ft	Winter[d] Design Dry-Bulb 99%	97.5%	Summer[e] Design Dry-Bulb and Mean Coincident Wet-Bulb 1%	2.5%	5%	Mean Daily Range	Design Wet-Bulb 1%	2.5%	5%	Humidity Ratio 2.5%	Ave. Winter Temp.*
Grand Island AP	41	0	98	2	1841	−8	−3	97/72	94/71	91/71	28	75	74	73	.0122	36.0
Hastings	40	4	98	3	1932	−7	−3	97/72	94/71	91/71	27	75	74	73	.0122	
Kearney	40	4	99	1	2146	−9	−4	96/71	93/70	90/70	28	74	73	72	.0116	
Lincoln CO (S)	40	5	96	5	1150	−5	−2	99/75	95/74	92/74	24	78	77	76	.0139	38.8
McCook	40	1	100	4	2565	−6	−2	98/69	95/69	91/69	28	74	72	71	.0107	
Norfolk	42	0	97	3	1532	−8	−4	97/74	93/74	90/73	30	78	77	75	.0147	34.0
North Platte AP (S)	41	1	100	4	2779	−8	−4	97/69	94/69	90/69	28	74	72	71	.0112	35.5
Omaha AP	41	2	95	5	978	−8	−3	94/76	91/75	88/74	22	78	77	75	.0157	35.6
Scottsbluff AP	41	5	103	4	3950	−8	−3	95/65	92/65	90/64	31	70	68	67	.0091	35.9
Sidney AP	41	1	103	0	4292	−8	−3	95/65	92/65	90/64	31	70	68	67	.0094	
NEVADA																
Carson City	39	1	119	5	4675	4	9	94/60	91/59	89/58	42	63	61	60	.0053	
Elko AP	40	5	115	5	5075	−8	−2	94/59	92/59	90/58	42	63	62	60	.0053	34.0
Ely AP (S)	39	1	114	5	6257	−10	−4	89/57	87/56	85/55	39	60	59	58	.0051	33.1
Las Vegas AP (S)	36	1	115	1	2162	25	28	108/66	106/65	104/65	30	71	70	69	.0048	53.5
Lovelock AP	40	0	118	3	3900	8	12	98/63	96/63	93/62	42	66	65	64	.0067	
Reno AP (S)	39	3	119	5	4404	5	10	95/61	92/60	90/59	45	64	62	61	.0057	
Reno CO	39	3	119	5	4490	6	11	96/61	93/60	91/59	45	64	62	61	.0055	
Tonopah AP	38	0	117	1	5426	5	10	94/60	92/59	90/58	40	64	62	61	.0055	39.3
Winnemucca AP	40	5	117	5	4299	−1	3	96/60	94/60	92/60	42	64	62	61	.0053	36.7
NEW HAMPSHIRE																
Berlin	44	3	71	1	1110	−14	−9	87/71	84/69	81/68	22	73	71	70	.0123	
Claremont	43	2	72	2	420	−9	−4	89/72	86/70	83/69	24	74	73	71	.0123	
Concord AP	43	1	71	3	339	−8	−3	90/72	87/70	84/69	26	74	73	71	.0121	33.0
Keene	43	0	72	2	490	−12	−7	90/72	87/70	83/69	24	74	73	71	.0121	
Laconia	43	3	71	3	505	−10	−5	89/72	86/70	83/69	25	74	73	71	.0123	
Manchester, Grenier AFB	43	0	71	3	253	−8	−3	91/72	88/71	85/70	24	75	74	72	.0127	
Portsmouth, Pease AFB	43	1	70	5	127	−2	2	89/73	85/71	83/70	22	75	74	72	.0130	
NEW JERSEY																
Atlantic City CO	39	3	74	3	11	10	13	92/74	89/74	86/72	18	78	77	75	.0146	43.2
Long Branch	40	2	74	0	20	10	13	93/74	90/73	87/72	18	78	77	75	.0135	
Newark AP	40	4	74	1	11	10	14	94/74	91/73	88/72	20	77	76	75	.0133	42.8
New Brunswick	40	3	74	3	86	6	10	92/74	89/73	86/72	19	77	76	75	.0137	
Paterson	40	5	74	1	100	6	10	94/74	91/73	88/72	21	77	76	75	.0133	
Phillipsburg	40	4	75	1	180	1	6	92/73	89/72	86/71	21	76	75	74	.0129	
Trenton CO	40	1	74	5	144	11	14	91/75	88/74	85/73	19	78	76	75	.0148	42.4
Vineland	39	3	75	0	95	8	11	91/75	89/74	86/73	19	78	76	75	.0146	
NEW MEXICO																
Alamagordo, Holloman AFB	32	5	106	1	4070	14	19	98/64	96/64	94/64	30	69	68	67	.0074	
Albuquerque AP (S)	35	0	106	4	5310	12	16	96/61	94/61	92/61	27	66	65	64	.0065	45.0
Artesia	32	5	104	2	3375	13	19	103/67	100/67	97/67	30	72	71	70	.0085	
Carlsbad AP	32	2	104	2	3234	13	19	103/67	100/67	97/67	28	72	71	70	.0082	
Clovis AP	34	3	103	1	4279	8	13	95/65	93/65	91/65	28	69	68	67	.0092	
Farmington AP	36	5	108	1	5495	1	6	95/63	93/62	91/61	30	67	65	64	.0075	
Gallup	35	3	108	5	6465	0	5	90/59	89/58	86/58	32	64	62	61	.0062	
Grants	35	1	107	5	6520	−1	4	89/59	88/58	85/57	32	64	62	61	.0061	
Hobbs AP	32	4	103	1	3664	13	18	101/66	99/66	97/66	29	71	70	69	.0080	
Las Cruces	32	2	107	0	3900	15	20	99/64	96/64	94/64	30	69	68	67	.0074	
Los Alamos	35	5	106	2	7410	5	9	89/60	87/60	85/60	32	62	61	60	.0084	
Raton AP	36	5	104	3	6379	−4	1	91/60	89/60	87/60	34	65	64	63	.0074	38.1
Roswell, Walker AFB	33	2	104	3	3643	13	18	100/66	98/66	96/66	33	71	70	69	.0082	47.5
Santa Fe CO	35	4	106	0	7045	6	10	90/61	88/61	86/61	28	63	62	61	.0087	
Silver City AP	32	4	108	2	5373	5	10	95/61	94/60	91/60	30	66	64	63	.0058	48.0
Socorro AP	34	0	106	5	4617	13	17	97/62	95/62	93/62	30	67	66	65	.0065	
Tucumcari AP	35	1	103	4	4053	8	13	99/66	97/66	95/65	28	70	69	68	.0087	
NEW YORK																
Albany AP (S)	42	5	73	5	277	−6	−1	91/73	88/72	85/70	23	75	74	72	.0135	34.6
Albany CO	42	5	73	5	19	−4	1	91/73	88/72	85/70	20	75	74	72	.0132	37.2
Auburn	43	0	76	3	715	−3	2	90/73	87/71	84/70	22	75	73	72	.0129	
Batavia	43	0	78	1	900	1	5	90/72	87/71	84/70	22	75	73	72	.0132	
Binghamton AP	42	1	76	0	1590	−2	1	86/71	83/69	81/68	20	73	72	70	.0128	33.9
Buffalo AP	43	0	78	4	705r	2	6	88/71	85/70	83/69	21	74	73	72	.0126	34.5
Cortland	42	4	76	1	1129	−5	0	88/71	85/71	82/70	23	74	73	71	.0137	
Dunkirk	42	3	79	2	590	4	9	88/73	85/72	83/71	18	75	74	72	.0142	
Elmira AP	42	1	76	5	860	−4	1	89/71	86/71	83/70	24	74	73	71	.0134	
Geneva (S)	42	5	77	0	590	−3	2	90/73	87/71	84/70	22	75	73	72	.0129	

Weather data and design conditions
Figure 5-10 (continued)

Climatic Conditions for the United States[a]

Col. 1	Col. 2		Col. 3		Col. 4	Winter[d] Col. 5		Summer[e] Col. 6			Col. 7	Col. 8			Col. 9	Col. 10
	Lati-tude[b]		Longi-tude[c]		Eleva-tion[c]	Design Dry-Bulb		Design Dry-Bulb and Mean Coincident Wet-Bulb			Mean Daily	Design Wet-Bulb			Humidity Ratio	Ave. Winter
State and Station	°	'	°	'	Ft	99%	97.5%	1%	2.5%	5%	Range	1%	2.5%	5%	2.5%	Temp.*
Glen Falls	42	2	73	4	321	−11	−5	88/72	85/71	82/69	23	74	73	71	.0134	
Gloversville	43	1	74	2	790	−8	−2	89/72	86/71	83/69	23	75	74	72	.0134	
Hornell	42	2	77	4	1325	−4	0	88/71	85/70	82/69	24	74	73	72	.0132	
Ithaca (S)	42	3	76	3	950	−5	0	88/71	85/71	82/70	24	74	73	71	.0137	
Jamestown	42	1	79	2	1390	−1	3	88/70	86/70	83/69	20	74	72	71	.0129	
Kingston	42	0	74	0	279	−3	2	91/73	88/72	85/70	22	76	74	73	.0135	
Lockport	43	1	78	4	520	4	7	89/74	86/72	84/71	21	76	74	73	.0139	
Massena AP	45	0	75	0	202r	−13	−8	86/70	83/69	80/68	20	73	72	70	.0120	
Newburg-Stewart AFB	41	3	74	1	460	−1	4	90/73	88/72	85/70	21	76	74	73	.0135	
NYC-Central Park (S)	40	5	74	0	132	11	15	92/74	89/73	87/72	17	76	75	74	.0137	42.8
NYC-Kennedy AP	40	4	73	5	16	12	15	90/73	87/72	84/71	16	76	75	74	.0134	41.4
NYC-La Guardia AP	40	5	73	5	19	11	15	92/74	89/73	87/72	16	76	75	74	.0137	43.1
Niagra Falls AP	43	1	79	0	596	4	7	89/74	86/72	84/71	20	76	74	73	.0139	
Olean	42	1	78	3	1420	−2	2	87/71	84/71	81/70	23	74	73	71	.0142	
Oneonta	42	3	75	0	1150	−7	−4	86/71	83/69	80/68	24	73	72	70	.0125	
Oswego CO	43	3	76	3	300	1	7	86/73	83/71	80/70	20	75	73	72	.0138	
Plattsburg AFB	44	4	73	3	165	−13	−8	86/70	83/69	80/68	22	73	72	70	.0120	
Poughkeepsie	41	4	73	5	103	0	6	92/74	89/74	86/72	21	77	75	74	.0146	
Rochester AP	43	1	77	4	543	1	5	91/73	88/71	85/70	22	75	73	72	.0127	35.4
Rome-Griffiss AFB	43	1	75	3	515	−11	−5	88/71	85/70	83/69	22	75	73	71	.0126	
Schenectady (S)	42	5	74	0	217	−4	1	90/73	87/72	84/70	22	75	74	72	.0134	35.4
Suffolk County AFB	40	5			57	7	10	86/72	83/71	80/70	16	76	74	73	.0129	
Syracuse AP	43	1	76	1	424	−3	2	90/73	87/71	84/70	20	75	73	72	.0129	35.2
Utica	43	1	75	2	714	−12	−6	88/73	85/71	82/70	22	75	73	71	.0134	
Watertown	44	0	76	0	497	−11	−6	86/73	83/71	81/70	20	75	73	72	.0138	
NORTH CAROLINA																
Ashville AP	35	3	82	3	217r	10	14	89/73	87/72	85/71	21	75	74	72	.0134	46.7
Charlotte AP	35	0	81	0	735	18	22	95/74	93/74	91/74	20	77	76	76	.0140	50.4
Durham	36	0	78	5	406	16	20	94/75	92/75	90/75	20	78	77	76	.0151	
Elizabeth City AP	36	2	76	1	10	12	19	93/78	91/77	89/76	18	80	78	78	.0168	
Fayetteville, Pope AFB	35	1	79	0	95	17	20	95/76	92/76	90/75	20	79	78	77	.0156	
Goldsboro, Seymour-Johnson AFB	35	2	78	0	88	18	21	94/77	91/76	89/75	18	79	78	77	.0159	
Greensboro AP (S)	36	1	80	0	887	14	18	93/74	91/73	89/73	21	77	76	75	.0139	47.5
Greenville	35	4	77	2	25	18	21	93/77	91/76	89/75	19	79	78	77	.0159	
Henderson	36	2	78	2	510	12	15	95/77	92/76	90/76	20	79	78	77	.0160	
Hickory	35	4	81	2	1165	14	18	92/73	90/72	88/72	21	75	74	73	.0133	
Jacksonville	34	5	77	3	24	20	24	92/78	90/78	88/77	18	80	79	78	.0179	
Lumberton	34	4	79	0	132	18	21	95/76	92/76	90/75	20	79	78	77	.0156	
New Bern AP	35	1	77	0	17	20	24	92/78	90/78	88/77	18	80	79	78	.0179	
Raleigh-Durham AP (S)	35	5	78	5	433	16	20	94/75	92/75	90/75	20	78	77	76	.0151	49.4
Rocky Mount	36	0	77	5	81	18	21	94/77	91/76	89/75	19	79	78	77	.0159	
Wilmington AP	34	2	78	0	30	23	26	93/79	91/78	89/77	18	81	80	79	.0177	54.6
Winston-Salem AP	36	1	80	1	967	16	20	94/74	91/73	89/73	20	76	75	74	.0139	48.4
NORTH DAKOTA																
Bismark AP (S)	46	5	100	5	1647	−23	−19	95/68	91/68	88/67	27	73	71	70	.0102	26.6
Devil's Lake	48	1	98	5	1471	−25	−21	91/69	88/68	85/66	25	73	71	69	.0109	22.4
Dickinson AP	46	5	102	5	2595	−21	−17	94/68	90/66	87/65	25	71	69	68	.0110	
Fargo AP	46	5	96	5	900	−22	−18	92/73	89/71	85/69	25	76	74	72	.0127	24.8
Grands Forks AP	48	0	97	2	832	−26	−22	91/70	87/70	84/68	25	74	72	70	.0124	
Jamestown AP	47	0	98	4	1492	−22	−18	94/70	90/69	87/68	26	74	74	71	.0112	
Minot AP	48	2	101	2	1713	−24	−20	92/68	89/67	86/65	25	72	70	68	.0099	
Williston	48	1	103	4	1877	−25	−21	91/68	88/67	85/65	25	72	70	68	.0104	25.2
OHIO																
Akron-Canton AP	41	0	81	3	1210	1	6	89/72	86/71	84/70	21	75	73	72	.0134	38.1
Ashtabula	42	0	80	5	690	4	9	88/73	85/72	83/71	18	75	74	72	.0142	
Athens	39	2	82	1	700	0	6	95/75	92/74	90/73	22	78	76	74	.0142	
Bowling Green	41	3	83	4	675	−2	2	92/73	89/73	86/71	23	76	75	73	.0141	
Cambridge	40	0	81	4	800	1	7	93/75	90/74	87/73	23	78	76	75	.0150	

Weather data and design conditions
Figure 5-10 (continued)

Climatic Conditions for the United States[a]

State and Station	Latitude[b] °	′	Longitude[c] °	′	Elevation[c] Ft	Winter[d] Design Dry-Bulb 99%	97.5%	Summer[e] Design Dry-Bulb and Mean Coincident Wet-Bulb 1%	2.5%	5%	Mean Daily Range	Design Wet-Bulb 1%	2.5%	5%	Humidity Ratio 2.5%	Ave. Winter Temp.*
Chillicothe	39	2	83	0	638	0	6	95/75	92/74	90/73	22	78	76	74	.0142	
Cincinnati CO	39	1	84	4	761	1	6	92/73	90/72	88/72	21	77	75	74	.0133	45.1
Cleveland AP (S)	41	2	81	5	777r	1	5	91/73	88/72	86/71	22	76	74	73	.0138	37.2
Columbus AP (S)	40	0	82	5	812	0	5	92/73	90/73	87/72	24	77	75	74	.0142	39.7
Dayton AP	39	5	84	1	997	−1	4	91/73	89/72	86/71	20	76	75	73	.0136	39.8
Defiance	41	2	84	2	700	−1	4	94/74	91/73	88/72	24	77	76	74	.0136	
Findlay AP	41	0	83	4	797	2	3	92/74	90/73	87/72	24	77	76	74	.0142	
Fremont	41	2	83	1	600	−3	1	90/73	88/73	85/71	24	76	75	73	.0143	
Hamilton	39	2	84	3	650	0	5	92/73	90/72	87/71	22	76	75	73	.0130	
Lancaster	39	4	82	4	920	0	5	93/73	91/73	88/72	23	77	75	74	.0139	
Lima	40	4	84	0	860	−1	4	94/74	91/73	88/72	24	77	76	74	.0139	
Mansfield AP	40	5	82	3	1297	0	5	90/73	87/72	85/72	22	76	74	73	.0143	36.9
Marion	40	4	83	1	920	0	5	93/74	91/73	88/72	23	77	76	74	.0139	
Middletown	39	3	84	3	635	0	5	92/73	90/72	87/71	22	76	75	73	.0130	
Newark	40	1	82	3	825	−1	5	94/73	92/73	89/73	23	77	75	74	.0137	
Norwalk	41	1	82	4	720	−3	1	90/73	88/73	85/71	22	76	75	73	.0143	
Portsmouth	38	5	83	0	530	5	10	95/76	92/74	89/73	22	78	77	75	.0142	
Sandusky CO	41	3	82	4	606	1	6	93/73	91/72	88/71	21	76	74	73	.0128	39.1
Springfield	40	0	83	5	1020	−1	3	91/74	89/73	87/72	21	77	76	74	.0144	
Steubenville	40	2	80	4	992	1	5	89/72	86/71	84/70	22	74	73	72	.0134	
Toledo AP	41	4	83	5	676r	−3	1	90/73	88/73	85/71	25	76	75	73	.0143	36.4
Warren	41	2	80	5	900	0	5	89/71	87/71	85/70	23	74	73	71	.0132	
Wooster	40	5	82	0	1030	1	6	89/72	86/71	84/70	22	75	73	72	.0134	
Youngstown AP	41	2	80	4	1178	−1	4	88/71	86/71	84/70	23	74	73	71	.0134	36.8
Zanesville AP	40	0	81	5	881	1	7	93/75	90/74	87/73	23	78	76	75	.0150	
OKLAHOMA																
Ada	34	5	96	4	1015	10	14	100/74	97/74	95/74	23	77	76	75	.0134	
Altus AFB	34	4	99	2	1390	11	16	102/73	100/73	98/73	25	77	76	75	.0122	
Ardmore	34	2	97	1	880	13	17	100/74	98/74	95/74	23	77	77	76	.0132	
Bartlesville	36	5	96	0	715	6	10	101/73	98/74	95/74	23	77	77	76	.0128	
Chickasha	35	0	98	0	1085	10	14	101/74	98/74	95/74	24	78	77	76	.0132	
Enid-Vance AFB	36	2	98	0	1287	9	13	103/74	100/74	97/74	24	79	77	76	.0131	
Lawton AP	34	3	98	2	1108	12	16	101/74	99/74	96/74	24	78	77	76	.0133	
Mc Alester	34	5	95	5	760	14	19	99/74	96/74	93/74	23	77	76	75	.0136	
Muskogee AP	35	4	95	2	610	10	15	101/74	98/75	95/75	23	79	78	77	.0137	
Norman	35	1	97	3	1109	9	13	99/74	96/74	94/74	24	77	76	75	.0136	
Oklahoma City AP (S)	35	2	97	4	1280	9	13	100/74	97/74	95/73	23	78	77	76	.0138	48.3
Ponca City	36	4	97	0	996	5	9	100/74	97/74	94/74	24	77	76	76	.0134	
Seminole	35	2	96	4	865	11	15	99/74	96/74	94/73	23	77	76	75	.0136	
Stillwater (S)	36	1	97	1	884	8	13	100/74	96/74	93/74	24	77	76	75	.0136	
Tulsa AP	36	1	95	5	650	8	13	101/74	98/75	95/75	22	79	78	77	.0137	47.7
Woodward	36	3	99	3	1900	6	10	100/73	97/73	94/73	26	78	76	75	.0132	
OREGON																
Albany	44	4	123	1	224	18	22	92/67	89/66	86/65	31	69	67	66	.0084	
Astoria AP (S)	46	1	123	5	8	25	29	75/65	71/62	68/61	16	65	63	62	.0098	45.6
Baker AP	44	5	117	5	3368	−1	6	92/63	89/61	86/60	30	65	63	61	.0066	
Bend	44	0	121	2	3599	−3	4	90/62	87/60	84/59	33	64	62	60	.0115	
Corvallis (S)	44	3	123	2	221	18	22	92/67	89/66	86/65	31	69	67	66	.0084	
Eugene AP	44	1	123	1	364	17	22	92/67	89/66	86/65	31	69	67	66	.0086	45.6
Grants Pass	42	3	123	2	925	20	24	99/69	96/68	93/67	33	71	69	68	.0088	
Klamath Falls AP	42	1	121	4	4091	4	9	90/61	87/60	84/59	36	63	61	60	.0066	
Medford AP (S)	42	2	122	5	1298	19	23	98/68	94/67	91/66	35	70	68	67	.0087	43.2
Pendleton AP	45	4	118	5	1492	−2	5	97/65	93/64	90/62	29	66	65	63	.0068	42.6
Portland AP	45	4	122	4	21	17	23	89/68	85/67	81/65	23	69	67	66	.0100	45.6
Portland CO	45	3	122	4	57	18	24	90/68	86/67	82/65	21	69	67	66	.0098	47.4
Roseburg AP	43	1	123	2	505	18	23	93/67	90/66	87/65	30	69	67	66	.0084	46.3
Salem AP	45	0	123	0	195	18	23	92/68	88/66	84/65	31	69	68	66	.0086	45.4
The Dalles	45	4	121	1	102	13	19	93/69	89/68	85/66	28	70	68	67	—	
PENNSYLVANIA																
Allentown AP	40	4	75	3	376	4	9	92/73	88/72	86/72	22	76	75	73	.0135	38.9
Altoona CO	40	2	78	2	1468	0	5	90/72	87/71	84/70	23	74	73	72	.0135	
Butler	40	4	80	0	1100	1	6	90/73	87/72	85/71	22	75	74	73	.0140	
Chambersburg	40	0	77	4	640	4	8	93/75	90/74	87/73	23	77	76	75	.0147	
Erie AP	42	1	80	1	732	4	9	88/73	85/72	83/71	18	75	74	72	.0142	36.8
Harrisburg AP	40	1	76	5	335	7	11	94/75	91/74	88/73	21	77	76	75	.0145	41.2
Johnstown	40	2	78	5	1214	−3	2	86/70	83/70	80/68	23	72	71	70	.0133	
Lancaster	40	1	76	2	255	4	8	93/75	90/74	87/73	22	77	76	75	.0147	
Meadville	41	4	80	1	1065	0	4	88/71	85/70	83/69	21	73	72	71	.0128	
New Castle	41	0	80	2	825	2	7	91/73	88/72	86/71	23	75	74	73	.0138	

Weather data and design conditions
Figure 5-10 (continued)

Climatic Conditions for the United States[a]

State and Station	Lati-tude[b] °	'	Longi-tude[c] °	'	Eleva-tion[c] Ft	Winter[d] Design Dry-Bulb 99%	97.5%	Design Dry-Bulb and Mean Coincident Wet-Bulb 1%	2.5%	5%	Mean Daily Range	Design Wet-Bulb 1%	2.5%	5%	Humidity Ratio 2.5%	Ave. Winter Temp.*
Philadelphia AP	39	5	75	2	7	10	14	93/75	90/74	87/72	21	77	76	75	.0144	41.8
Pittsburgh AP	40	3	80	1	1137	1	5	89/72	86/71	84/70	22	74	73	72	.0134	38.4
Pittsburgh CO	40	3	80	0	749r	3	7	91/72	88/71	86/70	19	74	73	72	.0127	42.2
Reading CO	40	2	76	0	226	9	13	92/73	89/72	86/72	19	76	75	73	.0129	42.4
Scranton/Wilkes-Barre	41	2	75	4	940	1	5	90/72	87/71	84/70	19	74	73	72	.0132	37.2
State College (S)	40	5	77	5	1175	3	7	90/72	87/71	84/70	23	74	73	72	.0132	
Sunbury	40	5	76	5	480	2	7	92/73	89/72	86/70	22	75	74	73	.0132	
Uniontown	39	5	79	4	1040	5	9	91/74	88/73	85/72	22	76	75	74	.0146	
Warren	41	5	79	1	1280	-2	4	89/71	86/71	83/70	24	74	73	72	.0137	
West Chester	40	0	75	4	440	9	13	92/75	89/74	86/72	20	77	76	75	.0149	
Williamsport AP	41	1	77	0	527	2	7	92/73	89/72	86/70	23	75	74	73	.0132	38.5
York	40	0	76	4	390	8	12	94/75	91/74	88/73	22	77	76	75	.0145	
RHODE ISLAND																
Newport (S)	41	3	71	2	20	5	9	88/73	85/72	82/70	16	76	75	73	.0136	
Providence AP	41	4	71	3	55	5	9	89/73	86/72	83/70	19	75	74	73	.0136	38.8
SOUTH CAROLINA																
Anderson	34	3	82	4	764	19	23	94/74	92/74	90/74	21	77	76	75	.0146	
Charleston AFB (S)	32	5	80	0	41	24	27	93/78	91/78	89/77	18	81	80	79	.0177	56.4
Charleston CO	32	5	80	0	9	25	28	94/78	92/78	90/77	13	81	80	79	.0175	57.9
Columbia AP	34	0	81	1	217	20	24	97/76	95/75	93/75	22	79	78	77	.0141	54.0
Florence AP	34	1	79	4	146	22	25	94/77	92/77	90/76	21	80	79	78	.0165	54.5
Georgetown	33	2	79	2	14	23	26	92/79	90/78	88/77	18	81	80	79	.0179	
Greenville AP	34	5	82	1	957	18	22	93/74	91/74	89/74	21	77	76	75	.0148	51.6
Greenwood	34	1	82	1	671	18	22	95/75	93/74	91/74	21	78	77	76	.0140	
Orangeburg	33	3	80	5	244	20	24	97/76	95/75	93/75	20	79	78	77	.0141	
Rock Hill	35	0	81	0	470	19	23	96/75	94/74	92/74	20	78	77	76	.0138	
Spartanburg AP	35	0	82	0	816	18	22	93/74	91/74	89/74	20	77	76	75	.0148	
Sumter-Shaw AFB	34	0	80	3	291	22	25	95/77	92/76	90/75	21	79	78	77	.0160	
SOUTH DAKOTA																
Aberdeen AP	45	3	98	3	1296	-19	-15	94/73	91/72	88/70	27	77	75	73	.0134	
Brookings	44	2	96	5	1642	-17	-13	95/73	92/72	89/71	25	77	75	73	.0132	
Huron AP	44	3	98	1	1282	-18	-14	96/73	93/72	90/71	28	77	75	73	.0130	28.8
Mitchel	43	5	98	0	1346	-15	-10	96/72	93/71	90/70	28	76	75	73	.0120	
Pierre AP	44	2	100	2	1718r	-15	-10	99/71	95/71	92/69	29	75	74	72	.0117	
Rapid City AP (S)	44	0	103	0	3165	-11	-7	95/66	92/65	89/65	28	71	69	67	.0085	33.4
Sioux Falls AP	43	4	96	4	1420	-15	-11	94/73	91/72	88/71	24	76	75	73	.0134	30.6
Watertown AP	45	0	97	0	1746	-19	-15	94/73	91/72	88/71	26	76	75	73	.0134	
Yankton	43	0	97	2	1280	-13	-7	94/73	91/72	88/71	25	77	76	74	.0134	
TENNESSEE																
Athens	33	3	84	4	940	13	18	95/74	92/73	90/73	22	77	76	75	.0137	
Bristol-Tri City AP	36	3	82	2	1519	9	14	91/72	89/72	87/71	22	75	75	73	.0139	46.2
Chattanooga AP	35	0	85	1	670	13	18	96/75	93/74	91/74	22	78	77	76	.0140	50.3
Clarksville	36	4	87	2	470	6	12	95/76	93/74	90/74	21	78	77	76	.0140	
Columbia	35	4	87	0	690	10	15	97/75	94/74	91/74	21	78	77	76	.0138	
Dyersburg	36	0	89	3	334	10	15	96/78	94/77	91/76	21	81	80	78	.0164	
Greenville	35	5	82	5	1320	11	16	92/73	90/72	88/72	22	76	75	74	.0137	
Jackson AP	35	4	88	5	413	11	16	98/76	95/75	92/75	21	79	78	77	.0144	
Knoxville AP	35	5	84	0	980	13	19	94/74	92/73	90/73	21	77	76	75	.0137	49.2
Memphis AP	35	0	90	0	263	13	18	98/77	95/76	93/76	21	80	79	78	.0153	50.5
Murfreesboro	35	5	86	2	608	9	14	97/75	94/74	91/74	22	78	77	76	.0138	
Nashville AP (S)	36	1	86	4	577	9	14	97/75	94/74	91/74	21	78	77	76	.0138	48.9
Tullahoma	35	2	86	1	1075	8	13	96/74	93/73	91/73	22	77	76	75	.0135	
TEXAS																
Abilene AP	32	3	99	4	1759	15	20	101/71	99/71	97/71	22	75	74	74	.0111	53.9
Alice AP	27	4	98	0	180	31	34	100/78	98/77	95/77	20	82	81	79	.0151	
Amarillo AP	35	1	101	4	3607	6	11	98/67	95/67	93/67	26	71	70	70	.0097	47.0
Austin AP	30	2	97	4	597	24	28	100/74	98/74	97/74	22	78	77	77	.0162	59.1
Bay City	29	0	96	0	52	29	33	96/77	94/77	92/77	16	80	79	79	.0161	
Beaumont	30	0	94	0	18	27	31	95/79	93/78	91/78	19	81	80	80	.0172	
Beeville	28	2	97	4	225	30	33	99/78	97/77	95/77	18	82	81	79	.0154	
Big Springs AP (S)	32	2	101	3	2537	16	20	100/69	97/69	95/69	26	74	73	72	.0102	
Brownsville AP (S)	25	5	97	3	16	35	39	94/77	93/77	92/77	18	80	79	79	.0163	67.7
Brownwood	31	5	99	0	1435	18	22	101/73	99/73	96/73	22	77	76	75	.0124	
Bryan AP	30	4	96	2	275	24	29	98/76	96/76	94/76	20	79	78	78	.0154	

Weather data and design conditions
Figure 5-10 (continued)

Climatic Conditions for the United States[a]

State and Station	Lati-tude[b] °	'	Longi-tude[c] °	'	Eleva-tion[c] Ft	Design Dry-Bulb 99%	97.5%	Design Dry-Bulb and Mean Coincident Wet-Bulb 1%	2.5%	5%	Mean Daily Range	Design Wet-Bulb 1%	2.5%	5%	Humidity Ratio 2.5%	Ave. Winter Temp.*
						Col. 5		Col. 6			Col. 7	Col. 8			Col. 9	Col. 10
Corpus Christi AP	27	5	97	3	43	31	35	95/78	94/78	92/78	19	80	80	79	.0170	64.6
Corsicana	32	0	96	3	425	20	25	100/75	98/75	96/75	21	79	78	77	.0137	
Dallas AP	32	5	96	5	481	18	22	102/75	100/75	97/75	20	78	78	77	.0132	55.3
Del Rio, Laughlin AFB	29	2	101	0	1072	26	31	100/73	98/73	97/73	24	79	77	76	.0123	
Denton	33	1	97	1	655	17	22	101/74	99/74	97/74	22	78	77	76	.0126	
Eagle Pass	28	5	100	3	743	27	32	101/73	99/73	98/73	24	78	78	77	.0118	
El Paso AP (S)	31	5	106	2	3918	20	24	100/64	98/64	96/64	27	69	68	68	.0070	52.9
Fort Worth AP (S)	32	5	97	0	544r	17	22	101/74	99/74	97/74	22	78	77	76	.0126	55.1
Galveston AP	29	2	94	5	5	31	36	90/79	89/79	88/78	10	81	80	80	.0191	62.2
Greenville	33	0	96	1	575	17	22	101/74	99/74	97/74	21	78	77	76	.0126	
Harlingen	26	1	97	4	37	35	39	96/77	94/77	93/77	19	80	79	79	.0161	
Houston AP	29	4	95	2	50	27	32	96/77	94/77	92/77	18	80	79	79	.0161	61.0
Houston CO	29	5	95	2	158r	28	33	97/77	95/77	93/77	18	80	79	79	.0158	62.0
Huntsville	30	4	95	3	494	22	27	100/75	98/75	96/75	20	78	78	77	.0137	
Killeen-Gray AFB	31	0	97	4	1021	20	25	99/73	97/73	95/73	22	77	76	75	.0125	
Lamesa	32	5	102	0	2965	13	17	99/69	96/69	94/69	26	73	72	71	.0107	
Laredo AFB	27	3	99	3	503	32	36	102/73	101/73	99/74	23	78	78	77	.0113	66.0
Longview	32	2	94	4	345	19	24	99/76	97/76	95/76	20	80	79	78	.0148	
Lubbock AP	33	4	101	5	3243	10	15	98/69	96/69	94/69	26	73	72	71	.0107	48.8
Lufkin AP	31	1	94	5	286	25	29	99/76	97/76	94/76	20	80	79	78	.0148	
Mc Allen	26	1	98	1	122	35	39	97/77	95/77	94/77	21	80	79	79	.0158	
Midland AP (S)	32	0	102	1	2815r	16	21	100/69	98/69	96/69	26	73	72	71	.0103	53.8
Mineral Wells AP	32	5	98	0	934	17	22	101/74	99/74	97/74	22	78	77	76	.0129	
Palestine CO	31	5	95	4	580	23	27	100/76	98/76	96/76	20	79	79	78	.0150	
Pampa	35	3	101	0	3230	7	12	99/67	96/67	94/67	26	71	70	70	.0091	
Pecos	31	2	103	3	2580	16	21	100/69	98/69	96/69	27	73	72	71	.0100	
Plainview	34	1	101	4	3400	8	13	98/68	96/68	94/68	26	72	71	70	.0102	
Port Arthur AP	30	0	94	0	16	27	31	95/79	93/78	91/78	19	81	80	80	.0172	60.5
San Angelo, Goodfellow AFB	31	2	100	2	1878	18	22	101/71	99/71	97/70	24	75	74	73	.0111	
San Antonio AP (S)	29	3	98	3	792	25	30	99/72	97/73	96/73	19	77	76	76	.0125	50.6
Sherman Perrin AFB	33	4	96	4	763	15	20	100/75	98/75	95/74	22	78	77	76	.0141	60.1
Snyder	32	4	101	0	2325	13	18	100/70	98/70	96/70	26	74	73	72	.0108	
Temple	31	1	97	2	675	22	27	100/74	99/74	97/74	22	78	77	77	.0126	
Tyler AP	32	2	95	2	527	19	24	99/76	97/76	95/76	21	80	79	78	.0148	
Vernon	34	1	99	2	1225	13	17	102/73	100/73	97/73	24	77	76	75	.0119	
Victoria AP	28	5	97	0	104	29	32	98/78	96/77	94/77	18	82	81	79	.0156	62.1
Waco AP	31	4	97	0	500	21	26	101/75	99/75	97/75	22	78	78	77	.0135	57.2
Wichita Falls AP	34	0	98	3	994	14	18	103/73	101/73	98/73	24	77	76	75	.0116	53.0
UTAH																
Cedar City AP	37	4	113	1	5613	-2	5	93/60	91/60	89/59	32	65	63	62	.0065	
Logan	41	4	111	5	4775	-3	2	93/62	91/62	88/60	33	65	64	63	.0069	
Moab	38	5	109	3	3965	6	11	100/60	98/60	96/60	30	65	64	63	.0041	
Ogden AP	41	1	112	0	4455	1	5	93/63	91/61	88/61	33	66	65	64	.0067	
Price	39	4	110	5	5580	-2	5	93/60	91/60	89/59	33	65	63	62	.0065	
Provo	40	1	111	4	4470	1	6	98/62	96/62	94/61	32	66	65	64	.0062	
Richfield	38	5	112	0	5300	-2	5	93/60	91/60	89/59	34	65	63	62	.0065	
St George CO	37	1	113	4	2899	14	21	103/65	101/65	99/64	33	70	68	67	.0065	
Salt Lake City AP (S)	40	5	112	0	4220	3	8	97/62	95/62	92/61	32	66	65	64	.0065	38.4
Vernal AP	40	3	109	3	5280	-5	0	91/61	89/60	86/59	32	64	63	62	.0069	
VERMONT																
Barre	44	1	72	3	1120	-16	-11	84/71	81/69	78/68	23	73	71	70	.0130	
Burlington AP (S)	44	3	73	1	331	-12	-7	88/72	85/70	82/69	23	74	72	71	.0126	29.4
Rutland	43	3	73	0	620	-13	-8	87/72	84/70	81/69	23	74	72	71	.0128	
VIRGINIA																
Charlottsville	38	1	78	3	870	14	18	94/74	91/74	88/73	23	77	76	75	.0148	
Danville AP	36	3	79	2	590	14	16	94/74	92/73	90/73	21	77	76	75	.0134	
Fredericksburg	38	2	77	3	50	10	14	96/76	93/75	90/74	21	78	77	76	.0145	
Harrisonburg	38	3	78	5	1340	12	16	93/72	91/72	88/71	23	75	74	73	.0134	
Lynchburg AP	37	2	79	1	947	12	16	93/74	90/74	88/73	21	77	76	75	.0150	46.0
Norfolk AP	36	5	76	1	26	20	22	93/77	91/76	89/76	18	79	78	77	.0159	49.2
Petersburg	37	1	77	3	194	14	17	95/76	92/76	90/75	20	79	78	77	.0156	

Weather data and design conditions
Figure 5-10 (continued)

Climatic Conditions for the United States[a]

Col. 1	Col. 2		Col. 3		Col. 4	Col. 5 (Winter[d])		Col. 6 (Summer[e])			Col. 7	Col. 8			Col. 9	Col. 10
State and Station	Latitude[b]		Longitude[c]		Elevation[c]	Design Dry-Bulb		Design Dry-Bulb and Mean Coincident Wet-Bulb			Mean Daily Range	Design Wet-Bulb			Humidity Ratio	Ave. Winter Temp.*
	°	'	°	'	Ft	99%	97.5%	1%	2.5%	5%	Range	1%	2.5%	5%	2.5%	
Richmond AP	37	3	77	2	162	14	17	95/76	92/76	90/75	21	79	78	77	.0156	47.3
Roanoke AP	37	2	80	0	1174r	12	16	93/72	91/72	88/71	23	75	74	73	.0131	46.1
Staunton	38	2	78	5	1480	12	16	93/72	91/72	88/71	23	75	74	73	.0143	
Winchester	39	1	78	1	750	6	10	93/75	90/74	88/74	21	77	76	75	.0150	
WASHINGTON																
Aberdeen	47	0	123	5	12	25	28	80/65	77/62	73/61	16	65	63	62	.0084	
Bellingham AP	48	5	122	3	150	10	15	81/67	77/65	74/63	19	68	65	63	.0104	
Bremerton	47	3	122	4	162	21	25	82/65	78/64	75/62	20	66	64	63	.0095	
Ellensburg AP	47	0	120	3	1729	2	6	94/65	91/64	87/62	34	66	65	63	.0070	
Everett-Paine AFB	47	5	122	2	598	21	25	80/65	76/64	73/62	20	67	64	63	.0102	
Kennewick	46	0	119	1	392	5	11	99/68	96/67	92/66	30	70	68	67	.0078	
Longview	46	1	123	0	12	19	24	88/68	85/67	81/65	30	69	67	66	.0100	
Moses Lake, Larson AFB	47	1	119	2	1183	1	7	97/66	94/65	90/63	32	67	66	64	.0070	
Olympia AP	47	0	122	5	190	16	22	87/66	83/65	79/64	32	67	66	64	.0090	44.2
Port Angeles	48	1	123	3	99	24	27	72/62	69/61	67/60	18	64	62	61	.0096	
Seattle-Boeing Fld	47	3	122	2	14	21	26	84/68	81/66	77/65	24	69	67	65	.0102	
Seattle CO (S)	47	4	122	2	14	22	27	85/68	82/66	78/65	19	69	67	65	.0113	46.9
Seattle-Tacoma AP (S)	47	3	122	2	386	21	26	84/65	80/64	76/62	22	66	64	63	.0093	44.2
Spokane AP (S)	47	4	117	3	2357	−6	2	93/64	90/63	87/62	28	65	64	62	.0073	36.5
Tacoma-Mc Chord AFB	47	1	122	3	350	19	24	86/66	82/65	79/63	22	68	66	64	.0095	
Walla Walla AP	46	1	118	2	1185	0	7	97/67	94/66	90/65	27	69	67	66	.0077	43.8
Wenatchee	47	2	120	2	634	7	11	99/67	96/66	92/64	32	68	67	65	.0070	
Yakima AP	46	3	120	3	1061	−2	5	96/65	93/65	89/63	36	68	66	65	.0073	39.1
WEST VIRGINIA																
Beckley	37	5	81	1	2330	−2	4	83/71	81/69	79/69	22	73	71	70	.0139	
Bluefield AP	37	2	81	2	2850	−2	4	83/71	81/69	79/69	22	73	71	70	.0142	
Charleston AP	38	2	81	4	939	7	11	92/74	90/73	87/72	20	76	75	74	.0142	44.8
Clarksburg	39	2	80	2	977	6	10	92/74	90/73	87/72	21	76	75	74	.0142	
Elkins AP	38	5	79	5	1970	1	6	86/72	84/70	82/70	22	74	72	71	.0137	40.1
Huntington CO	38	2	82	3	565r	5	10	94/76	91/74	89/73	22	78	77	75	.0145	45.0
Martinsburg AP	39	2	78	0	537	6	10	93/75	90/74	88/74	21	77	76	75	.0147	
Morgantown AP	39	4	80	0	1245	4	8	90/74	87/73	85/73	22	76	75	74	.0149	
Parkersburg CO	39	2	81	3	615r	7	11	93/75	90/74	88/73	21	77	76	75	.0147	43.5
Wheeling	40	1	80	4	659	1	5	89/72	86/71	84/70	21	74	73	72	.0131	
WISCONSIN																
Appleton	44	2	88	2	742	−14	−9	89/74	86/72	83/71	23	76	74	72	.0139	
Ashland	46	3	90	5	650	−21	−16	85/70	82/68	79/66	23	72	70	68	.0112	
Beloit	42	3	89	0	780	−7	−3	92/75	90/75	86/74	24	78	77	75	.0159	
Eau Claire AP	44	5	91	3	888	−15	−11	92/75	89/73	86/71	23	77	75	73	.0144	
Fond du Lac	43	5	88	3	760	−12	−8	89/74	86/72	84/71	23	76	74	72	.0143	
Green Bay AP	44	3	88	1	683	−13	−9	88/74	85/72	83/71	23	76	74	72	.0142	30.3
La Crosse AP	43	5	91	2	652	−13	−9	91/75	88/73	85/72	22	77	75	74	.0143	31.5
Madison AP (S)	43	1	89	2	858	−11	−7	91/74	88/73	85/71	22	77	75	73	.0146	30.9
Manitowoc	44	1	87	4	660	−11	−7	89/74	86/72	83/71	21	76	74	72	.0139	
Marinette	45	0	87	4	605	−15	−11	87/73	84/71	82/70	20	75	73	71	.0136	
Milwaukee AP	43	0	87	5	672	−8	−4	90/74	87/73	84/71	21	76	74	73	.0145	32.6
Racine	42	4	87	4	640	−6	−2	91/75	88/73	85/72	21	77	75	74	.0143	
Sheboygan	43	4	87	4	648	−10	−6	89/75	86/73	83/72	20	77	75	74	.0148	
Stevens Point	43	0	89	3	1079	−15	−11	92/75	89/73	86/71	23	77	75	73	.0144	
Waukesha	43	0	88	1	860	−9	−5	90/74	87/73	84/71	22	76	74	73	.0149	
Wausau AP	44	6	89	4	1196	−16	−12	91/74	88/72	85/70	23	76	74	72	.0138	
WYOMING																
Casper AP	42	5	106	3	5319	−11	−5	92/58	90/57	87/57	31	63	61	60	.0046	33.4
Cheyene AP	41	1	104	5	6126	−9	−1	89/58	86/58	84/57	30	63	62	60	.0064	34.2
Cody AP	44	3	109	0	5090	−19	−13	89/60	86/60	83/59	32	64	63	61	.0073	
Evanston	41	2	111	0	6860	−9	−3	86/55	84/55	82/54	32	59	58	57	.0053	
Lander AP (S)	42	5	108	4	5563	−16	−11	91/61	88/61	85/60	32	64	63	61	.0077	31.4
Laramie AP (S)	41	2	105	3	7266	−14	−6	84/56	81/56	79/55	28	61	60	59	.0067	
Newcastle	43	5	104	1	4480	−17	−12	91/64	87/63	84/63	30	69	68	66	.0090	
Rawlins	41	5	107	1	6736	−12	−4	86/57	83/57	81/56	40	62	61	60	.0144	
Rock Springs AP	41	4	109	0	6741	−9	−3	86/55	84/55	82/54	32	59	58	57	.0051	
Sheridan AP	44	5	107	0	3942	−14	−8	94/62	91/62	88/61	32	66	65	63	.0071	32.5
Torrington	42	0	104	1	4098	−14	−8	94/62	91/62	88/61	30	66	65	63	.0071	

Weather data and design conditions
Figure 5-10 (continued)

Infiltration Load Factor

Btu/hr = Building volume x Air changes x Btu/cu.ft./hr x Temp difference

where: Btu/cu. ft./hr = 0.075 x 0.24 = 0.018

Allowance Factors, Latent Cooling Loads

Dry climates = 1.2   All others = 1.3 ⟵

Estimated Fuel Consumption

$$E = \frac{He \times D \times 24 \times C}{(Ti - To) \times k \times V}$$

where: E = Estimated fuel used
He = Heat loss in Btu/hr
D = Degree days
24 = Conversion of Btu/hr loss to Btu/day loss
Ti = Inside design temperature
To = Outside design temperature
k = Annual fuel utilization efficiency (AFUE) of furnace
V = Heating value of fuel in Btu per unit used
C = Correction factor, Figure 5-20B

Cfm Sizing

$$\text{Heating cfm} = \frac{\text{Heat loss}}{(Ta - Ti) \times 1.1}$$

$$\text{Cooling cfm} = \frac{\text{Heat gain}}{(Ti - Tc) \times 1.1}$$

Area of a Circle

$$P_i \times \frac{d^2}{4}$$

UA Calculations

$$U \times \text{Area}$$

Fan Law

$$\frac{cfm_1}{cfm_2} = \sqrt{\frac{P_1}{P_2}}$$

Formula summary
Figure 5-11

Overall Coefficients[a] of Heat Transmission (*U*-Factor) of Windows and Skylights, Btu/(hr · ft² · F)

Description	Exterior Vertical Panels				Exterior Horizontal Panels (Skylights)	
	Summer**		Winter*		Summer [j]	Winter [i]
	No Indoor Shade	Indoor Shade***	No Indoor Shade	Indoor Shade***		
Flat Glass [b]						
Single Glass	1.04	0.81	1.10	0.83	0.83	1.23
Insulating Glass, Double [c]						
3/16 in. air space [d]	0.65	0.58	0.62 ←	0.52	0.57	0.70
1/4 in. air space [d]	0.61	0.55	0.58	0.48	0.54	0.65
1/2 in. air space [e]	0.56	0.52	0.49	0.42	0.49	0.59
1/2 in. air space, low emittance coating [f]						
e = 0.20	0.38	0.37	0.32	0.30	0.36	0.48
e = 0.40	0.45	0.44	0.38	0.35	0.42	0.52
e = 0.60	0.51	0.48	0.43	0.38	0.46	0.56
Insulating Glass, Triple [c]						
1/4 in. air space [d]	0.44	0.40	0.39	0.31		
1/2 in. air space [g]	0.39	0.36	0.31	0.26		
Storm Windows						
1 in. to 4 in. air spaces [d]	0.50	0.48	0.50	0.42		
Plastic Bubbles [k]						
Single Walled					0.80	1.15
Double Walled					0.46	0.70

Adjustment Factors for Various Window and Sliding Patio Door Types (Multiply *U*-Values in Part A by These Factors)

Description	Single Glass	Double or Triple Glass	Storm Windows
Windows			
All Glass [h]	1.00	1.00	1.00
Wood Sash; 80% Glass	0.90	0.95	0.90
Wood Sash; 60% Glass	0.80	0.85	0.80
Metal Sash; 80% Glass	1.00	1.20 [m]	1.20 [m]
Sliding Patio Doors			
Wood Frame	0.95	1.00	—
Metal Frame	1.00	1.10 [m]	—

[a] See Table 3.14B for adjustments for various windows and sliding patio doors.
[b] Emittance of uncoated glass surface = 0.84.
[c] Double and triple refer to number of lights of glass.
[d] 0.125-in. glass.
[e] 0.25-in. glass
[f] Coating on either glass surface facing air space; all other glass surfaces uncoated.
[g] Window design: 0.25-in. glass, 0.125-in. glass, 0.25-in. glass.
[h] Refers to windows with negligible opaque areas.
[i] For heat flow up.
[j] For heat flow down.
[k] Based on area of opening, not total surface area.

[m] Values will be less than these when metal sash and frame incorporate thermal breaks. In some thermal break designs *U* values will be equal to or less than those for the glass. Window manufacturers should be consulted for specific data.
*15 mph outdoor air velocity; 0 F outdoor air; 70 F inside air temp natural convection.
**7.5 mph outdoor air velocity; 89 F outdoor air; 75 F inside air natural convection; solar radiation 248.3 Btu/(hr · ft²)
***Values apply to tightly closed venetian and vertical blinds, draperies, and roller shades.
The reciprocal of the above *U*-factors is the thermal resistance, *R*, for each type of glazing. If tightly drawn drapes (heavy close weave), closed Venetian blinds, or closely fitted roller shades are used internally, the additional *R* is approximately 0.29 (hr · ft² · F)/Btu. If miniature louvered solar screens are used in close proximity to the outer fenestration surface, the additional *R* is approximately 0.24 (hr · ft² · F)/Btu.

Courtesy: American Society of Heating, Refrigerating and Air-Conditioning Engineers, Inc. (ASHRAE)
1981 Fundamentals Handbook, all rights reserved.

Common U-values
Figure 5-12A

Adjusted U-Value for Some Insulated Walls and Roofs with Wood Framing Members (Winter Conditions)

Construction	Wood Members, Nominal Sizes	U-Value At Wood Section	% Wood Framing For Surface	Air Space	R-7	R-11	R-19	R-30	R-38
Roof/Ceiling, Combined — Asphalt shingles, felt membrane, plywood sheathing, gypsum wallboard, with foil backing	2-in.×4-in. 16-in. o.c.	0.145	*	0.213	0.086	0.074			
			9.4	0.207	0.092	0.081			
			15	0.203	0.095	0.085			
	2-in.×6-in. 16-in. o.c.	0.106	*	0.213	0.086	0.064	0.046		
			9.4	0.203	0.088	0.068	0.052		
			15	0.197	0.089	0.070	0.055 ←		
Ceiling — Metal lath & plaster, joists, no floor above	Joists 2-in.×8-in. 16-in. o.c.	0.093	*	0.592	0.115	0.079	0.048	0.032	0.025
			9.4	0.545	0.113	0.080	0.052	0.038	0.029
			15	0.517	0.112	0.081	0.055	0.041	0.031
Frame Wall — Wood siding, sheathing, framing, gypsum wall board or plaster (approximately the same values can be used for brick veneer and for metal siding with insulated frame walls)	2-in.×4-in. 16-in. o.c.	0.128	*	0.225	0.087	0.069			
			9.4	0.216	0.091	0.074			
			15	0.210	0.093	0.078			
			20	0.206	0.095	0.081			
	2-in.×4-in. 16-in. o.c. with insulated sheathing of R = 5	0.087	*	0.123	0.066	0.055			
			9.4	0.120	0.068	0.058 ←			
			15	0.118	0.069	0.060			
			20	0.116	0.070	0.061			
	2-in.×6-in. 24-in. o.c.	0.097	*	0.225	0.087	0.069	0.045		
			6.3	0.217	0.088	0.071	0.048		
			10	0.212	0.088	0.072	0.050		
			15	0.206	0.089	0.073	0.053		
Masonry Wall — 8-in. concrete block, furring, dry wall with foil backing	2-in.×1-in., 16-in. o.c. (Flat)	0.240	*	0.167	0.098				
			9.4	0.174	0.111				
	2-in.×2-in. 16-in. o.c.	0.196	*	0.176	0.098	0.070			
			9.4	0.178	0.107	0.082			

*U-Values for the section between wood members.

Coefficients of Transmission (U) for Slab Doors
Btu per (hr·ft²-F)

Thickness[a]	Winter — Solid Wood, No Storm Door	Winter — Storm Door[b] Wood	Winter — Storm Door[b] Metal	Summer — No Storm Door
1-in.	0.64	0.30	0.39	0.61
1.25-in.	0.55	0.28	0.34	0.53
1.5-in.	0.49	0.27	0.33	0.47
2-in.	0.43	0.24	0.29	0.42
Steel Door				
1.75 in.				
A[c]	0.59 ←	—	—	0.58
B[d]	0.19	—	—	0.18
C[e]	0.47	—	—	0.46

[a] Nominal thickness.
[b] Values for wood storm doors are for approximately 50% glass; for metal storm door values apply for any percent of glass.
[c] A = Mineral fiber core (2 lb/ft^3).
[d] B = Solid urethane foam core with thermal break.
[e] C = Solid polystyrene core with thermal break.

Common U-values
Figure 5-12A (continued)

Thermal Properties of Typical Building and Insulating Materials — (Design Values)[a]

Conductivity and conductance are expressed in Btu per (hour) (square foot) (degree Fahrenheit temperature difference). Conductivities (k) are per inch thickness, and conductances (C) are for thickness or construction stated, not per inch thickness. All values are for a mean temperature of 75 F, except as noted by an asterisk (*) which have been reported at 45 F.

Description	Density (lb/ft³)	Conductivity (k)	Conductance (C)	Resistance[b] (R) Per inch thickness (1/k)	Resistance[b] (R) For thickness listed (1/C)	Specific Heat, Btu/(lb) (deg F)	Wt lb ft²	Heat Capacity Btu ft²·F	Heat Capacity Btu ft³·F
BUILDING BOARD[c]									
Boards, Panels, Subflooring, Sheathing									
Woodboard Panel Products									
Asbestos-cement board	120	4.0	—	0.25	—	0.24	—	—	28.8
Asbestos-cement board ....0.125 in.	120	—	33.00	—	0.03		1.25	0.30	28.8
Asbestos-cement board ....0.25 in.	120	—	16.50	—	0.06		2.50	0.60	28.8
Gypsum or plaster board ......0.375 in.	50	—	3.10	—	0.32	0.26	1.56	0.41	13.0
Gypsum or plaster board ......0.5 in.	50	—	2.22	—	0.45		2.08	0.54	13.0
Gypsum or plaster board ......0.625 in.	50	—	1.78	—	0.56		2.60	0.68	13.0
Plywood (Douglas Fir)	34	0.80	—	1.25	—	0.29	—	—	9.86
Plywood (Douglas Fir) ....0.25 in.	34	—	3.20	—	0.31		0.71	0.21	9.86
Plywood (Douglas Fir) ....0.375 in.	34	—	2.13	—	0.47		1.06	0.31	9.86
Plywood (Douglas Fir) ....0.5 in.	34	—	1.60	—	0.62		1.42	0.41	9.86
Plywood (Douglas Fir) ....0.625 in.	34	—	1.29	—	0.77		1.77	0.51	9.86
Plywood or wood panels ......0.75 in.	34	—	1.07	—	0.93	0.29	2.13	0.62	9.86
Vegetable Fiber Board									
Sheathing, regular density .....0.5 in.	18	—	0.76	—	1.32	0.31	0.75	0.23	5.58
................0.78125 in.	18	—	0.49	—	2.06		1.17	0.36	5.58
Sheathing intermediate density ....0.5 in.	22	—	0.82	—	1.22	0.31	0.92	0.28	6.82
Nail-base sheathing ....0.5 in.	25	—	0.88	—	1.14	0.31	1.04	0.32	7.75
Shingle backer ....0.375 in.	18	—	1.06	—	0.94	0.31	0.56	0.17	5.58
Shingle backer ....0.3125 in.	18	—	1.28	—	0.78		0.47	0.15	5.58
Sound deadening board ....0.5 in.	15	—	0.74	—	1.35	0.30	0.62	0.19	4.50
Tile and lay-in panels, plain or acoustic	18	0.40	—	2.50	—	0.14	—	—	2.52
................0.5 in.	18	—	0.80	—	1.25		0.75	0.11	2.52
................0.75 in.	18	—	0.53	—	1.89		1.13	0.16	2.52
Laminated paperboard	30	0.50	—	2.00	—	0.33	—	—	9.90
Homogeneous board from repulped paper	30	0.50	—	2.00	—	0.28	—	—	8.40
Hardboard									
Medium density	50	0.73	—	1.37	—	0.31	—	—	17.60
High density, service temp. service underlay	55	0.82	—	1.22	—	0.32	—	—	17.60
High density, std. tempered	63	1.00	—	1.00	—	0.32	—	—	20.16
Particleboard									
Low density	37	0.54	—	1.85	—	0.31			11.47
Medium density	50	0.94	—	1.06	—	0.31			15.50
High density	62.5	1.18	—	0.85	—	0.31			19.38
Underlayment ....0.625 in.	40	—	1.22	—	0.82	0.29	2.08	0.6	11.60
Wood subfloor ....0.75 in.		—	1.06	—	0.94	0.33	2.00	0.6	9.60
BUILDING MEMBRANE									
Vapor—permeable felt	—	—	16.70	—	0.06		—	—	—
Vapor—seal, 2 layers of mopped 15-lb felt	—	—	8.35	—	0.12		—	—	—
Vapor—seal, plastic film	—	—	—	—	Negl.		—	—	—
FINISH FLOORING MATERIALS									
Carpet and fibrous pad	—	—	0.48	—	2.08	0.34	—	—	—
Carpet and rubber pad	—	—	0.81	—	1.23	0.33	—	—	—
Cork tile ....0.125 in.	—	—	3.60	—	0.28	0.48	—	—	—
Terrazzo ....1 in.	—	—	12.50	—	0.08	0.19	11.7	2.22	26.60
Tile—asphalt, linoleum, vinyl, rubber	—	—	20.00	—	0.05	0.30	—	—	—
vinyl asbestos						0.24	—	—	—
ceramic						0.19	—	—	—
Wood, hardwood finish ....0.75 in.			1.47		0.68		2.81	0.84	13.50
INSULATING MATERIALS									
BLANKET AND BATT									
Mineral Fiber, fibrous form processed from rock, slag, or glass									
approx.[e] 2-2.75 in.	0.3-2.0	—	0.143	—	7[d]	0.17-0.23	.12-.40	.02-.09	0.1-0.46
approx.[e] 3-3.5 in.	0.3-2.0	—	0.091	—	11[d]		.16-.54	.03-.12	0.1-0.46
approx.[e] 3.50-6.5	0.3-2.0	—	0.053	—	19[d]		.30-98	.05-.23	0.1-0.46
approx.[e] 6-7 in.	0.3-2.0	—	0.045	—	22[d]		.30-1.10	.05-.25	0.1-0.46
approx.[d] 8.5 in.	0.3-2.0	—	0.033	—	30[d]		.40-1.42	.07-.32	0.1-0.46
BOARD AND SLABS									
Cellular glass	8.5	0.38	—	2.63	—	0.24	—	—	2.64
Glass fiber, organic bonded	4-9	0.25	—	4.00	—	0.23	—	—	.9-2.1
Expanded rubber (rigid)	4.5	0.22	—	4.55	—	0.40	—	—	1.8
Expanded polystyrene extruded Cut cell surface	1.8	0.25	—	4.00	—	0.29	—	—	0.52
Expanded polystyrene extruded Smooth skin surface	2.2	0.20	—	5.00	—	0.29	—	—	0.64
Expanded polystyrene extruded Smooth skin surface	3.5	0.19	—	5.26	—		—	—	1.02
Expanded polystyrene, molded beads	1.0	0.28	—	3.57	—	0.29	—	—	0.29
Expanded polyurethane[f] (R-11 exp.)	1.5	0.16	—	6.25	—	0.38	—	—	0.57
(Thickness 1 in. or greater)	2.5								

Thermal properties of materials
Figure 5-12B

Thermal Properties of Typical Building and Insulating Materials — (Design Values)[a]

Description	Customary Unit								
	Density (lb/ft³)	Conductivity (k)	Conductance (C)	Resistance[b] (R)		Specific Heat, Btu/(lb) (deg F)	Wt lb ft²	Heat Capacity	
				Per inch thickness (1/k)	For thickness listed (1/C)			Btu ft²·F	Btu ft³·F
Mineral fiber with resin binder	15	0.29	—	3.45	—	0.17	—	—	9.15
Mineral fiberboard, wet felted									
Core or roof insulation	16–17	0.34	—	2.94	—		—	—	2.2-2.4
Acoustical tile	18	0.35	—	2.86	—	0.19	—	—	3.42
Acoustical tile	21	0.37	—	2.70	—		—	—	2.94
Mineral fiberboard, wet molded									
Acoustical tile[g]	23	0.42	—	2.38	—	0.14	—	—	3.22
Wood or cane fiberboard									
Acoustical tile[g] .... 0.5 in.	—	—	0.80	—	1.25	0.31	—	—	—
Acoustical tile[g] .... 0.75 in.	—	—	0.53	—	1.89		—	—	—
Interior finish (plank, tile)	15	0.35	—	2.86	—	0.32	—	—	4.80
Wood shredded (cemented in preformed slabs)	22	0.60	—	1.67	—	0.31	—	—	6.82
LOOSE FILL									
Cellulosic insulation (milled paper or wood pulp)	2.3-3.2	0.27-0.32	—	3.13-3.70	—	0.33	—	—	.76-1.06
Sawdust or shavings	8.0-15.0	0.45	—	2.22	—	0.33	—	—	2.64-4.45
Wood fiber, softwoods	2.0-3.5	0.30	—	3.33	—	0.33	—	—	.66-1.16
Perlite, expanded	5.0-8.0	0.37	—	2.70	—	0.26	—	—	1.3 -2.08
Mineral fiber (rock, slag or glass)									
approx.[e] 3.75-5 in.	0.6-2.0	—	—		11	0.17	0.2 - .71	.04-.12	0.1
approx.[e] 6.5-8.75 in.	0.6-2.0	—	—		19		.51-1.27	.06-.22	0.1- .34
approx.[e] 7.5-10 in.	0.6-2.0	—	—		22		.45-1.46	.07-.25	0.1- .34
approx.[e] 10.25-13.75 in.	0.6-2.0	—	—		30		.60-2.02	.1 -.34	0.1- .34
Vermiculite, exfoliated	7.0-8.2	0.47	—	2.13	—	3.20	—	—	1.4-1.64
	4.0-6.0	0.44	—	2.27	—		—	—	0.8-1.2
ROOF INSULATION[h]									
Preformed, for use above deck									
Different roof insulations are available in different thicknesses to provide the design C values listed.[h]			0.72 to		1.39 to		—	—	—
Consult individual manufacturers for actual *thickness of their material.*			0.12		8.33		—	—	—
MASONRY MATERIALS									
CONCRETES									
Cement mortar	116	5.0	—	0.20	—		—	—	23.2
Gypsum-fiber concrete 87.5% gypsum, 12.5% wood chips	51	1.66	—	0.60	—	0.21	—	—	10.71
Lightweight aggregates including expanded shale, clay or slate; expanded slags; cinders; pumice; vermiculite; also cellular concretes	120	5.2	—	0.19	—		—	—	24.0
	100	3.6	—	0.28	—		—	—	20.0
	80	2.5	—	0.40	—		—	—	16.0
	60	1.7	—	0.59	—		—	—	12.0
	40	1.15	—	0.86	—		—	—	8.0
	30	0.90	—	1.11	—		—	—	6.0
	20	0.70		1.43			—	—	4.0
Perlite, expanded	40	0.93		1.08			—	—	12.8
	30	0.71		1.41			—	—	9.6
	20	0.50		2.00		0.32	—	—	6.4
Sand and gravel or stone aggregate (oven dried)	140	9.0	—	0.11		0.22	—	—	30.8
Sand and gravel or stone aggregate (not dried)	140	12.0	—	0.08			—	—	28.0
Stucco	116	5.0	—	0.20			—	—	23.2
MASONRY UNITS									
Brick, common[i]	120	5.0	—	0.20	—	0.19	—	—	22.8
Brick, face[i]	130	9.0	—	0.11	—		—	—	24.7
Clay tile, hollow:									
1 cell deep .... 3 in.	—	—	1.25	—	0.80	0.21	15.0	3.2	12.6
1 cell deep .... 4 in.	—	—	0.90	—	1.11		16.0	3.4	10.1
2 cells deep .... 6 in.	—	—	0.66	—	1.52		25.0	5.25	10.5
2 cells deep .... 8 in.	—	—	0.54	—	1.85		30.0	6.3	9.5
2 cells deep .... 10 in.	—	—	0.45	—	2.22		35.0	7.4	8.8
3 cells deep .... 12 in.	—	—	0.40	—	2.50		40.0	8.4	8.4
Concrete blocks, three oval core:									
Sand and gravel aggregate .... 4 in.	—	—	1.40	—	0.71	0.22	23.0	5.1	15.2
.... 8 in.	—	—	0.90	—	1.11		43.0	9.4	14.1
.... 12 in.	—	—	0.78	—	1.28		63.0	13.9	13.9
Cinder aggregate .... 3 in.	—	—	1.16	—	0.86	0.21	17.0	3.6	14.3
.... 4 in.	—	—	0.90	—	1.11		20.0	4.2	12.6
.... 8 in.	—	—	0.58	—	1.72		37.0	7.9	11.8
.... 12 in.	—	—	0.53	—	1.89		53.0	11.1	11.1
Lightweight aggregate .... 3 in.	—	—	0.79	—	1.27	0.21	15.0	2.6	12.6
(expanded shale, clay, slate .... 4 in.	—	—	0.67	—	1.50		17.0	3.6	10.9
or slag; pumice) .... 8 in.	—	—	0.50	—	2.00		32.0	6.7	10.1
.... 12 in.	—	—	0.44	—	2.27		43.0	9.0	9.0
Concrete blocks, rectangular core.[*][j]									
Sand and gravel aggregate									
2 core, 8 in. 36 lb.[k*]	—	—	0.96	—	1.04	0.22	43.1	9.5	14.2
Same with filled cores[j*]	—	—	0.52	—	1.93	0.22	—	—	—

**Thermal properties of materials
Figure 5-12B (continued)**

Thermal Properties of Typical Building and Insulating Materials — (Design Values)[a]

Description	Density (lb/ft³)	Conductivity (k)	Conductance (C)	Resistance[b] (R) Per inch thickness (1/k)	Resistance[b] (R) For thickness listed (1/C)	Specific Heat, Btu/(lb) (deg F)	Wt lb ft²	Heat Capacity Btu ft²·F	Heat Capacity Btu ft³·F
Lightweight aggregate (expanded shale, clay, slate or slag, pumice):									
3 core, 6 in. 19 lb. k*	—	—	0.61	—	1.65	0.21	22.8	4.8	9.6
Same with filled cores¹*	—	—	0.33	—	2.99				
2 core, 8 in. 24 lb. k*	—	—	0.46	—	2.18		28.8	6.0	9.1
Same with filled cores¹*	—	—	0.20	—	5.03				
3 core, 12 in. 38 lb. k*	—	—	0.40	—	2.48		45.6	9.6	9.6
Same with filled cores¹*	—	—	0.17	—	5.82				
Stone, lime or sand.	—	12.50	—	0.08	—	0.19			28.5
Gypsum partition tile:						0.19			
3 × 12 × 30 in. solid	—	—	0.79	—	1.26		11.0	2.1	8.6
3 × 12 × 30 in. 4-cell	—	—	0.74	—	1.35		9.0	1.7	6.7
4 × 12 × 30 in. 3-cell	—	—	0.60	—	1.67		13.0	2.5	7.2
PLASTERING MATERIALS									
Cement plaster, sand aggregate	116	5.0	—	0.20	—	0.20	—	—	23.2
Sand aggregate ... 0.375 in.	—	—	13.3	—	0.08	0.20	3.63	0.72	23.2
Sand aggregate ... 0.75 in.	—	—	6.66	—	0.15	0.20	7.25	1.45	23.2
Gypsum plaster:									
Lightweight aggregate ... 0.5 in.	45	—	3.12	—	0.32		1.88	0.38	9.0
Lightweight aggregate ... 0.625 in.	45	—	2.67	—	0.39		2.34	0.47	9.0
Lightweight agg. on metal lath ... 0.75 in.	—	—	2.13	—	0.47		—	0.57	
Perlite aggregate	45	1.5	—	0.67	—	0.32	—	—	14.4
Sand aggregate	105	5.6	—	0.18	—	0.20	—	—	21.0
Sand aggregate ... 0.5 in.	105	—	11.10	—	0.09		4.38	0.88	21.0
Sand aggregate ... 0.625 in.	105	—	9.10	—	0.11		5.47	1.09	21.0
Sand aggregate on metal lath ... 0.75 in.	—	—	7.70	—	0.13		—	1.32	
Vermiculite aggregate	45	1.7	—	0.59	—		—	—	9.0
ROOFING									
Asbestos-cement shingles	120	—	4.76	—	0.21	0.24	—	—	28.8
Asphalt roll roofing	70	—	6.50	—	0.15	0.36	—	—	25.2
Asphalt shingles	70	—	2.27	—	0.44	0.30	—	—	25.2
Built-up roofing ... 0.375 in.	70	—	3.00	—	0.33	0.35	2.19	0.73	24.5
Slate ... 0.5 in.	—	—	20.00	—	0.05	0.30	—	—	
Wood shingles, plain and plastic film faced	—	—	1.06	—	0.94	0.31	—	—	
SIDING MATERIALS (On Flat Surface)									
Shingles									
Asbestos-cement	120	—	4.75	—	0.21		—	—	28.8
Wood, 16 in., 7.5 exposure	—	—	1.15	—	0.87	0.31	—	—	—
Wood, double, 16-in., 12-in. exposure	—	—	0.84	—	1.19	0.28	—	—	—
Wood, plus insul. backer board, 0.3125 in.	—	—	0.71	—	1.40	0.31	—	—	—
Siding									
Asbestos-cement, 0.25 in., lapped	—	—	4.76	—	0.21	0.24	—	—	29.5
Asphalt roll siding	—	—	6.50	—	0.15	0.35	—	—	24.5
Asphalt insulating siding (0.5 in. bed.)	—	—	0.69	—	1.46	0.35	—	—	—
Wood, drop, 1 × 8 in.	—	—	1.27	—	0.79	0.28	—	—	—
Wood, bevel, 0.5 × 8 in., lapped	—	—	1.23	—	0.81	0.28	—	—	—
Wood, bevel, 0.75 × 10 in., lapped	—	—	0.95	—	1.05	0.28	—	—	—
Wood, plywood, 0.375 in., lapped	—	—	1.59	—	0.59	0.29	—	—	—
Wood, medium density siding, 0.4375 in.	40	1.49	—	0.67	—	0.28	—	—	11.5
Aluminum or Steel[m], over sheathing Hollow-backed	—	—	1.61	—	0.61	0.29	—	—	—
Insulating-board backed nominal 0.375 in.	—	—	0.55	—	1.82	0.32	—	—	—
Insulating-board backed nominal 0.375 in., foil backed.		—	0.34		2.96	0.32			
Architectural glass	—	—	10.00	—	0.10	0.20			
WOODS									
Maple, oak, and similar hardwoods	45	1.10	—	0.91	—	0.30	—	—	13.5
Fir, pine, and similar softwoods	32	0.80	—	1.25	—	0.33	—	—	10.6
Fir, pine, and similar softwoods ... 0.75 in.	32	—	1.06	—	0.94	0.33	2.0	0.66	10.6
... 1.5 in.		—	0.53	—	1.89		4.0	1.32	10.6
... 2.5 in.		—	0.32	—	3.12		6.7	2.20	10.6
... 3.5 in.		—	0.23	—	4.35		9.3	3.08	10.6

[a] Representative values for dry materials were selected by ASHRAE TC4.4, Insulation and Moisture Barriers. They are intended as design (not specification) values for materials in normal use. For properties of a particular product, use the value supplied by the manufacturer or by unbiased tests

[b] Resistance values are the reciprocals of C before rounding off C to two decimal places.

[c] Also see Insulating Materials, Board.

[d] Does not include paper backing and facing, if any. Where insulation forms a boundary (reflective or otherwise) of an air space, see Tables 3.3 and 3.4 for the insulating value of air space for the appropriate effective emittance and temperature conditions of the space.

[e] Conductivity varies with fiber diameter. Insulation is produced by different densities; therefore, there is a wide variation in thickness for the same R-value among manufacturers. No effort should be made to relate any specific R-value to any specific thickness.

[f] Values are for aged board stock. For change in conductivity with age of expanded urethane, see Chapter 19, Factors Affecting Thermal Conductivity. 1977 Fundamentals Volume.

[g] Insulating values of acoustical tile vary, depending on density of the board and on type, size, and depth of perforations.

[h] The U. S. Department of Commerce, *Simplified Practice Recommendation for Thermal Conductance Factors for Preformed Above-Deck Roof Insulation*, No. R 257-55, recognizes the specification of roof insulation on the basis of the C-values shown. Roof insulation is made in thicknesses to meet these values.

[i] Face brick and common brick do not always have these specific densities. When density is different from that shown, there will be a change in thermal conductivity.

[j] Data on rectangular core concrete blocks differ from the above data on oval core blocks, due to core configuration, different mean temperatures, and possibly differences in unit weights. Weight data on the oval core blocks tested are not available.

[k] Weights of units approximately 7.625 in. high and 15.75 in. long. These weights are given as a means of describing the blocks tested, but conductance values are all for 1 ft² of area.

[l] Vermiculite, perlite, or mineral wool insulation. Where insulation is used, vapor barriers or other precautions must be considered to keep insulation dry.

[m] Values for metal siding applied over flat surfaces vary widely, depending on amount of ventilation of air space beneath the siding; whether air space is reflective or nonreflective; and on thickness, type, and application of insulating backing-board used. Values given are averages for use as design guides, and were obtained from several guarded hotbox tests (ASTM C236) or calibrated hotbox (BSS 77) on hollow-backed types and types made using backing-boards of wood fiber, foamed plastic, and glass fiber. Departures of ±50% or more from the values given may occur.

Thermal properties of materials
Figure 5-12B (continued)

Heat Loss of Concrete Floors at or Near Grade Level per Foot of Exposed Edge (less than 3 ft below grade)

Outdoor Design Temperature, F	Heat Loss per Foot of Exposed Edge, Btu/(hr·ft)		
	R = 5.0 Edge Insulation	R = 2.5 Edge Insulation	No Edge Insulation [a]
−20 to −30	50	60	75
−10 to −20	45	55	65
0 to −10	40 ←	50	60
+10 to 0	35	45	55
+20 to +10	30	40	50

[a] This construction not recommended; shown for comparison only.

Floor Heat Loss to be Used When Warm Air Perimeter Heating Ducts Are Embedded in Slabs[a] [Btu/hr per (linear foot of heated edge)]

Outdoor Design Temperature, F	Edge Insulation		
	R = 2.5 Vertical Extending Down 18 in. Below Floor Surface	R = 2.5 L-Type Extending at Least 12 in. Deep and 12 in. Under	R = 5 L-Type Extending at Least 12 in. Down and 12 in. Under
−20	105	100	85
−10	95	90	75
0	85	80	65
10	75	70	55
20	62	57	45

[a] Factors include loss downward through inner area of slab.

Heat Loss for Below-Grade Walls with Insulation on Inside Surface (For walls extending more than 3 ft below grade) Average Btu/(hr · ft² · F)

Distance Wall Extends Below-Grade,* ft	Insulation Over Full Surface			Wall Insulated to a Depth of Two Feet Below Grade		
	R-4	R-8	R-13	R-4	R-8	R-13
4	0.110	0.075	0.057	0.136	0.102	0.090
5	0.102	0.071	0.054	0.128	0.100	0.091
6	0.095	0.067	0.052	0.120	0.097	0.089
7	0.089	0.064	0.050	0.112	0.093	0.086

* For a depth below-grade of 3 feet or less, treat as a slab on grade.

Heat Loss Through Basement Floors (For Floors more than 3 ft below grade) Btu/(hr · ft² · F)

Depth of Foundation Wall Below Grade,* ft	Width of House, ft			
	20	24	28	32
4	0.035	0.032	0.027	0.024
5	0.032	0.029	0.026	0.023
6	0.030	0.027	0.025	0.022
7	0.029	0.026	0.023	0.021

*For a depth below-grade of 3 feet or less, treat as a slab on grade.

Heat loss tables
Figure 5-13

Residential Heating and Cooling Loads

Construction Type	Description
Tight ↑	New buildings where there is close supervision of workmanship and special precautions are taken to prevent infiltration. Descriptions for tight windows and doors are given in Tables 5.6 and 5.7.
Medium	Building is constructed using conventional construction procedures. Medium fitting windows and doors are described in Tables 5.6 and 5.7.
Loose	Buildings constructed with poor workmanship or older buildings where joints have separated. Loose windows and doors are described in Tables 5.6 and 5.7.

Design Infiltration Rate, Winter (Heating) (Air Changes/hr) Wind speed = 15 mph

Type of Construction	Winter Outdoor Design Temperature, deg F									
	50	40	30	20	10	0	−10	−20	−30	−40
Tight	0.4	0.5	0.6	0.6	0.7	0.8	0.8	0.9	0.9	1.0
Medium	0.6	0.7	0.8	0.9	1.0	1.1	1.2	1.2	1.3	1.4
Loose	0.8	0.9	1.0	1.2	1.3	1.4	1.5	1.6	1.8	1.9

Design Infiltration Rate, Summer (Cooling) (Air Changes/hr) Wind Speed = 7.5 mph

Type of Construction	Summer Outdoor Design Temperature, deg F					
	85	90	95	100	105	110
Tight	0.3	0.3 ←	0.3	0.4	0.4	0.4
Medium	0.4	0.4	0.5	0.5	0.5	0.6
Loose	0.4	0.5	0.6	0.6	0.7	0.8

Infiltration rates
Figure 5-14

Duct Heat Gain and Loss Allowance

Duct Description	Cooling — Heat Gain, %	Heating — Heat Loss, %			
		Design Heat Loss Btu/(hr·ft²)	Outside Design Temp. F		
			Below 0	0 to +15	Above +15
Located in attic or crawl space and with duct insulation* of R-5 or more	10 ←	Less than 35	20 ←	15	10
		35 and over	15	10	5
Unconditioned basement	5	Less than 35	20	15	10
		35 and over	10	10	5

*Check appropriate standards for insulation required.

Note: Since new residential structures are being built to tighter thermal standards and older structures are being improved thermally, the duct heat loss or gain can become a more significant fraction of the total load. Care must be exercised to reduce duct air leakage as well as duct heat losses or gains. This is particularly true when runs are made through attic spaces, crawl spaces, and unconditioned spaces such as garages and through outside wall stud spaces.

Courtesy: American Society of Heating, Refrigerating and Air-Conditioning Engineers, Inc. (ASHRAE) 1981 Fundamentals Handbook, all rights reserved.

Duct heat gain and loss
Figure 5-15

Design Equivalent Temperature Differences

Design Temperature, deg F	85		90			95			100		105	110
Daily Temperature Range [a]	L	M	L	M	H	L	M	H	M	H	H	H
WALLS AND DOORS												
1. Frame and veneer-on-frame	17.6	13.6	22.6	18.6	13.6	27.6	23.6	18.6	28.6	23.6	28.6	33.6
2. Masonry walls, 8-in. block or brick	10.3	6.3	15.3	11.3	6.3	20.3	16.3	11.3	21.3	16.3	21.3	26.3
3. Partitions, frame	9.0	5.0	14.0	10.0	5.0	19.0	15.0	10.0	20.0	15.0	20.0	25.0
masonry	2.5	0	7.5	3.5	0	12.5	8.5	3.5	13.5	8.5	13.5	18.5
4. Wood doors	17.6	13.6	22.6	18.6	13.6	27.6	23.6	18.6	28.6	23.6	28.6	33.6
CEILINGS AND ROOFS [b]												
1. Ceilings under naturally vented attic or vented flat roof—dark	38.0	34.0	43.0	39.0	34.0	48.0	44.0	39.0	49.0	44.0	49.0	54.0
—light	30.0	26.0	35.0	31.0	26.0	40.0	36.0	31.0	41.0	36.0	41.0	46.0
2. Built-up roof, no ceiling—dark	38.0	34.0	43.0	39.0	34.0	48.0	44.0	39.0	49.0	44.0	49.0	54.0
—light	30.0	26.0	35.0	31.0	26.0	40.0	36.0	31.0	41.0	36.0	41.0	46.0
3. Ceilings under unconditioned rooms	9.0	5.0	14.0	10.0	5.0	19.0	15.0	10.0	20.0	15.0	20.0	25.0
FLOORS												
1. Over unconditioned rooms	9.0	5.0	14.0	10.0	5.0	19.0	15.0	10.0	20.0	15.0	20.0	25.0
2. Over basement, enclosed crawl space or concrete slab on ground	0	0	0	0	0	0	0	0	0	0	0	0
3. Over open crawl space	9.0	5.0	14.0	10.0	5.0	19.0	15.0	10.0	20.0	15.0	20.0	25.0

[a] Daily Temperature Range
L (Low) Calculation Value: 12 deg F.   M (Medium) Calculation Value: 20 deg F.   H (High) Calculation Value: 30 deg F.
Applicable Range: Less than 15 deg F.   Applicable Range: 15 to 25 deg F.   Applicable Range: More than 25 deg F.
[b] Ceiling and Roofs: For roofs in shade, 18-hr average = 11 deg temperature differential. At 90 deg F design and medium daily range, equivalent temperature differential for light-colored roof equals $11 + (0.71)(39 - 11) = 31$ deg F.

Courtesy: American Society of Heating, Refrigerating and Air-Conditioning Engineers, Inc. (ASHRAE) 1981 Fundamentals Handbook, all rights reserved.

Design equivalent temperatures
Figure 5-16

Step 1: Select Design Conditions
The weather data in Figure 5-10, from the U.S. Weather Bureau, is based on station averages over a 30-year period. Our sample residence is in Milwaukee, Wisconsin. Looking in Column 5, we find the winter outside design temperature for Milwaukee is -8°F. Note that the column is headed 99%. This means that in a normal winter, 22 hours — 1% of a 2200-hour base period — would be at or below -8°F. You also have a value in the 97.5% temperature column of -4°F. This means that in a normal winter, the temperature is at or below -4°F for 2.5% of the base period, or 54 hours. We'll be conservative and select the 99% value of -8°F.

Column 6 gives the summer conditions. We'll select the 1% value. The 1% indicates that the temperatures met or exceeded the values given 1% of the time for a base period of 2,928 hours, June through September. For Milwaukee, Column 6 shows a 90°F dry bulb (Fdb) and a 74°F wet bulb (Fwb).

The temperature values from Figure 5-10 were figured using an average wind velocity of 7.5 mph for summer and 15 mph for winter. These are commonly-used values.

Higher wind velocities can increase the heating and cooling loads through increased *infiltration* — the air that the wind blows into a building. Later, you'll see how the calculation of the infiltration load accounts for this. Overall, high-quality, tight construction is recommended to reduce the infiltration effects of high speed winds. Consult the ASHRAE Fundamentals Handbook if your job is located in an area where unusual weather conditions have to be evaluated when choosing outside design temperatures.

We also need to record the summer Mean Daily Range of 21°F from Column 7 to find the cooling loads.

We write all of this temperature data on the load calculation worksheet, Figure 5-8, near the top, under Design Conditions.

We gave a detailed discussion of comfort conditions in Chapter 4. We've selected an inside design of 67°Fdb in winter and 78°Fdb in summer. Relative humidity (rh) in winter for this northern location shouldn't be too high or condensation may occur on the windows. From 15 to 20% is a good range. Humidity in summer should be 50 to 55%. Keep in mind that our sample residence is located in Milwaukee, Wisconsin. You may want

to refer to the ASHRAE Fundamentals Handbook section on comfort conditions before making a determination of inside design data for jobs in your area.

Both inside and outside design conditions may be mandated by local codes. Be sure to check them for your area before making a final choice.

Step 2: Determine U-Values
Heat is lost through a building's exterior walls, roof, windows, doors and foundation. The rate at which the heat moves from the warm area to the outside depends on the type of material through which it passes. These rates have been found through experiments and have been given numerical values. Multiplying these heat transfer constants (U-values) by a temperature difference will give us our load factors.

You can look up the U-values in Figure 5-12A after evaluating the building materials. Write the U-values on the worksheet in the column under UA Calculations.

Windows— You'll find window and door U-values in the tables in Figure 5-12A. The windows in our house are wood frames with double insulating glass, 3/16'' air space. The plans don't call for any special shading so we'll assume they'll have no indoor shade. We'll call them Type A and write in a description in the Window Schedule at the top of our worksheet. We'll call them Type A in case we have other windows of different construction with different U-values on this job.

Find the window sizes and the direction they face from the floor plan and building elevations. Fill in this information on the appropriate lines of the worksheet. Note that we've written in the total area of all the same-type windows facing the same direction in the Entire House column. Next we fill in the U-value from Figure 5-12A. Our U-value is 0.62.

Doors— All three of our doors are called Type A — steel doors with insulated cores. One door leads into the garage. The garage is unheated so we treat it as having outside conditions for design purposes. Our door U-value is 0.59.

Walls— The construction of the walls is shown in Figure 5-4, Section 6-A10. We have a 4'' brick veneer, 1/2'' air space, 1'' polyurethane board and a 2 x 4 wood stud frame 16'' o.c., 3½'' fiberglass insulation and 1/2'' gypsum board interior finish. A wall of this type in the Wall and Roof table in Figure 5-12A has a U-value of 0.058. We'll round it off to 0.06.

When the U-value for a complete wall thickness isn't available, you can calculate overall U-values if you know the construction components. The R-values for common building materials are shown in Figure 5-12B. Tables are also available from materials manufacturers. The total resistance of our wall would be the sum of the resistances of the components, from outside to inside: 4'' brick plus 1/2'' air space plus 1'' polyurethane plus 3½'' fiberglass batt plus 1/2'' gypsum board. Using the separate R-values, we can find the total R-value:

$$.44 + .6 + 6.25 + 11 + .45 = 18.74$$

Since the overall U-value is equal to 1 divided by the R-value, the U-value for the space between the studs is:

$$\frac{1}{18.74} = .053$$

The resistance through the wall at the studs won't be as high. The total R-value for 2 x 4 wood studs equals about 7.90. One divided by 7.90 is 0.126, so our U-value is 0.126. We know that the space between the studs has a U-value of 0.053. For each average square foot of wall, 0.093 is studs and 0.907 is filled with batt insulation. The calculated U-value of the wall, then, equals:

.907 x .053 = .048 (U-value of insulated wall)
.093 x .126 = .012 (U-value of wall at stud)
                .060 (U-value of total wall)

This is close to the 0.058 value from Figure 5-12A.

Checking the elevation of our residence, we find that the brick veneer extends only half-way up the exterior. But, as noted in the table in Figure 5-12A, the remainder of the wood-framed wall has about the same U-values.

The residence has an unheated garage, so we'll assume the conditions in the garage are the same as

outside. The interior garage wall, for heat loss or gain purposes, will be similar to the outside wall of the residence. We'll use the same U-value of 0.058.

Note also that the kitchen and mechanical room walls won't need a U-value since they're not exposed. These two rooms will have a ceiling loss, however. Even so, we won't heat them with a separate heating supply since the loss will be relatively small. Also, the kitchen will pick up heat from the living room. The mechanical room is seldom occupied for long periods of time and will get some heat from its equipment. We'll consider these rooms again when we size the warm air system in the next chapter.

Perimeter heat loss— Heat will also be lost through the concrete slab along the edge at the outside wall. Figure 5-5, Section 1-A11, shows an 8'' (average) solid block wall with 1'' of foam insulation, R-5, along the outside perimeter of the building, extending below the frost line. Figure 5-13 tells us that this type of construction will result in a loss of 40 Btu/hr/linear foot of exposed edge for our outside design temperature, -8°F.

Note that in the two left-hand tables in Figure 5-13 dealing with heat loss per linear foot, the values are not based *per degree* as they are in tables dealing with Btu/hr loss for square feet areas. This means that the value given for a linear foot heat loss *does not* have to be multiplied by the temperature difference. Instead, we enter the value of 40 Btu/hr/linear foot directly into the Load Factors — Heating column on the worksheet, Figure 5-8.

Basement— In our sample residence, the house is built on a slab. We'll assume no floor loss other than at the perimeter. If you have a house with a basement, you'll have to compute the heat loss for the foundation wall that's above and below grade. Enter this in the wall section of the worksheet. Also enter any floor loss as described in Figure 5-13.

Ceiling— The ceiling of our residence has insulation of R-19 (Figure 5-5). Looking up the construction closest to our building in Figure 5-12A, we find a U-value of 0.055. This figure goes on the worksheet.

Step 3: Take-off Building Dimensions
Our worksheet (Figure 5-8) has lines for areas and running (linear) feet for the building's parts. Using

the floor plan and elevations we can determine the dimensions to the nearest foot and calculate the areas. We'll round off the figures to the nearest foot, since this will be accurate enough for our design.

Exterior windows and doors, because they have a different U-value from the walls, are calculated separately. Note that we'll need the total running feet of exposed walls in the building to calculate perimeter loss.

Calculations are only needed for those insulated walls, ceilings, roofs and other building sections exposed to the outside. In our residence, for example, the right-hand wall isn't figured since it's not exposed. It's a party wall, dividing our residence from the other half of the duplex. We also don't figure any exterior side-wall area above the ceiling line, since this is where the insulation ends. On one of your jobs, you might have a two-story building where you wouldn't figure first floor ceilings. This may seem obvious, but it's easy to make a mistake. Check the plans carefully when doing your jobs.

Keep in mind that for our wall dimensions, the Gross Area line on our worksheet is the total footage. We have to subtract the area of the windows and doors, since these have a different U-value. These net wall areas are written on the Above Grade line of the worksheet. Note that there's room on the worksheet for more than one line of wall data in case your job has wall areas of different U-values.

As we mentioned earlier, although the kitchen and mechanical room won't have a wall loss or gain, they will have heat loss through the ceiling. So we need to calculate the ceiling area for these rooms. The hallway ceiling is also calculated, along with a small section of its wall around the door into the garage.

Step 4: Calculate Heating Load Factors
Under the worksheet's UA Calculations, there's a column for multiplying U-values by areas (U x A). These calculations determine the "UA" of the areas.

There's also room to subtotal the data. Take the sum of the U x A for the walls, doors and windows, and divide it by the total area. This gives a U_0-value, which is the average value for the entire wall area. We then add in any ceiling and floor areas and U x A values to calculate an overall U-average for our building. We use these values to determine if the construction meets energy conservation codes. Most codes limit how high a U-average may be.

The worksheet also has a column called Load Factors. To calculate the load factors for heating, multiply the U-values by the design temperature difference. Our temperature difference (TD) is 75°F. For example, here are the calculations for the windows:

$$TD = 67° - (-8°) = 75°F$$
$$Load\ factor = 75° \times .62 = 46.5$$

Doors, walls and ceilings are figured the same way.

Step 5: Calculate Cooling Load Factors
We'll calculate cooling load factors just like we did heating load factors, with two important differences: First, multiply the U-values by the design equivalent temperature differences for the walls, doors, ceilings and roofs instead of by the summer design temperature difference. Look these up in Figure 5-16. Second, take the design cooling load factors for windows directly from the table in Figure 5-17.

Equivalent temperature differences (ETD) are used to account for the interaction of the mass of the building's walls, roof and floors with the changing outside air temperature and solar loads. On a hot day, for example, the surface of a dark roof will absorb and hold heat at a higher rate than the actual outside air temperature. This places a greater cooling load on the HVAC system. This wouldn't be taken into consideration if you figured the design using typical design temperatures.

To find the equivalent temperatures we need, we first take the Mean Daily Range from the Design Conditions section in Figure 5-8. This is 21°F. Footnote A in Figure 5-16 tells us that this is an M (Medium) Calculation Value.

Remember that for our building, the outside design temperature for summer was 90°F. Inside was 78°F. So our computed design TD is 90 minus 78, or 12°F.

Look at Column M under 90°F in Figure 5-16 to find the wall design equivalent temperature of 18.6°F. We'll multiply this by our wall U-value of 0.06 to get a cooling load factor of 1.1. Column M also gives us the ceiling ETD of 39°F, which we multiply by the ceiling U-value of 0.055 to get a cooling load factor of 2.145. Round up to 2.2 as this is more conservative and gives a larger calculated cooling load. Design conditions for each job will vary, so read the footnotes in Figure 5-16 carefully when you use it.

For window heat gain, consider not only the area but also the direction each window faces and its amount of inside shading. Look at Figure 5-17. Write these load factors directly into the cooling load factor column of the worksheet.

Since our doors are steel, we can't use the Equivalent Temperature table in Figure 5-16. Instead we'll use a U-value from Figure 5-12A and multiply it by the summer temperature difference to get the cooling factor.

In residences, the heat gain from inhabitants' body heat is figured at a load factor of 225 Btu/hr per person. It's industry practice to calculate two people for each bedroom in the house. This load is generally assigned to the living room. The load from lights and appliances in the kitchen is averag-

Shade Line Factors

Direction Window Faces	N Latitude, Deg						
	25	30	35	40	45	50	55
E/W	0.8	0.8	0.8	0.8	0.8	0.8	0.8
SE/SW	1.9	1.6	1.4	1.3	1.1	1.0	0.9
S	10.1	5.4	3.6	2.6	2.0	1.7	1.4

Note: Distance shadow line falls below the edge of the overhang equals shade line factor multiplied by width of overhang. Values are averages for 5 hr of greatest solar intensity on August 1.

Outdoor Design Temp	Regular Single Glass						Regular Double Glass						Heat Absorbing Double Glass						Clear Triple Glass		
	85	90	95	100	105	110	85	90	95	100	105	110	85	90	95	100	105	110	85	90	95
No Awnings or inside Shading																					
North	23	27	31	35	39	44	19	21	24	26	28	30	12	14	17	19	21	23	17	19	20
NE and NW	56	60	64	68	72	77	46	48	51	53	55	57	27	29	32	34	36	38	42	43	44
East and West	81	85	89	93	97	102	68	70	73	75	77	79	42	44	47	49	51	53	62	63	64
SE and SW	70	74	78	82	86	91	59	61	64	66	68	70	35	37	40	42	44	46	53	55	56
South	40	44	48	52	56	61	33	35	38	40	42	44	19	21	24	26	28	30	30	31	33
Horiz. Skylight	160	164	168	172	176	181	139	141	144	146	148	150	89	91	94	96	98	100	126	127	129
Draperies or Venetian Blinds																					
North	15	19	23	27	31	36	12	14	17	19	21	23	9	11	14	16	18	20	11	12	14
NE and NW	32	36	40	44	48	53	27	29	32	34	36	38	20	22	25	27	29	31	24	26	27
East and West	48	52	56	60	64	69	42	44	47	49	51	53	30	32	35	37	39	41	38	39	41
SE and SW	40	44	48	52	56	61	35	37	40	42	44	46	24	26	29	31	33	35	32	33	34
South	23	27	31	35	39	44	20	22	25	27	29	31	15	17	20	22	24	26	18	19	21
Roller Shades Half-Drawn																					
North	18	22	26	30	34	39	15	17	20	22	24	26	10	12	15	17	19	21	13	14	15
NE and NW	40	44	48	52	56	61	38	40	43	45	47	49	24	26	29	31	33	35	34	35	35
East and West	61	65	69	73	77	82	54	56	59	61	63	65	35	37	40	42	44	46	49	49	50
SE and SW	52	56	60	64	68	73	46	48	51	53	55	57	30	32	35	37	39	41	41	42	43
South	29	33	37	41	45	50	27	29	32	34	36	38	18	20	23	25	27	29	25	26	26
Awnings																					
North	20	24	28	32	36	41	13	15	18	20	22	24	10	12	15	17	19	21	11	12	13
NE and NW	21	25	29	33	37	42	14	16	19	21	23	25	11	13	16	18	20	22	12	13	14
East and West	22	26	30	34	38	43	14	16	19	21	23	25	12	14	17	19	21	23	12	13	14
SE and SW	21	25	29	33	37	42	14	16	19	21	23	25	11	13	16	18	20	22	12	13	14
South	21	24	28	32	36	41	13	15	18	20	22	24	11	13	16	18	20	22	11	12	13

Design cooling load factors through glass
Figure 5-17

ed out at 1,200 Btu/hr. Write this data in the appropriate columns on the datasheet. In commercial jobs, lights and equipment can add a heavier load. Large numbers of building occupants also add heat. Chapter 8 shows how to figure these commercial loads.

Before continuing the steps, take the load factors and areas from the worksheet and transfer them to the datasheet (Figure 5-9). Calculate the Btu/hr losses and gains for each room by multiplying the load factors by the areas. Also figure linear edge (perimeter) losses.

Note that the calculation sheet in Figure 5-9 is laid out so you can figure the loads two ways: You can calculate the load for each room, or you can calculate like loads for the total areas they represent in the entire house. You should calculate the values both ways, to cross-check your work. You'll also need room losses and gains to size the ductwork.

After entering the data in the appropriate columns in Figure 5-9, we can get subtotals for the entire job. Before we can find the total Room Sensible Heat Losses, we have to account for heat loss and gain from air infiltration and from our supply ducts.

Step 6: Calculate Infiltration Factors

The infiltration rate for our building is determined from the tables in Figure 5-14. With tight construction, at a winter temperature of -8°F (closest to -10°F in the table), it's equal to 0.8 air changes per hour. This means that for our house, 80% of the volume of the air inside the house will be displaced each hour by outside air. This outside air must be heated. Summer air will also infiltrate and need to be cooled.

The first step is to find the infiltration heating load factor (HLF). This will give us a value that we can apply to different areas of our residence to account for the extra Btu's needed to heat the infiltrated air. The formula, from Figure 5-11, is:

Btu/hr = building volume x air changes x Btu/cubic foot/hr x TD

This formula is made up of the variables that represent our building volumes and air changes, and the math constants that represent the heat we need. Here are the variables we already know:

HLF = Btu/hr factor
Temperature inside (Ti) - Temperature outside (To) =
$67° - (-8°) = 75°$

Air changes/hr = .8

Math constants = .075 and .24 (based on Btu's needed to heat 1 cu. ft. of air/hr.)

Substituting them into the formula:

$$HLF = \text{room volume} \times .8 \times .075 \times .24 \times 75°$$

Before we can plug in the building volumes, keep in mind that we want the answer to be a factor that we can apply to all the rooms or areas on our worksheets. These are designated on the worksheet in square feet, or area. But volume equals area times height. The height of all rooms in our sample residence is 8 feet. So we can plug that height into our HLF formula and use the different areas as they come up on the worksheet to get the volumes and hence the infiltration heat loss for each. But, like the number of air changes and the temperature difference, this height of 8 feet is a variable. It only applies to this job. You may have jobs where it's different. Don't forget to change these variables as your job requires. Be especially careful if your height varies from room to room.

For our job then:

$$HLF = (\text{area}) \times 8' \times .8 \times .075 \times .24 \times 75°$$
$$HLF = (\text{area}) \times 8.64$$

We'll write the HLF of 8.64 in the Load Factor column at the bottom of Figure 5-9. The Btu/hr loss due to infiltration is found by multiplying the HLF by the area we're figuring. For example, for our whole building:

$$HLF = 8.64 \times 937 \text{ sq. ft.} = 8,096 \text{ Btu/hr}$$

Note that the HLF we used was based on tight construction. Poorer building types will need adjustment, as shown in Figure 5-14. Also, when we calculate heat losses for each room, we choose to ignore infiltration for our interior rooms. Although in theory there may be some loss through the roof areas, it's small enough to be handled by our dimensional heat losses. But if you have a building in a windy location or with interior rooms with very large roof areas, you may want to calculate infiltration through the roof areas.

The cooling load caused by infiltration is handled by the same formula used for heat load infiltration. But the values for the design temperature and air changes/hour are different:

$$To - Ti = 90° - 78° = 12°F$$
Air changes = .3 (from Figure 5-14)

Substituting these values into the formula gives us:

$$CLF = (area) \times 8' \times .3 \times .075 \times .24 \times 12°F$$
$$CLF = (area) \times .518$$

Enter the CLF on the datasheet. Multiply the total square footage of the residence by 0.518 to get a CLF heat gain of:

$$937 \times .518 = 485 \text{ Btu/hr}$$

Step 7: Calculate Duct Loss and Gain
Besides loads due to infiltration, the total room sensible loads will include some heat lost and gained from the duct system. Look at Figure 5-15, duct heat gain and loss, for this factor.

For our residence, we'll take the subtotal Btu/hr loss of 23,546 and divide it by our total residence square footage of 937. That gives us an average per foot loss of 25.1 Btu/hr/sq. ft. Then we refer to the table in Figure 5-15. Since our design heat loss is less than 35 Btu/hr/sq. ft., and we'll run our ducts with insulation in the attic, we find the duct heat loss will be 20%. Write this in the lower left of the datasheet. Taking 20% times our subtotal of 23,546 Btu/hr gives 4,709 Btu/hr loss in the ductwork. This gives us a total Btu/hr residence loss of 28,255.

Figure duct heat gain the same way as the duct heat loss. After subtotaling the Btu/hr gain for the residence, we go to the table in Figure 5-15. Here we see that the Btu/hr gain isn't that critical. Instead, a value is chosen only on where the duct will run and how it will be insulated. For our residence, then, duct heat gain is 10%.

Step 8: Summarize Worksheet Data
After totaling the calculations, we summarize the data at the upper left of the worksheet, Figure 5-8, in the Load Calculation Summary for the Entire House. By dividing our total heat loss by the design temperature difference of 75°F, we get a Btu/hr/degree F value of 377. We can also compare by using the Btu/hr loss per square foot (28,255/937 equals 30.1). These values are useful in comparing like-size buildings as to energy efficiency.

Latent Cooling Load— For our Cooling Load Summary, we have a sensible heat gain (SHG) of 7,905 Btu/hr. The total load is the sensible load plus an allowance factor for latent heat gain (LHG). This allowance factor, from Figure 5-11, is 1.2 for dry climates and 1.3 for all others. Since our building is in Milwaukee, we'll use the 1.3 allowance:

$$1.3 \times 7905 = 10,277 \text{ total Btu/hr gain}$$

You can find another cooling load comparison by using the ton rating. This method of measuring is often used in cooling and refrigeration design and is based on the rate of cooling produced when one ton of ice melts during 24 hours. One ton of refrigeration equals 12,000 Btu/hr cooling capacity. For our residence:

$$10,277 / 12,000 = .86 \text{ tons}$$
$$937 / .86 = 1,089 \text{ sq. ft./ton}$$

Computer Programs
By now you've discovered that heat loss and heat gain calculations aren't difficult, but they do require careful input and accurate arithmetic. They take time. You can use a computer program to reduce calculation time and insure accuracy. Relatively inexpensive programs are available to run on small personal computers. On larger projects they can save time and money. They also make it easy to find the difference in heating and cooling loads when you change building components. For example, changing from double to triple pane windows will change the loads. A computer will do these calculations quickly.

An input list of program data for our residence, which we calculated manually in the above steps, is shown in Figure 5-18. The output after the program was run is shown in Figure 5-19.

The program we used is based on a modified ASHRAE method of load calculation. To check their accuracy, the manual calculations in Figures 5-8 and 5-9 should be within plus/minus 7% of the results from the computer program. This computer program runs on an IBM-PC, Zenith, Eagle and other IBM-compatible machines. The program and input instructions are sold by Blitz Publishing Company, 1600 Verona Street, Middleton, Wisconsin 53562.

```
04 CCFH
Duplex B Milwaukee, Wis Design conditions winter outside -8, 15mph wind,
inside 67F, summer outside 90DB, 74WB, inside 78F. U factors wall U=.058,
roof ceiling U=.055 glass and doors U=.62 infiltration by the air change
method.
 1 22 27 31 33 34 33 31 27 22 17 13
 2118 72 36 34 34 33 31 27 22 15  6
 3213195147 80 37 33 31 27 22 15  6
 4184202194167127 60 33 27 22 15  6
 5 471181341561651561341118 47 18  6
 6 22 27 33 60116167195202184138 54
 7 22 27 31 33 37 80147195214190 91
 8 22 26 31 33 34 34 36 72118131 74
 6 14 21 27 34 36 38 35 32 25 18
0.700.000.00.0557.44 .60
46.54.350.000.004.13 40      77
L/D  2112 8 12 8      118            121      21 12 12      15
CLOST 9 3 8  3 8                        9  3  3      8
BR   11113 8 24 8    116                11 13 23      15      70
BATH  6 8 8  6 8                        6  8  6      8
HALL  8 4 8                             8  4         8
BR   21014 8 24 8              116      10 14 24      15      70
HALL 17 4 8                             17  4         8
KITH  8 8 8                             8  8          8
NE/LA 8 8 8                             8  8          8
ENTRY 7 6 8  6 8              126       7  6  6      3
COATS 7 5 8                             7  5          8
STOR  7 7 8 12 8                        7  7 12      15
FINIS
```

Computer program input
Figure 5-18

Estimating Fuel Consumption

It's possible to make an estimate of the fuel used to heat a residence or commercial building. The degree-day method is widely used.

A *degree-day* is the difference between 65°F and the average of the high and low temperatures on a given day. The higher the number of degree-days, the more fuel used for heating.

For example, if the high temperature is 42°F and the low temperature 22°F on a particular day, the average is 32°F. Subtract this from 65°F. This particular day will have 65 minus 32, or 33 degree-days.

If we wish to compare a building in different years or months when the occupancy levels are about the same, degree-days compensate for different weather conditions over time. Although a degree-day reading is generally accurate, it ignores factors such as excessive infiltration caused by high winds or extra heat from the sun. The degree-day data is collected for the heating season. Figure 5-20A shows monthly and yearly degree-days for many cities in the country. Degree-day data is also available from your local weather bureau or utility company.

You'll find a detailed degree-day method in the ASHRAE Systems Handbook. But if we know the heat loss of the residence, the type of fuel, and the system heating efficiency, we can use the equation from Figure 5-11 to estimate the fuel used:

$$E = \frac{He \times D \times 24 \times C}{(Ti-To) \times k \times V}$$

In our example, we'll use natural gas at 1,000 Btu/cubic foot (Figure 5-21). Our residence has a heat loss of 28,255 Btu/hr. We'll choose a seasonal system efficiency of 65%, typical for gas and oil systems. (Manufacturers supply efficiencies for their heating and cooling units, as do the Underwriter's Laboratories and the Federal Department of Energy.) In Milwaukee, the annual degree-days equal 7,635 (Figure 5-20).

$$E = \frac{28,255 \times 7,635 \times 24 \times .7}{(67-(-8)) \times .65 \times 1,000}$$

$$E = 74,342 \text{ cubic feet per year}$$

If gas costs $.60 per 100 cubic feet, our estimated average annual bill will be:

$$\frac{74,342}{100} = 743 \times \$.60 = \$446$$

```
Duplex B Milwaukee, Wis Design conditions winter outside -8, 15mph wind,
inside 67F, summer outside 90DB, 74WB, inside 78F. U factors wall U=.058,
roof ceiling U=.055 glass and doors U=.62 infiltration by the air change
method.
   8 22, 27, 31, 33, 34, 33, 31, 27, 22, 17, 13,
   8118, 72, 36, 34, 34, 33, 31, 27, 22, 15,  6,
   8213,195,147, 80, 37, 33, 31, 27, 22, 15,  6,
   8184,202,194,167,127, 60, 33, 27, 22, 15,  6,
   8 47,118,134,156,165,156,134,118, 47, 18,  6,
   8 22, 27, 33, 60,116,167,195,202,184,138, 54,
   8 22, 27, 31, 33, 37, 80,147,195,214,190, 91,
   8 22, 26, 31, 33, 34, 34, 36, 72,118,131, 74,
   6, 14, 21, 27, 34, 36, 38, 35, 32, 25, 18,
       .70    .00    .00    .05    7.44    .60
      46.50   4.35   .00    .00    4.13   40.00    .00   77.00
```

RM=ROOM NUMBER L=ROOM LENGTH W=ROOM WIDTH H=ROOM HEIGHTROOM DIMENSIONS TO
NEARST FOOT AREA=ROOM AREA SQ.FT. VOL=ROOM VOLUME CUBIC FT. LOSS=HEAT LOSS BTU/HR
EK=KITCHEN EXHAUST CFM ET=TOILET EXHAUST CFM HGPM=GPM HEATING WATER FLOW BASED ON
20 DEGREE DROP GAIN= HEAT GAIN BTU/HR CGPM=GPM COOLING WATER FLOW   BASED ON 20
DEGREE RISE HCFM=HEATING AIR FLOW BASED ON 6   AIR CHANGES CCFM=COOLING AIR FLOW
BASED ON 20 DEGREE RISE

RM	L	W	H	AREA	VOL	LOSS	EK	ET	HCFM	GAIN	MAX CCFM	MIN CCFM
L/D	21.	12.	8.	252.	2016.	7773.	0.	0.	201.6	3708.	165.	41.
CLOST	9.	3.	8.	27.	216.	575.	0.	0.	21.6	73.	3.	1.
BR 1	11.	13.	8.	143.	1144.	5399.	0.	0.	114.4	1516.	58.	43.
BATH	6.	8.	8.	48.	384.	1073.	0.	0.	38.4	134.	6.	2.
HALL	8.	4.	8.	32.	256.	416.	0.	0.	25.6	67.	3.	0.
BR 2	10.	14.	8.	140.	1120.	5376.	0.	0.	112.0	3203.	133.	42.
HALL	17.	4.	8.	68.	544.	884.	0.	0.	54.4	142.	6.	1.
KITH	8.	8.	8.	64.	512.	832.	0.	0.	51.2	134.	6.	1.
NE/LA	8.	8.	8.	64.	512.	832.	0.	0.	51.2	134.	6.	1.
ENTRY	7.	6.	8.	42.	336.	1858.	0.	0.	33.6	2861.	127.	15.
COATS	7.	5.	8.	35.	280.	455.	0.	0.	28.0	73.	3.	1.
STOR	7.	7.	8.	49.	392.	1915.	0.	0.	39.2	170.	8.	4.

```
TOTALS, HEAT LOSS =   27388,  VOL =    7712, EXH =    0,   BTU/SQ.FT.= 28,   TOTAL AREA=   964,
  BTU/SQ.FT. SHELL= 8.76   WINDOW AREA SQ.FT.= 97.SHELL AREA SQ.FT.= 1660,
PERCENT OF SENSIBLE TO TOTAL = 96.16
MAXIMUM TOTAL HEAT GAIN IS   11051, AT 3 PM,  GLASS GAIN=   7374,   TOTAL WATTS BTU=        0,
  PEOPLE  GAIN=    1334,   OUTSIDE AIR CFM =     20,   SQ.FT./TON=1047,
```

Computer program output
Figure 5-19

Any method of determining fuel consumption must be used with care. The results are only as good as the input data.

Remember these methods are *estimates* and not guarantees of fuel consumption. This is because of the large differences in weather from year to year and the variation in the way people use their residences.

A computer program for personal computers is an easy way to calculate fuel consumption and cost. You can change data easily to compare different design temperatures, equipment efficiency and changing fuel costs. Figure 5-22 shows the input and output for the sample residence. You can purchase this program from Blitz Publishing Company, at the address just given.

Now we have all the data we need to begin designing the warm air system for our sample residence. That's the subject of the next chapter.

Average Winter Temperature and Yearly Degree Days for Cities in the United States and Canada[a,b,c] (Base 65F)

State	Station		Avg. Winter Temp.[d] F	Degree-Days Yearly Total	State	Station		Avg. Winter Temp. F	Degree-Days Yearly Total
Ala.	Birmingham	A	54.2	2551	Fla.	Miami Beach	C	72.5	141
	Huntsville	A	51.3	3070	(Cont'd)	Orlando	A	65.7	766
	Mobile	A	59.9	1560		Pensacola	A	60.4	1463
	Montgomery	A	55.4	2291		Tallahassee	A	60.1	1485
Alaska	Anchorage	A	23.0	10864		Tampa	A	66.4	683
	Fairbanks	A	6.7	14279		West Palm Beach	A	68.4	253
	Juneau	A	32.1	9075	Ga.	Athens	A	51.8	2929
	Nome	A	13.1	14171		Atlanta	A	51.7	2961
Ariz.	Flagstaff	A	35.6	7152		Augusta	A	54.5	2397
	Phoenix	A	58.5	1765		Columbus	A	54.8	2383
	Tucson	A	58.1	1800		Macon	A	56.2	2136
	Winslow	A	43.0	4782		Rome	A	49.9	3326
	Yuma	A	64.2	974		Savannah	A	57.8	1819
Ark.	Fort Smith	A	50.3	3292		Thomasville	C	60.0	1529
	Little Rock	A	50.5	3219	Hawaii	Lihue	A	72.7	0
	Texarkana	A	54.2	2533		Honolulu	A	74.2	0
Calif.	Bakersfield	A	55.4	2122		Hilo	A	71.9	0
	Bishop	A	46.0	4275	Idaho	Boise	A	39.7	5809
	Blue Canyon	A	42.2	5596		Lewiston	A	41.0	5542
	Burbank	A	58.6	1646		Pocatello	A	34.8	7033
	Eureka	C	49.9	4643	Ill.	Cairo	C	47.9	3821
	Fresno	A	53.3	2611		Chicago(O'Hare)	A	35.8	6639
	Long Beach	A	57.8	1803		Chicago(Midway)	A	37.5	6155
	Los Angeles	A	57.4	2061		Chicago	C	38.9	5882
	Los Angeles	C	60.3	1349		Moline	A	36.4	6408
	Mt. Shasta	C	41.2	5722		Peoria	A	38.1	6025
	Oakland	A	53.5	2870		Rockford	A	34.8	6830
	Red Bluff	A	53.8	2515		Springfield	A	40.6	5429
	Sacramento	A	53.9	2502	Ind.	Evansville	A	45.0	4435
	Sacramento	C	54.4	2419		Fort Wayne	A	37.3	6205
	Sandberg	C	46.8	4209		Indianapolis	A	39.6	5699
	San Diego	A	59.5	1458		South Bend	A	36.6	6439
	San Francisco	A	53.4	3015	Iowa	Burlington	A	37.6	6114
	San Francisco	C	55.1	3001		Des Moines	A	35.5	6588
	Santa Maria	A	54.3	2967		Dubuque	A	32.7	7376
Colo.	Alamosa	A	29.7	8529		Sioux City	A	34.0	6951
	Colorado Springs	A	37.3	6423		Waterloo	A	32.6	7320
	Denver	A	37.6	6283	Kans.	Concordia	A	40.4	5479
	Denver	C	40.8	5524		Dodge City	A	42.5	4986
	Grand Junction	A	39.3	5641		Goodland	A	37.8	6141
	Pueblo	A	40.4	5462		Topeka	A	41.7	5182
Conn.	Bridgeport	A	39.9	5617		Wichita	A	44.2	4620
	Hartford	A	37.3	6235	Ky.	Covington	A	41.4	5265
	New Haven	A	39.0	5897		Lexington	A	43.8	4683
Del.	Wilmington	A	42.5	4930		Louisville	A	44.0	4660
D.C.	Washington	A	45.7	4224	La.	Alexandria	A	57.5	1921
Fla.	Apalachicola	C	61.2	1308		Baton Rouge	A	59.8	1560
	Daytona Beach	A	64.5	879		Lake Charles	A	60.5	1459
	Fort Myers	A	68.6	442		New Orleans	A	61.0	1385
	Jacksonville	A	61.9	1239		New Orleans	C	61.8	1254
	Key West	A	73.1	108		Shreveport	A	56.2	2184
	Lakeland	C	66.7	661	Me.	Caribou	A	24.4	9767
	Miami	A	71.1	214		Portland	A	33.0	7511
					Md.	Baltimore	A	43.7	4654

[a] Data for United States cities from a publication of the United States Weather Bureau, *Monthly Normals of Temperature, Precipitation and Heating Degree Days,* 1962, are for the period 1931 to 1960 inclusive. These data also include information from the 1963 revisions to this publication, where available.

[b] Data for airport station, A, and city stations, C, are both given where available.

[c] Data for Canadian cities were computed by the Climatology Division, Department of Transport from normal monthly mean temperatures, and the monthly values of heating days data were obtained using the National Research Council computer and a method devised by H. C. S. Thom of the United States Weather Bureau. The heating days are based on the period from 1931 to 1960.

[d] For period October to April, inclusive.

Degree-days
Figure 5-20A

Average Winter Temperature and Yearly Degree Days for Cities in the United States and Canada (Base 65F)

State	Station		Avg. Winter Temp, F	Degree-Days Yearly Total
	Baltimore	C	46.2	4111
	Frederich	A	42.0	5087
Mass.	Boston	A	40.0	5634
	Nantucket	A	40.2	5891
	Pittsfield	A	32.6	7578
	Worcester	A	34.7	6969
Mich.	Alpena	A	29.7	8506
	Detroit(City)	A	37.2	6232
	Detroit(Wayne)	A	37.1	6293
	Detroit(Willow Run)	A	37.2	6258
	Escanaba	C	29.6	8481
	Flint	A	33.1	7377
	Grand Rapids	A	34.9	6894
	Lansing	A	34.8	6909
	Marquette	C	30.2	8393
	Muskegon	A	36.0	6696
	Sault Ste. Marie	A	27.7	9048
Minn.	Duluth	A	23.4	10000
	Minneapolis	A	28.3	8382
	Rochester	A	28.8	8295
Miss.	Jackson	A	55.7	2239
	Meridian	A	55.4	2289
	Vicksburg	C	56.9	2041
Mo.	Columbia	A	42.3	5046
	Kansas City	A	43.9	4711
	St. Joseph	A	40.3	5484
	St. Louis	A	43.1	4900
	St. Louis	C	44.8	4484
	Springfield	A	44.5	4900
Mont.	Billings	A	34.5	7049
	Glasgow	A	26.4	8996
	Great Falls	A	32.8	7750
	Havre	A	28.1	8700
	Havre	C	29.8	8182
	Helena	A	31.1	8129
	Kalispell	A	31.4	8191
	Miles City	A	31.2	7723
	Missoula	A	31.5	8125
Neb.	Grand Island	A	36.0	6530
	Lincoln	C	38.8	5864
	Norfolk	A	34.0	6979
	North Platte	A	35.5	6684
	Omaha	A	35.6	6612
	Scottsbluff	A	35.9	6673
	Valentine	A	32.6	7425
Nev.	Elko	A	34.0	7433
	Ely	A	33.1	7733
	Las Vegas	A	53.3	2709
	Reno	A	39.3	6332
	Winnemucca	A	36.7	6761
N.H.	Concord	A	33.0	7383
	Mt. Washington Obsv.		15.2	13817
N.J.	Atlantic City	A	43.2	4812
	Newark	A	42.8	4589
	Trenton	C	42.4	4980
N.M.	Albuquerque	A	45.0	4348
	Clayton	A	42.0	5158
	Raton	A	38.1	6228
	Roswell	A	47.5	3793
	Silver City	A	48.0	3705
N.Y.	Albany	A	34.6	6875
	Albany	C	37.2	6201
	Binghamton	A	33.9	7286
	Binghamton	C	36.6	6451
	Buffalo	A	34.5	7062
	New York (Cent. Park)	C	42.8	4871
	New York (LaGuardia)	A	43.1	4811
	New York (Kennedy)	A	41.4	5219
	Rochester	A	35.4	6748
	Schenectady	C	35.4	6650
	Syracuse	A	35.2	6756
N. C.	Asheville	C	46.7	4042
	Cape Hatteras		53.3	2612
	Charlotte	A	50.4	3191
	Greensboro	A	47.5	3805
	Raleigh	A	49.4	3393
	Wilmington	A	54.6	2347
	Winston-Salem	A	48.4	3595
N. D.	Bismarck	A	26.6	8851
	Devils Lake	C	22.4	9901
	Fargo	A	24.8	9226
	Williston	A	25.2	9243
Ohio	Akron-Canton	A	38.1	6037
	Cincinnati	C	45.1	4410
	Cleveland	A	37.2	6351
	Columbus	A	39.7	5660
	Columbus	C	41.5	5211
	Dayton	A	39.8	5622
	Mansfield	A	36.9	6403
	Sandusky	C	39.1	5796
	Toledo	A	36.4	6494
	Youngstown	A	36.8	6417
Okla.	Oklahoma City	A	48.3	3725
	Tulsa	A	47.7	3860
Ore.	Astoria	A	45.6	5186
	Burns	C	35.9	6957
	Eugene	A	45.6	4726
	Meacham	A	34.2	7874
	Medford	A	43.2	5008
	Pendleton	A	42.6	5127
	Portland	A	45.6	4635
	Portland	C	47.4	4109
	Roseburg	A	46.3	4491
	Salem	A	45.4	4754
Pa.	Allentown	A	38.9	5810
	Erie	A	36.8	6451
	Harrisburg	A	41.2	5251
	Philadelphia	A	41.8	5144
	Philadelphia	C	44.5	4486
	Pittsburgh	A	38.4	5987
	Pittsburgh	C	42.2	5053
	Reading	C	42.4	4945
	Scranton	A	37.2	6254
	Williamsport	A	38.5	5934
R. I.	Block Island	A	40.1	5804
	Providence	A	38.8	5954
S. C.	Charleston	A	56.4	2033
	Charleston	C	57.9	1794
	Columbia	A	54.0	2484
	Florence	A	54.5	2387
	Greenville-Spartanburg	A	51.6	2980

Degree-days
Figure 5-20A (continued)

Average Winter Temperature and Yearly Degree Days for Cities in the United States and Canada (Base 65F)

State	Station		Avg. Winter Temp, F	Degree-Days Yearly Total
S. D.	Huron	A	28.8	8223
	Rapid City	A	33.4	7345
	Sioux Falls	A	30.6	7839
Tenn.	Bristol	A	46.2	4143
	Chattanooga	A	50.3	3254
	Knoxville	A	49.2	3494
	Memphis	A	50.5	3232
	Memphis	C	51.6	3015
	Nashville	A	48.9	3578
	Oak Ridge	C	47.7	3817
Tex.	Abilene	A	53.9	2624
	Amarillo	A	47.0	3985
	Austin	A	59.1	1711
	Brownsville	A	67.7	600
	Corpus Christi	A	64.6	914
	Dallas	A	55.3	2363
	El Paso	A	52.9	2700
	Fort Worth	A	55.1	2405
	Galveston	A	62.2	1274
	Galveston	C	62.0	1235
	Houston	A	61.0	1396
	Houston	C	62.0	1278
	Laredo	A	66.0	797
	Lubbock	A	48.8	3578
	Midland	A	53.8	2591
	Port Arthur	A	60.5	1447
	San Angelo	A	56.0	2255
	San Antonio	A	60.1	1546
	Victoria	A	62.7	1173
	Waco	A	57.2	2030
	Wichita Falls	A	53.0	2832
Utah	Milford	A	36.5	6497
	Salt Lake City	A	38.4	6052
	Wendover	A	39.1	5778
Vt.	Burlington	A	29.4	8269
Va.	Cape Henry	C	50.0	3279
	Lynchburg	A	46.0	4166
	Norfolk	A	49.2	3421
	Richmond	A	47.3	3865
	Roanoke	A	46.1	4150
Wash.	Olympia	A	44.2	5236
	Seattle-Tacoma	A	44.2	5145
	Seattle	C	46.9	4424
	Spokane	A	36.5	6655
	Walla Walla	C	43.8	4805
	Yakima	A	39.1	5941
W. Va.	Charleston	A	44.8	4476
	Elkins	A	40.1	5675
	Huntington	A	45.0	4446
	Parkersburg	C	43.5	4754
Wisc.	Green Bay	A	30.3	8029
	La Crosse	A	31.5	7589
	Madison	A	30.9	7863
	Milwaukee	A	32.6	7635 ←
Wyo.	Casper	A	33.4	7410
	Cheyenne	A	34.2	7381
	Lander	A	31.4	7870
	Sheridan	A	32.5	7680

Prov.	Station		Avg. Winter Temp, F	Degree-Days Yearly Total
Alta.	Banff	C	—	10551
	Calgary	A	—	9703
	Edmonton	A	—	10268
	Lethbridge	A	—	8644
B. C.	Kamloops	A	—	6799
	Prince George*	A	—	9755
	Prince Rupert	C	—	7029
	Vancouver*	A	—	5515
	Victoria*	A	—	5699
	Victoria	C	—	5579
Man.	Brandon*	A	—	11036
	Churchill	A	—	16728
	The Pas	C	—	12281
	Winnipeg	A	—	10679
N. B.	Fredericton*	A	—	8671
	Moncton	C	—	8727
	St. John	C	—	8219
Nfld.	Argentia	A	—	8440
	Corner Brook	C	—	8978
	Gander	A	—	9254
	Goose*	A	—	11887
	St. John's*	A	—	8991
N. W. T.	Aklavik	C	—	18017
	Fort Norman	C	—	16109
	Resolution Island	C	—	16021
N. S.	Halifax	C	—	7361
	Sydney	A	—	8049
	Yarmouth	A	—	7340
Ont.	Cochrane	C	—	11412
	Fort William	A	—	10405
	Kapuskasing	C	—	11572
	Kitchener	C	—	7566
	London	A	—	7349
	North Bay	C	—	9219
	Ottawa	C	—	8735
	Toronto	C	—	6827
P.E.I.	Charlottetown	C	—	8164
	Summerside	C	—	8488
Que.	Arvida	C	—	10528
	Montreal*	A	—	8203
	Montreal	C	—	7899
	Quebec*	A	—	9372
	Quebec	C	—	8937
Sasks	Prince Albert	A	—	11630
	Regina	A	—	10806
	Saskatoon	C	—	10870
Y. T.	Dawson	C	—	15067
	Mayo Landing	C	—	14454

*The data for these normals were from the full ten-year period 1951-1960, adjusted to the standard normal period 1931-1960.

Degree-days
Figure 5-20A (continued)

Degree days	Correction factors
0	.85
1,000	.80
2,000	.75
3,000	.70
4,000	.65
5,000	.60
6,000	.64
7,000	.68
8,000	.72

**Correction factors, degree-days
Figure 5-20B**

Fuel Oil	1 gal.	=	140,000 Btu's
Natural Gas	1 cu. ft.	=	1,000 Btu's

(1 CCF = 100 cu. ft. = 1 therm = 100,000 Btu's)

LP Gas	1 gal.	=	91,600 Btu's
Electricity	1 Kwh	=	3,413 Btu's
Coal	1 ton	=	24,000,000 Btu's
Wood	1 cord	=	21,000,000 Btu's

**Btu values for different fuels
Figure 5-21**

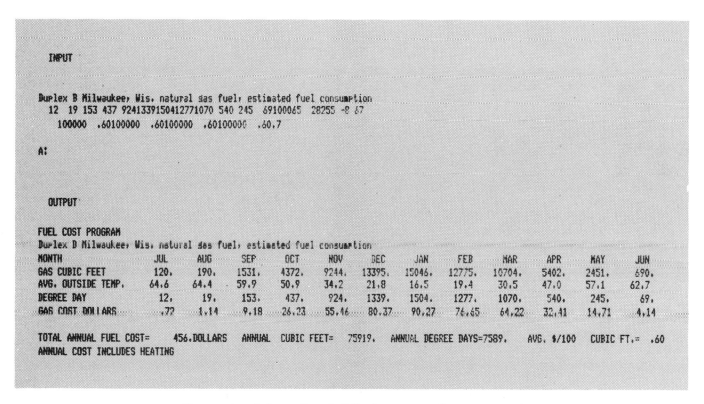

**Computer data, estimated fuel consumption program
Figure 5-22**

Designing a Warm Air System

With the heating and cooling loads and the other data now in hand, we can continue our HVAC design. This chapter covers warm air systems, an industry mainstay.

In designing the system, we'll use the residence from Chapter 5. Follow these steps:

1) Size, select and locate on the floor plan the air supply outlets, returns and exhaust air outlets

2) Size, select and locate the furnace and cooling equipment

3) Size and locate the ductwork

4) Locate the thermostat

5) Select fuel piping

6) Determine electrical requirements

7) Consider the chimney, breeching and combustion air

8) Review the design checklist

9) Review the plans and specifications checklist

Step 1: Air Supply, Return and Exhaust
The location and sizing of the air intakes and outlets are critical in a high-quality warm air system. Review the following definitions so you're sure to keep them straight when we begin calculating.

Free area— The total area of the openings in the outlet or inlet through which air can pass.

Feet per minute (fpm)— The measure of the velocity of the air stream. It's measured with a velocity meter calibrated in feet per minute.

Face velocity— The average face velocity of air passing through the face of the outlet or return.

Cubic feet per minute (cfm)— The measure of volume of air. When testing systems, find the cfm by multiplying the face velocity times the free area in square feet, following the register manufacturer's recommendations.

Terminal velocity— The point at which discharged air from an outlet decreases to a given speed, generally accepted as 50 feet per minute minimum.

Throw— The distance, measured in feet, that the air stream travels from the outlet to the point of terminal velocity. It's measured vertically from perimeter diffusers and horizontally from registers and ceiling diffusers.

Spread— The measurement in feet of the maximum width of the air pattern at the point of terminal velocity.

Drop— Generally associated with cooling where air is discharged horizontally from high sidewall outlets. Measured at the point of terminal velocity, it's the distance in feet that the air has fallen below the level of the outlet.

Static pressure (st pr)— The outward force of air within a duct, measured in inches of water. It's comparable to the air pressure in your car's tires.

Velocity pressure (vel pr)— The forward-moving force of air within the duct. Also measured in inches of water, it's comparable to the pressure of air rushing from a punctured tire.

Total pressure (tot pr)— The sum of the static pressure and the velocity pressure. Measured in inches of water, it's related to the sound level of an outlet. Anything that increases the total pressure, such as undersizing outlets or speeding up the blower, will also increase the noise level. For design purposes, inlets and outlets should not exceed 0.05 total pressure loss or drop.

Registers— Outlets which can deliver air in a concentrated stream into a given area.

Diffusers— Outlets which send out a widespread, fan-shaped pattern of air and which cause local recirculation of the air near the diffuser.

Perimeter installations— Heating or cooling installations where the diffusers or registers are installed to supply air to blanket outside walls. Returns are usually located in one or more central locations.

Sizing Supplies and Returns

You need to find the amount of air needed to offset the heat loss or gain without excess drafts or noise. First, calculate the air volume using these formulas (from Figure 5-11 in Chapter 5):

$$\text{Heating cfm} = \frac{\text{Heat loss}}{(Ta - Ti) \times 1.1}$$

$$\text{Cooling cfm} = \frac{\text{Heat gain}}{(Ti - Tc) \times 1.1}$$

For heating, 1.1 is a math constant reflecting the heat needed to heat the air volume in cfm. *Ta* is the temperature of the air from the furnace — for most jobs, 120 to 130°F. *Ti* is the room design temperature in winter, 67°F. *Tc* is the temperature of the air from the air conditioning unit. For cooling cfm, we use an air temperature from the unit of 55°F, typical for most units. Our room design temperature is 78°F. Take the heat gains and losses for each room from the calculations sheet, Figure 5-9.

Recall that for some of our interior rooms, we had a heat loss only through the ceiling areas and heat gains due to people and appliances. Now we'll assign all our losses and gains to rooms which have air supplies and returns:

Bedroom 1

$$\text{Heating cfm} = \frac{5646}{(130 - 67) \times 1.1} = 82$$

$$\text{Cooling cfm} = \frac{1060}{(78 - 55) \times 1.1} = 42$$

Bedroom 2

$$\text{Heating cfm} = \frac{4547}{(130 - 67) \times 1.1} = 66$$

$$\text{Cooling cfm} = \frac{1632}{(78 - 55) \times 1.1} = 65$$

Living room (plus kitchen and 1/2 hall)

$$\text{Heating cfm} = \frac{7082 + 262 + 701}{(130 - 67) \times 1.1} = 116$$

$$\text{Cooling cfm} = \frac{2258 + 114 + 1475}{(78 - 55) \times 1.1} = 152$$

Entry (plus 1/2 hall and mech room)

$$\text{Heating cfm} = \frac{3452 + 295 + 700}{(130 - 67) \times 1.1} = 64$$

$$\text{Cooling cfm} = \frac{530 + 113 + 174}{(78 - 55) \times 1.1} = 32$$

Bathroom (minimum of 50 cfm)

$$\text{Heating cfm} = \frac{1282}{(130 - 67) \times 1.1} = 19$$

$$\text{Cooling cfm} = \frac{202}{(78 - 55) \times 1.1} = 8$$

Storage (minimum of 50 cfm)

$$\text{Heating cfm} = \frac{1387}{(130 - 67) \times 1.1} = 20$$

$$\text{Cooling cfm} = \frac{215}{(78 - 55) \times 1.1} = 9$$

When you figure small rooms, you may come up with less than 50 cfm of air required. But don't put less than 50 cfm into the room. Less than 50 cfm will result in poor air movement and stale air.

Use the larger cfm needed to select the supply outlet size in a room. You can round off the cfm to the nearest increment of 10.

Now here's something you may want to consider for your jobs. Keep in mind that although it will get colder than the -8°F outside design temperature only 1% of the time for the sample residence, all our calculations so far have been based on minimum amounts. In other words, we haven't done any oversizing of the system so far. Oversizing when it's not necessary will only increase job and operating costs. But many designers like to build in a little extra capacity to fall back on, especially if it won't affect the job cost. Most furnaces for residences of 1,500 square feet or less will have more than enough capacity to handle slight increases in cfm. So here's what we're going to do.

The two bedrooms in our residence have the most exposed wall area. So for sizing air supplies in the two bedrooms, we'll go to the next larger size, and add 10 cfm to each room. In the entry, an area of high infiltration because of the two doors, we'll add two sizes, or 20 cfm.

Keep in mind that increasing the cfm may mean an increase in the size of ducts and registers. If you're close to the maximum in a system, it may also push you up to a larger, more costly, furnace. You'll have to weigh any possible increase in job costs against the potential customer benefits from a slight oversizing of the system.

Now we total the heating cfm for all the rooms:

Bedroom # 1	=	82	=	90	+	10 =	100
Bedroom # 2	=	66	=	70	+	10 =	80
Living Room	=	152				=	160
Entry	=	64	=	70	+	20 =	90
Bathroom						=	50
Storage						=	50

$$- \ 530 \text{ total cfm}$$

The air supply temperatures used, Ta and Tc, are typical of the industry. Different ones may be needed for your jobs, depending on the conditions. Consult with equipment manufacturers or an engineer if you have a job that's out of the ordinary.

When heating losses begin to exceed 10,000 Btu/hr per room, or in large rooms with several windows, it's best to use more than one outlet. The total cfm can be divided by the number of outlets desired. Too many outlets, however, will add needlessly to the job cost with little increase in efficiency. Too few will cause poor heat distribution and excessive noise.

Size return air grilles to limit the face velocity of the air to less than 500 feet per minute. Higher velocities may be noisy.

From our definitions earlier in this section, we know we can find the cfm by multiplying the face velocity by the free area, in square feet, of the register. If we turn the equation around, we can find the free area of a register by dividing the cfm by the face velocity. Bedroom 2 has an 80 cfm supply. We'll divide this by the maximum face velocity desired, 500 fpm, to find the free area of return:

Free area of return = 80/500 = 0.16 sq. ft.

Now convert 0.16 square feet to square inches: multiply 0.16 by 144, the number of square inches in a square foot:

0.16 × 144 sq. in. = 23 sq. in.

That's the amount of free area needed in bedroom 2.

Register manufacturers list register free area in

square inches in their tables, so choosing a return is simple once you've found the size you need. Once again, we're talking about a minimum size. Many designers use this rule of thumb: Make the return one register size larger than the supply. Never make it smaller.

Unless the room is very large, with many outlets, you'll usually only install one return, located on a wall opposite the supply. Good industry practice is to have a return located in every room or area. However, you can cut costs by placing the return in a central location, such as a hallway, if you use grilles in the room doors or undercut the doors to insure air flow.

Selecting the Outlets and Returns
You can use the cfm size to choose registers. But there are also other variables to consider. For best results, the register shouldn't exceed a total pressure loss of 0.050 inches, or a face velocity of 500 fpm. The horizontal throw should extend from 1/2 to 3/4 of the distance to the opposite wall. Register size, color and appearance may also have to be matched to the room decor. In small rooms, it's a good idea to use a register a size larger than the minimum to reduce drafts, which are more noticeable in smaller areas.

Figure 6-1 is data for registers and grilles manufactured by the Lima Corporation. Note that in the Series 10M table, the data for the different registers is divided by double lines. The data to the left of these lines is recommended as the best combination to insure good outlets and returns. The heating and cooling Btu lines give the maximum that each register can deliver at the listed cfm, TP, throw and velocity.

For our residence we've chosen these registers:

Bedroom 1	12'' x 6''
Bedroom 2	12'' x 6''
Living room	14'' x 6''
Entry	12'' x 6''
Bathroom	10'' x 6''
Storage	10'' x 6''

When ordering registers, make sure each supply outlet has a damper so the air flow can be adjusted and turned off.

Locating Outlets and Returns
After sizing and selecting the supplies and returns, we'll draw them as indicated on the plan shown in Figure 6-2.

The air supply in each room should be located close to the outside wall, under or near windows, to offset the heat loss or gain through the walls, windows, floor, and ceiling. Place it at the baseboard or in the floor. The baseboard level is the preferred industry practice unless the building construction requires otherwise. Locate return air grilles in the wall, either at floor level or near the ceiling. The high location is preferred for climates with a long cooling season, such as in Arizona. The floor locations are better for areas with a long heating season, such as New England. Label each location with the register size and the cfm of that outlet or return. List the type of register and the manufacturer in the fixture schedule or in the specs. Check this carefully to avoid mix-ups.

Ventilating
Small exhaust fans and kitchen hoods with fans are used to exhaust the bathroom and kitchen. They're operated from manual switches by the occupants. Small bathrooms need 50 or 60 cfm of exhaust over the toilet area. A kitchen hood needs from 100 to 150 cfm of exhaust over the stove. Fans, ductwork, and roof or wall outlets can all be chosen from manufacturer's catalogs.

For residential and many commercial jobs, exhaust air is supplied through the natural infiltration into the building. The air being drawn out by the exhaust has been heated through convection and adds very little heat loss. In rooms with tight doors, such as bathrooms, undercut the doors 1/2'' to 1'' to make sure exhaust air gets into the room. Where larger amounts of air are vented, install a door grille.

Where large amounts of fresh air are needed, you may have to install a separate intake. In some cases this air may have to be heated or cooled. In apartment or commercial buildings with many exhaust fans, it's most economical to run several fans into a common duct. Be sure to check local codes for limitations. Chapter 9 covers different kinds of commercial exhaust systems.

Humidifiers
In very dry climates, you'll probably want to install a humidifier in the system's ductwork. A humidistat controls the unit, allowing it to operate whenever the furnace fan operates.

In very cold climates, the humidity may need to be reduced. If there's too much moisture in the air, it can condense on cold surfaces, such as windows.

STEEL Sidewall registers, grilles

series 10M

This shallow multi-valve unit permits installation where stackhead space is limited and in ducts where a single valve would interfere with the air stream. Series 10M Registers can be used for combination heating and cooling systems when installed high on the wall or for low sidewall heating applications. ¼" lances.

series 10M, 10BIM

- Series 10 BIM not available in 8" x 6"
- For overall O.D. 10M add 1 1/2" to nominal listed size

Product Number and Size	FREE AREA SQ. IN.	Heating BTU h	3045	4565	6090	7610	9515	11415	13320	15220	17125	19025	20930	22830	24735
		Cooling BTU h	855	1280	1710	2135	2670	3200	3735	4270	4805	5340	5870	6405	6940
		C.F.M.	40	60	80	100	125	150	175	200	225	250	275	300	325
10M 8" x 6"	34	T.P. Loss	.007	.013	.021	.031	.050	.078	.101						
		Horiz. Throw (ft.)	3.5	5	6.5	8	11	15	19						
		Velocity	170	255	340	425	530	635	745						
10M, 10BIM 10" x 6"	43	T.P. Loss		.007	.012	.020	.028	.041	.056	.083	.109				
		Horiz. Throw (ft.)		4.5	6	7.5	10	12	15	18.5	21.5				
		Velocity		200	265	335	415	500	585	665	750				
10M, 10BIM 12" x 6"	53	T.P. Loss			.007	.012	.021	.029	.042	.060	.081	.095	.107		
		Horiz. Throw			4.5	6	8	10	12	14	16.5	20	23.5		
		Velocity			220	275	340	410	480	560	615	680	750		
10M, 10BIM 14" x 6"	62	T.P. Loss				.008	.015	.023	.033	.041	.053	.077	.092	.103	.119
		Horiz. Throw				5	6	9	10.5	12	14	16.5	18.5	21.5	24.5
		Velocity				230	290	350	410	465	525	580	640	695	755

Finish: White

series 10FG

Product Number and Size	Free Area Sq. In.	Maximum Recommended C.F.M.	Outside Dimensions Inches
10FG- 8 x 6	29	83	9½ x 7½
10FG-10 x 6	36	102	11½ x 7½
10FG-12 x 6	44	123	13½ x 7½
10FG-14 x 6	52	144	15½ x 7½
10FG-24 x 6	90	250	25¼ x 7¼
10FG-30 x 6	111	308	31¼ x 7¼

Finish: White

Series 10FG Sidewall Grille matches the sidewall registers on this page in both styling and finish. Horizontal ¼" louvers are deflected downward 23 degrees . . . (they can be inverted for high sidewall applications to prevent see-through). Vertical blades (spaced 2 to 3 inches apart) are resistance welded . . . there's no vibration or whistling.

Courtesy: Lima Corp.

**Register data and specifications
Figure 6-1**

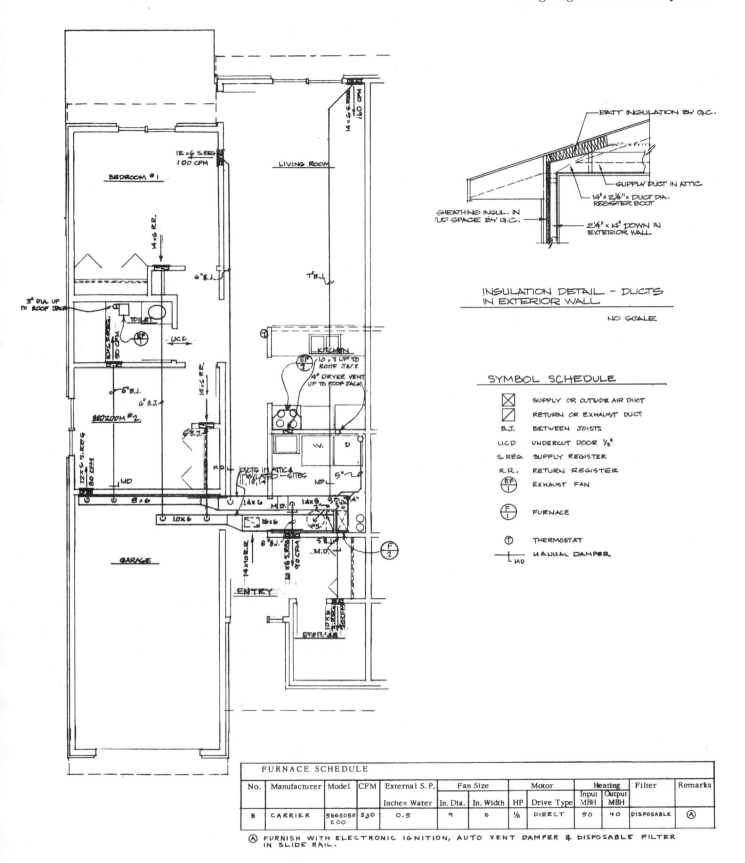

INSULATION DETAIL - DUCTS IN EXTERIOR WALL

NO SCALE

SYMBOL SCHEDULE

⊠	SUPPLY OR OUTSIDE AIR DUCT
⊡	RETURN OR EXHAUST DUCT
B.J.	BETWEEN JOISTS
U.C.D.	UNDERCUT DOOR ½"
S. REG.	SUPPLY REGISTER
R.R.	RETURN REGISTER
EF	EXHAUST FAN
F	FURNACE
T	THERMOSTAT
MD	MANUAL DAMPER

FURNACE SCHEDULE												
No.	Manufacturer	Model	CFM	External S.P.	Fan Size		Motor		Heating		Filter	Remarks
				Inches Water	In. Dia.	In. Width	HP	Drive Type	Input MBH	Output MBH		
B	CARRIER	58GS050 200	530	0.5	9	6	⅙	DIRECT	50	40	DISPOSABLE	Ⓐ

Ⓐ FURNISH WITH ELECTRONIC IGNITION, AUTO VENT DAMPER & DISPOSABLE FILTER IN SLIDE RAIL.

Sample residence, warm air system floor plan
Figure 6-2

Read the literature from window manufacturers on the insulating value of their windows and frames. With this information, you can calculate the inside surface temperature of the windows. Condensation won't occur if the dew point of the air is higher than the surface temperature. To find the dew point, measure the dry bulb and wet bulb temperatures and plot the values on a psychrometric chart. Chapter 9 has more information on humidity control systems.

Step 2: Furnace and Cooling Equipment Selection

For the heating system, look for a furnace that has:

- Efficient and quiet operation

- Sufficient capacity at design conditions to meet heating, cooling, and air flow needs

- Easy maintenance

- Reasonable cost

Figure 6-3 describes several models of Borg-Warner Furnaces, including the annual fuel utilization efficiency (AFUE), heating capacity, and other data. We can compare model capacity and calculate estimated fuel consumption to help us choose a model for our residence.

Efficient Operation

Heating equipment is tested and given efficiency ratings by the manufacturer, trade associations and the government.

The American Gas Association (AGA) and Underwriter's Laboratories (UL) test and certify that certain standards have been met. The Gas Appliance Manufacturers Association provides efficiency ratings. The Department of Energy (DOE) also tests and provides efficiency ratings for heating and cooling equipment.

In Figure 6-3, model number D10N04001, with a 40 MBH (thousand Btu/hr) input capacity and spark ignition, has an AFUE of 67.1% to 77.2%, for an average 72% AFUE. The standard pilot-light model has an AFUE of 61.3%. Using the estimated fuel consumption/degree-day equation from the previous section, we calculate the annual fuel cost will be $446. Since it's about 10% more efficient, the spark ignition model will save about $45 annually. An accessory price sheet (not shown)

lists the added cost for the spark ignition cost at $128. At a savings of $45 annually, payback will take about 2.8 years ($128 divided by $45). Since the life of the unit is well over three years, the spark ignition would be a good investment, especially since the cost of gas will likely continue to rise in the future.

You can make comparisons like this between various brands and accessories as part of your job design. Although we now favor model D10N04001 for our residence, we'll evaluate the remaining factors before making a final decision.

Sufficient Capacity at Design Conditions

The peak output of the furnace should never be less than the heat loss of the building. It's best to size it a little larger, but avoid inefficient operation. A grossly oversized furnace will cost too much to operate. The furnace also will need a fan to furnish the right amount of cfm for our house. Cooling equipment must be of the right capacity and compatible with the furnace.

Heating— In the data in Figure 6-3, the smallest furnace available is model D10N04001, with an input of 40,000 Btu/hr and a heating output of 30,000 Btu/hr. The output is the actual amount of heat available for the system after taking into account heat loss up the chimney and within the unit itself. Since our building heat loss is 28,255 Btu/hr, the model D10N04001 will do nicely.

Note that most manufacturers show reductions in the heat output of gas or oil fired equipment when used at elevations of over 2,000 feet. (See the Note in the furnace data.) The reduction is usually about 4% for every 1,000 feet above sea level. The reduction occurs because air is less dense at higher altitudes. When you design systems, you must increase the equipment size to adjust for this reduction in heat output. The fan will also need to circulate more cfm.

Fan volume— The furnace fan volume required is at least the total of all the air supply outlets in the system. The total cfm for the sample residence is 530.

But you must also consider pressure drops within the system when figuring fan volume. We'll explain this further in the next sections. For now, we can refer to the Blower Performance table in Figure 6-4.

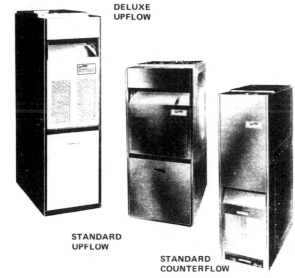

DELUXE
UPFLOW

STANDARD
UPFLOW

STANDARD
COUNTERFLOW

BORG-WARNER®

GAS-FIRED FURNACES

STANDARD MODELS —
Upflow & Counterflow
- **Standing Pilot (P2UG, P2CG) or**
- **Intermittent Spark Ignition (P2US, P2CS)**

DELUXE MODELS —
Upflow (P2UX)

DESCRIPTION

The compact furnaces above are styled for residential or commercial installation in a basement, closet, recreation room or garage. All furnaces are natural gas-fired. Standing pilot furnaces may easily be converted to propane gas by dealer or distributor personnel. Furnaces are completely factory assembled, wired and tested to assure dependable and economical operation.

FEATURES

DURABLE CASING - The casing is fabricated from heavy gauge steel. All panels are degreased, bonderized, and finished with baked enamel. Side panels are formed with surface indentations which provide added durability.

LASTING HEAT EXCHANGER - Die-formed from heavy gauge cold rolled steel, the standard models carry a ten-year limited warranty. Deluxe models are die-formed from heavy gauge aluminized steel and carry a twenty-year limited warranty. All heat exchangers are designed to provide an optimum heat transfer surface.

EFFICIENT IGNITION - Spark ignition pilot is provided on models P2US, P2CS, and P2UX for a more efficient operation by shutting off pilot as well as the main gas supply during off cycle.

OVERHEAT PROTECTION - All furnaces include a temperature actuated fan and limit control to guard against overheating. An auxiliary limit control is included with all counterflow furnaces to provide double protection.

VENT DAMPER - Shipped with all Deluxe Furnaces. Available as accessory on US & CS units.

VERSATILE TRANSFORMER AND FAN RELAY - The factory installed transformer is suitable for heating-only, cooling/heating, or add-on cooling applications. A fan relay is factory supplied on all furnaces.

LOW OPERATING SOUND LEVEL - Blower motor and blower assembly are rubber mounted to isolate vibration. Sound and temperature insulated heat exchanger compartment, casing indentations, and heat exchanger expansion joint keep metal expansion and contraction noises to a minimum. The blower compartment on Deluxe furnaces is accoustically insulated using 3-pound density insulation.

SAFETY DEVICES - All units use a redundant gas valve and a redundant limit on Spark Ignition units or energy cut-off device (ECO) on Standing Pilot units.

VARIABLE AIR HANDLING CAPABILITY - Multi-speed motors on direct-drive and variable pitch V-belt drive blowers adjust to deliver the proper cfm of air for a wide range of air requirements. Direct-drive furnaces allow operation of blower at a higher cfm for cooling and a lower cfm for heating.

EASY ACCESS - Lift-lock front panels provide easy access with no removal of screws required.

BLOWER DOOR SAFETY SWITCH - This switch disconnects power to the unit when the blower panel door is removed.

SPECIAL SERVICING FEATURES - (1) Slide-out blower assembly on all upflow models. (2) Easily removable blower assemblies on counterflow models.

FILTERS - High velocity cleanable filters are provided with furnace.

PRINTED CIRCUIT BOARD (Deluxe Models only) - Pre-wired control system has central terminal board for simplified field wiring. Plug-in blower relay permits different blower speeds for cooling and heating. Plug-in relay receptacle provided for control of electronic air cleaner and/or humidifier.

Gas furnace data
Figure 6-3

PHYSICAL DATA

Line	AFUE/AFUE with Damper[7] Std. Pilot P2UG	Std. Spark Ign. P2US	Deluxe Spark Ign. P2UX	Model No	Heating Capacity Input MBH[1]	SS[7] Out-Put MBH	Air Temp. Rise °F	CFM @ (Mean Air Temp. Rise)	Max. Outlet Air Temp. °F	Blower Diam.	Width	H.P.	Max. Over-Current Protection[8]	Unit Amp-acity
	P2UG	**P2US**	**P2UX**	**UPFLOW — HEATING/ADD-ON COOLING**										
1	61.3	67.1/77.2	—	D[3]10N[5]04001	40	30	15-45	990	145	10	4	1/3	10	9
2	67.5	71.9/81.2	—	D10N05501	55	44	25-55	1020	155	9	9	1/4		8
3	65.0	69.2/78.7	—	D08N06501	65	50	50-80	740	180	9	6	1/5	5	4
4	65.6	—	—	D10N06501	65	50	35-65	960	165		9	1/4		8
5	66.3	71.6/79.9	79.9	D12N06501	65	50	25-55	1200	155	10	6	1/3		9
6	61.3		—	D10N08001	80	59	45-75	910	175	9	9	1/4	10	8
7	61.3		—	D12N08001	80	59	35-65	1090	165	10	6	1/3		8
8	67.2	73.0/80.4	—	D10N09001	90	70	50-80	1030	180	9	9	1/4		8
9	68.5	73.0/80.4	80.4	D16N09001	90	70	50-80	1480	160	10	10	1/3		9
10	69.8	73.1/80.5	—	D20N09001	90	71	25-55	1670	155	11		1/2	20	17
11	66.1	70.8/79.1	79.1	D10N10501	105	80	60-90	1040	190	9	9	1/4	10	8
12	66.7	70.5/78.2	78.2	D16N10501	105	80	35-65	1560	165	10		1/3		9
13	67.4	70.5/78.2	—	D20N10501	105	80	35-65	1730	160	11		1/2	20	17
14	63.0	64.4/71.6	—	D16N12001	120	86	45-75	1330	175	10		1/3	10	9
15	—	69.6/79.2	79.2	D16N14001	140	108	50-80	1600	180	10	10	1/3	10	9
16	68.6		—	D20N14001	140	109	50-80	2070	165	11		1	20	17
17	—	68.8/72.1	—	D22N20001	200	144	50-80	2670	180			1	25	20
18	70.7		—	B[4]30N180[10]31	180	141	35-65	2280	165	13	15	1	115V 25 / 230V 15	15 / 10
	P2CG	**P2CS**	**—**	**COUNTERFLOW — HEATING/ADD-ON COOLING**										
19	65.2	72.0/80.7	—	D08N06501	65	51	45-75	800	175	9	9	1/5	5	4
20	67.5	73.7/81.1	—	D12N06501	65	51	25-55	1200	155					9
21	61.3	—	—	D12N08001	80	59	35-65	1090	165		8			8
22	71.4	74.3/81.2	—	D14N09001	90	73	35-65	1330	165	10	8	1/3	10	9
23	68.6	73.1/80.0	—	D14N10501	105	82	45-75	1330	175					
24	70.2		—	D17N10501	105	83	30-60	1730	160		10			
25	—	74.0/80.4	—	D16N14001	140	111	50-80	1600	180			1/2	15	13
26	70.7		—	D20N14001	140	111	35-65	2070	165	12	11	1	20	17

NOTES:

[1] For altitudes above 2000 ft, reduce capacities 4% for each 1000 ft. above sea level.

[2] Available external static pressure .50 IWG is at furnace outlet and ahead of cooling coil.

[3] D = Direct Drive multi-speed blower motors. Maximum Speed 1050 RPM.

[4] B = Belt Drive blower. Motor Pulley is adjustable 1.9 - 2.08D, blower pulley 6" PD. Belt pitch length is 40".

[5] N = Natural gas models shown. Accessory available for propane conversion of standing pilot models only.

[6] Side air return on upflow 140, 180, 200 use single side return filter frame 1FF0308 (Filters shipped in bottom of unit)

[7] SS = Steady State capacity and AFUE (Annual Fuel Utilization Efficiency) determined in accordance with DOE test procedures.

[8] Dual element fuse or circuit breaker

[9] Based on copper AWG60C, 3% voltage drop.

[10] P2UGB30N180 furnace is dual-voltage, 230/115-1-60 shipped for 230/1/60 operation. Field conversion to 115/1/60.

[11] All filters supplied with furnace are high velocity, cleanable type.

N.A. = Not Applicable

See Form 650.55-AD1 for blower performance.

DIMENSIONS—INCHES

UPFLOW FURNACE

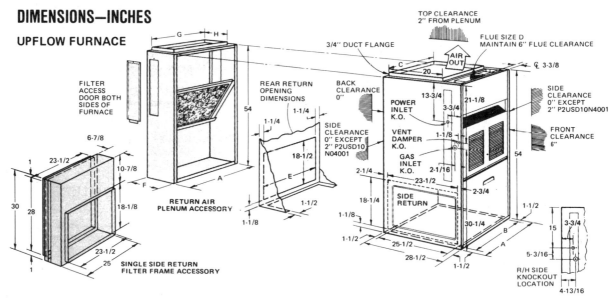

Gas furnace data
Figure 6-3 (continued)

LINE	Electrical Min Wire Size @ 75 Ft. Circuit[9] (One Way)	Filter Size Side[11] Supplied With Unit	Bottom	Add-On Cooling Tons	CFM @ .50" ESP[2]	Approx. Oper. Wt. Lbs.	A	B	C	D	E	F	G	H
				UPFLOW – HEATING/ADD-ON COOLING										
1		20 x 25	10 x 25	1, 1.5, 2	1180	105	12-1/4	9-1/4	11	4	9-3/4	16	11	14-3/4
2			14 x 25	2, 2.5	980	131								
3		20 x 25	14 x 25	1.5, 2	760	125	16-1/4	13-1/4	15	4	13-3/4		15	
4				2, 2.5	980	131								
5	14			2.5, 3	1280	141								
6		20 x 25	14 x 25	2, 2.5	980	131	16-1/4	13-1/4	15	4	13-3/4		15	
7				2.5, 3	1280	141								
8		20 x 25	20 x 25	2, 2.5	1025	155						16		14-3/4
9				3, 3.5, 4	1575	175								
10	10			4, 5	1905	200	22-1/4	19-1/4	21	5	19-3/4		21	
11	14	20 x 25	20 x 25	2, 2.5	1025	155	22-1/4	19-1/4	21	5	19-3/4		21	
12	14			3, 3.5, 4	1575	175								
13	10			4, 5	1905	200								
14	14			3, 3.5, 4	1575	175								
15	14	24 x 25[6]	24 x 25	3, 3.5, 4	1680	207	26-1/4	23-1/4	25	6	23-3/4		25	19-3/4
16	10			4, 5	1995	210						21		
17	12	30 x 25[6]	30 x 25	4, 5	2205	260	32-1/4	29-1/4	31	7	29-3/4		31	16-1/8
18	14			5, 7.5	2800	285								
				COUNTERFLOW – HEATING/ADD-ON COOLING										
19		(2) 14 x 20	N.A.	1, 1.5, 2	940	125	16-1/4	13-1/4	15	4	16-1/2	18-7/16	13	14-1/2
20				2, 2.5, 3	1260	141								
21		(2) 14 x 20	N.A.	2, 2.5, 3	1260	141	16-1/4	13-1/4	15	4	16-1/2	18-7/16	13	14-1/2
22	14			2, 2.5	1330	164								
23				3, 3.5	1330	164	22-1/4	19-1/4	21	5	22-1/2	24-7/16	19	20-1/2
24				3.5, 4	1550	190								
25	12	(2) 14 x 20	N.A.	3.5, 4	1570	210	26-1/4	23-1/4	25	6	26-1/2	28-7/16	23	24-1/2
26	10			4, 5	1865	210								

COUNTERFLOW FURNACE

TYPICAL WIRING AND PIPING

BASEMENT INSTALLATION

GROUND LEVEL INSTALLATION

Gas furnace data
Figure 6-3 (continued)

ACCESSORIES FOR FIELD INSTALLATION

SINGLE-SIDE-RETURN FILTER FRAME (Upflow) - For introducing air into only one side of upflow furnaces (Not required on 040, 065, or 080 furnaces). Filters are not supplied with this accessory. Two field-supplied filters are required, a 14 x 25 x 1 and a 16 x 25 x 1 inch filter.

RETURN-AIR PLENUM (Upflow) - For top return of air on upflow furnaces. Includes filter rack and two filter access doors. Filters must be field-supplied according to the sizes given in the accessory instructions.

BOTTOM CLOSURE (Upflow) - Must be used to close the bottom of the furnace when the application utilizes side or rear air return.

COMBUSTIBLE-FLOOR BASE (Counterflow) - For installation of counterflow furnaces on combustible floors.

PROPANE CONVERSION (Standard Furnaces with standing pilot ONLY) - Done by dealer or distributor prior to installation in areas being supplied with propane gas.

TWO-STAGE GAS VALVE (Upflow & Counterflow) - For installations requiring two steps of heat.

THERMOSTATS - Remote, wall-mounted, low-voltage thermostats, both standard and clock set-back models, are available for use in heating only and heating/cooling applications.

VENT DAMPER (P2US, P2CS, & P2UX Models ONLY) - Automatic damper in flue gas pipe opens and closes upon demand of room thermostat to prevent waste of heat during off cycle. (P2UX shipped with furnace as standard)

FIELD WIRING DIAGRAMS

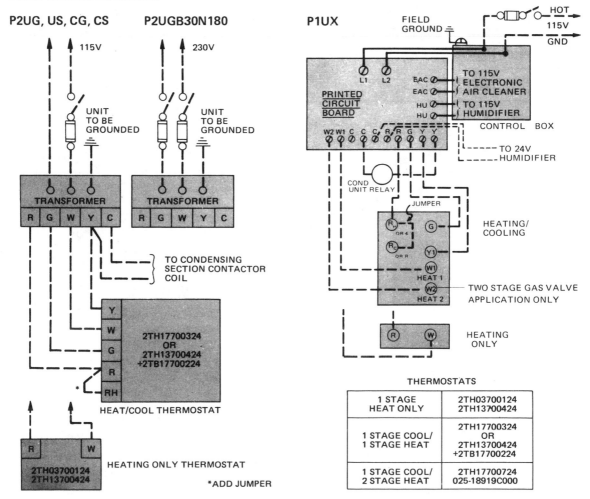

THERMOSTATS	
1 STAGE HEAT ONLY	2TH03700124 2TH13700424
1 STAGE COOL/ 1 STAGE HEAT	2TH17700324 OR 2TH13700424 +2TB17700224
1 STAGE COOL/ 2 STAGE HEAT	2TH17700724 025-18919C000

Gas furnace data
Figure 6-3 (continued)

BORG-WARNER

GAS-FIRED FURNACES
BLOWER PERFORMANCE DATA
UPFLOW & COUNTERFLOW MODELS

APPLICATION DATA	Supersedes: Form 650.54-AD1 (1279)	482	Form 650.55-AD1

STANDARD MODELS — UPFLOW & COUNTERFLOW
STANDING PILOT (P2UG, P2CG)

INTERMITTENT SPARK IGNITION (P2US, P2CS)
DELUXE MODELS — UPFLOW
INTERMITTENT SPARK IGNITION (P2UX)

The table below shows the Blower Performance Data for Borg-Warner Furnaces at various external static pressure conditions and blower speeds.

Note that the cfm values shown include allowances for the furnace's filter(s). Refer to Form 650.55-TG1 for Ratings, Physical Data, Wiring Data, Dimensions, etc.

TABLE 1 — UPFLOW FURNACES — DIRECT DRIVE BLOWER
[Allowance made for filter(s)]

Std.	Std. With Spk. Ign.	Deluxe	Model	Blower Speed	AVAILABLE EXTERNAL STATIC PRESSURE, IWG.					
					.10	.20	.30	.40	.50	.60
					CFM					
P2UG	P2US	—	D10N040	High	1310	1280	1250	1230	1180	1150
				Med	1250	1220	1190	1160	1130	1090
				Low	1040	1020	1000	980	960	930
P2UG	P2US	—	D08N065	High	975	925	875	820	760	695
				Med	845	820	775	735	685	640
				Low	715	705	695	675	650	595
P2UG	P2US	—	D10N055	High	1210	1160	1105	1045	980	910
P2UG	—	—	D10N065	Med	1095	1055	1010	965	910	845
P2UG	—	—	D10N080	Low	—	—	—	—	750	715
P2UG	P2US	P2UX	D12N065	High	1460	1430	1390	1340	1280	1205
P2UG	—	—	D12N080	Med	1280	1265	1245	1220	1180	1115
				Low	1090	1085	1070	1045	1010	945
P2UG	P2US	—	D10N090	High	1210	1170	1125	1080	1025	970
P2UG	P2US	P2UX	D10N105	Med	980	975	960	935	895	840
				Low	775	770	765	755	745	710
P2UG	P2US	P2UX	D16N090	High	1740	1710	1670	1625	1575	1525
P2UG	P2US	P2UX	D16N105	Med	1520	1510	1485	1460	1420	1370
P2UG	P2US	—	D16N120	Low	—	—	1245	1220	1180	1130
P2UG	P2US	—	D20N090	High	2155	2100	2040	1975	1905	1820
P2UG	P2US	—	D20N105	Med	1900	1875	1840	1790	1710	1595
				Low	1620	1605	1585	1555	1510	1460
—	P2US	P2UX	D16N140	High	1830	1800	1760	1730	1680	1650
				Med	1670	1650	1630	1600	1570	1530
				Low	1340	1335	1320	1300	1270	1230
P2UG	—	—	D20N140	High	2105	2085	2060	2030	1995	1950
				Med	1725	1710	1690	1670	1645	1605
				Low	1495	1485	1475	1460	1440	1415
—	P2US	—	D22N200	High	2360	2345	2315	2270	2205	2025
				Med	1940	1930	1915	1895	1860	1770
				Low	1745	1720	1695	1665	1630	1565

NOTE: Static pressures in above table which exceed A.G.A. certified ESP values (shown in Physical Data Table in Form 650.55-TG1 are listed for air conditioning (cooling) applications which may require higher static pressures.

Blower performance data
Figure 6-4

In the column for model D10N040, we see that the cfm for varying fan speeds and static pressures are all over 530. This is ample for our system if we stay within recommended pressures. So this unit is more than adequate for our heating load so far. If we were to suddenly find it had insufficient fan volume, it would probably mean we had made a mistake in our calculations somewhere. This is because the manufacturers of the units try to size the fans to fit the maximum application of the Btu capacity of the unit.

Cooling equipment— Since we intend to cool the residence, we need to check on the cooling capacity of our furnace. In Figure 6-3, the add-on cooling capacity is 1, 1.5, or 2 tons. *One ton of refrigeration is equal to a 12,000 Btu/hr cooling capacity.* In Chapter 5, we computed our total heat load or gain for the residence at 10,277 Btu/hr. A 1-ton add-on to model D10N04001 will handle our cooling needs.

Cooling equipment for the furnace consists of an evaporator coil installed in the furnace. Refrigeration piping connects it to a condensing unit located on a concrete pad outside the residence. Figure 6-5 has an Accessory Information table for evaporator coils that can be added to the various models of furnaces. A D10N04001 furnace uses an ''A'' type upflow coil, number M1UF024A. Figure 6-6 gives the data for the condensing units available.

The evaporator coil and condenser we select must be able to handle the 10,277 Btu/hr load at the summer outside design temperature of 90°Fdb. Since we know that the smallest furnace available in our literature is still slightly oversized, we'll begin by arbitrarily selecting the smallest available coil. Then we'll see if it's adequate.

The Cooling Capacity table in Figure 6-5 shows the model M1UF024A has a gross capacity rating of 655 cfm. It ranges from 12.2 MBH up to 39.1 MBH, depending on evaporator temperature and pressure. We can determine a suitable evaporator temperature and pressure by using our residence data. We need to cool the inside air to a design temperature of 78°Fdb. This means our return air will enter the coil at about 78°F or slightly higher, and at 50% relative humidity. This air has a wet bulb temperature of about 67°F (see psychrometric chart).The Cooling Capacity data for the coil shows that at 67°Fwb at 655 cfm, the M1UF024A has a capacity of 14.2 MBH at 50/84.7°F/psig (temperature and pressure).

Since our residence needs about 10.2 MBH, the coil will be large enough. But we also have to make sure our condenser can be sized properly. It must be chosen at the same 50/84.7 coil temp and pressure. We also need to use the outside design temperature of 90°Fdb. Look at Figure 6-6. At 95°Fdb and 50/84.7°F/psig, we have 15 MBH available from the condenser. At 75°Fdb, we have 18 MBH. By interpolating, we can find the MBH for our outside design temperature of 90°Fdb:

$$\frac{18\ MBH - 15\ MBH}{95F - 75F} = .15\ MBH$$

$$(90F - 75F) \times .15 = 2.25\ MBH$$

$$18\ MBH - 2.25\ MBH = 15.75\ MBH$$

Keep in mind that for every degree of increase in temperature, we have a *decrease* in MBH.

The condensing unit will be about 11% larger than required for our coil, but it's close enough to use.

The total MBH available will be found at the coil, rather than at the condenser. Note that the evaporator coil data table lists gross capacities. If we want to fine-tune our choice, we have to find the *net* capacity of the unit — the amount of cooling that's actually available for the system. According to the note in the table, we have to subtract the cooling that's needed to cool the blower motor located in the air stream. Here's how:

$$\frac{1}{3}\ hp\ motor \times 746\ watts/hp \times .003415 = .85\ MBH$$

A 1/3 hp motor produces 1/3 hp of mechanical work at the shaft. To get this, about 1/2 hp of electrical energy must be coming into the motor. A motor of this type typically operates at about 60% efficiency. So we have to take our 1/3 hp MBH loss and divide it by 0.60 to get a more accurate answer.

$$.85/.60 = 1.42\ MBH$$

By subtracting this from the rated capacity, we get the total amount of latent and sensible heat which can be removed by the coil and condenser. Since our evaporator capacity is 14.2 MBH:

14.2 − 1.42 = 12.78 net capacity MBH
Total building load = 10.2 MBH

We still have enough capacity, as 12.78 MBH is greater than 10.2 MBH. If we didn't, we'd have to go to the next size coil.

UPFLOW "A" COILS

YORK®

SPLIT SYSTEM EVAPORATOR COILS

FLATTOP™ DESIGN
"A" TYPE DESIGN
COIL CASINGS & ADAPTERS

COOLING: 12.5 thru 54.5 MBH with Matched Condensing Sections. (Rated at 80°F db, 67°F wb on evaporator coil, 95°F db on condenser coil.)

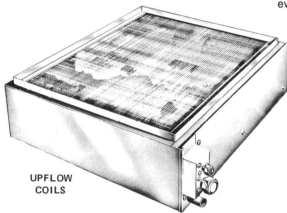

UPFLOW COILS

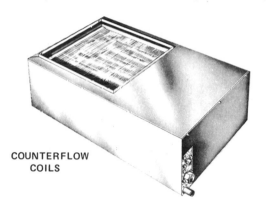

COUNTERFLOW COILS

DESCRIPTION

These evaporator coils are designed to be installed on Borg-Warner furnaces. The coils are used with Champion Series condensing sections. Possible split-system combinations to be made with these coils are shown in Form 550.08-TG1, 550.12-TG1 and 550.14-TG1.

FEATURES

INTERNALLY CLEAN — All evaporator coils are factory leak-tested, dehydrated, sealed and shipped with a holding charge. Protection plugs and rubber "O" rings seal out damaging dirt and moisture until piping connections are ready to be made. This minimizes installation time. No evacuation is necessary.

QUICK, RELIABLE COUPLINGS — Quik-Lok compression couplings in the condensing section and on the evaporator coil make installation of refrigerant lines quick and easy. Simply remove protection plugs and insert tubing through nut and ferrule into the coupling body. Then tighten. No brazing required.

LOW SILHOUETTE COILS — The low height of the York flattop is ideally suited to low ceiling or limited access applications. Upflow 7-3/16" high. Counterflow 9-3/8" high.

COMPACT CABINET COILS — For those high cooling capacity applications with compact furnace size. Three upflow models with "A" coil design and three counterflow models with flattop coil design.

COIL CASING — Upflow and counterflow models may be installed with furnace to permit easy application of flattop coil at a later date. Shipped knocked-down.

COIL ADAPTERS — Use with upflow coil casing for application of flattop coil which is 2" or 4" less than width of furnace or coil casing.

ACCESSORIES FOR FIELD INSTALLATION

ADAPTERS — Required to mate upflow coils to Borg-Warner gas-fired furnaces when coil does not match furnace opening.

COIL CASING — Upflow and counterflow for future flat-top coil installation.

COIL BAFFLES — Use with upflow coil casing for coil 2" or 4" less in width than the furnace or casing.

INTERCONNECTING TUBING REDUCERS — Required when refrigerant lines are incompatible with evaporator connections.

Evaporator coil data
Figure 6-5

COOLING CAPACITY—MBH— UPFLOW AND COUNTERFLOW EVAPORATOR COILS ONLY

Model	CFM	Temp. on Coil FWB	Evaporator Temperature and Corresponding Pressure-F/PSIG					Model	CFM	Temp. on Coil FWB	Evaporator Temperature and Corresponding Pressure-F/PSIG				
			30/55.2	35/61.9	40/69.0	45/76.6	50/84.7				30/55.2	35/61.9	40/69.0	45/76.6	50/84.7
M1UF024A	655	72	39.1	35.3	30.7	26.2	20.6	M1UF052A M1CF052A	1420	72	84.8	76.4	66.6	56.7	44.7
		67	31.9	28.3	24.0	19.2	14.2			67	69.2	61.4	52.0	41.6	30.7
		62	25.9	22.1	17.8	14.6	12.2			62	56.2	47.8	38.5	31.7	26.5
		57	21.1	19.0	16.8	14.6	12.2			57	45.8	41.1	36.4	31.7	26.5
M1UF032A M1CF032A	875	72	52.2	47.0	41.0	34.9	27.5	M1UF064A M1UA064A	1745	72	104.3	94.1	81.9	69.8	55.0
		67	42.6	37.8	32.0	25.6	18.9			67	85.1	75.5	64.0	51.2	37.8
		62	34.6	29.4	23.7	19.5	16.3			62	69.1	58.9	47.4	39.0	32.6
		57	28.2	25.3	22.4	19.5	16.3			57	56.3	50.6	44.8	39.0	32.6
M1UF044A M1CF044A M1UA044A	1200	72	71.7	64.7	56.3	48.0	37.8	M1UA067A	1830	72	109.2	98.5	85.8	73.0	57.6
		67	58.5	51.9	44.0	35.2	26.0			67	89.1	79.1	67.0	53.6	39.5
		62	47.5	40.5	32.6	26.8	22.4			62	72.4	61.6	49.6	40.9	34.2
		57	38.7	34.8	30.8	26.8	22.4			57	59.0	52.9	46.9	40.9	34.2

Note: Ratings are gross capacities—for net capacities, deduct evaporator blower motor heat, KW × .3415 = MBH (If not included in calculated load).

STATIC PRESSURE DROP-IWG

Model		Static Pressure Drop Thru Coil						Model		Static Pressure Drop Thru Coil					
		.10	.15	.20	.25	.30	.35*			.10	.15	.20	.25	.30	.35*
		UPFLOW FLATTOP COIL								UPFLOW "A" COIL					
M1UF024A	CFM	540	620	700	780	860	935	M1UA044A	CFM	750	870	1025	1170	1300	1430
M1UF032A		680	805	925	1050	1175	1300	M1UA064A		1000	1260	1540	1780	2000	2180
M1UF044A		1000	1145	1285	1430	1570	1715	M1UA067A		1200	1440	1750	2030	2300	2450
M1UF052A		1110	1290	1465	1645	1820	2000			COUNTERFLOW FLATTOP COIL					
M1UF064A		1475	1670	1805	2060	2250	2440	M1CF032A	CFM	680	805	925	1050	1175	1300
								M1CF044A		1000	1145	1285	1430	1570	1715
								M1CF052A		1110	1290	1465	1645	1820	2000

*For applications where a high latent load exists, do not exceed CFM which gives .30 IWG pressure drop thru coil.

ACCESSORY INFORMATION

COIL CASING — Upflow and counterflow casings are available for installation with the furnace for future use of flattop cooling coil. Casings are knock-down type, individually wrapped and shipped 5 to a carton. Each casing consists of 2 side panels, back panel, top panel (plus bottom panel on counterflow) and a front panel which is discarded when cooling coil is installed. Upflow casing is 22 gauge painted steel and counterflow casing is 18 gauge painted steel. When casings are assembled, the dimensions are the same as corresponding evaporator coil models.

Coil			Matching Furnace		Matching Coil Model
Type	Casing Model	Width Inches		Capacity MBH	
Upflow	1CC1101	12-1/4		40	M1UF024A
	1CC1102	16-1/4		55,65,80	M1UF032A
	1CC1103	22-1/4		90,105,120	M1UF044A
	1CC1104	26-1/4		140	M1UF052A
	1CC1105	32-1/4		180,200	M1UF064A
Counter Flow	1CC1106	16-1/4		65,80	M1CF032A
	1CC1107	22-1/4		105,120	M1CF044A
	1CC1108	26-1/4		140	M1CF052A

COIL ADAPTERS — For use with upflow coil casing to accommodate a coil which is either 2" or 4" less in width than the furnace or upflow casing. Consists of two unpainted baffles and two painted front panel filler pieces. One baffle and filler piece are on each side of the coil to insure proper condensate drainage, full air flow over coil and rated coil capacity for the smaller coil. Coil adapters are individually wrapped and shipped ten to a carton.

Courtesy: York Environmental Systems, Borg-Warner Corp., all rights reserved.

Evaporator coil data
Figure 6-5 (continued)

DIMENSION-INCHES

"A" COILS

UPFLOW COILS

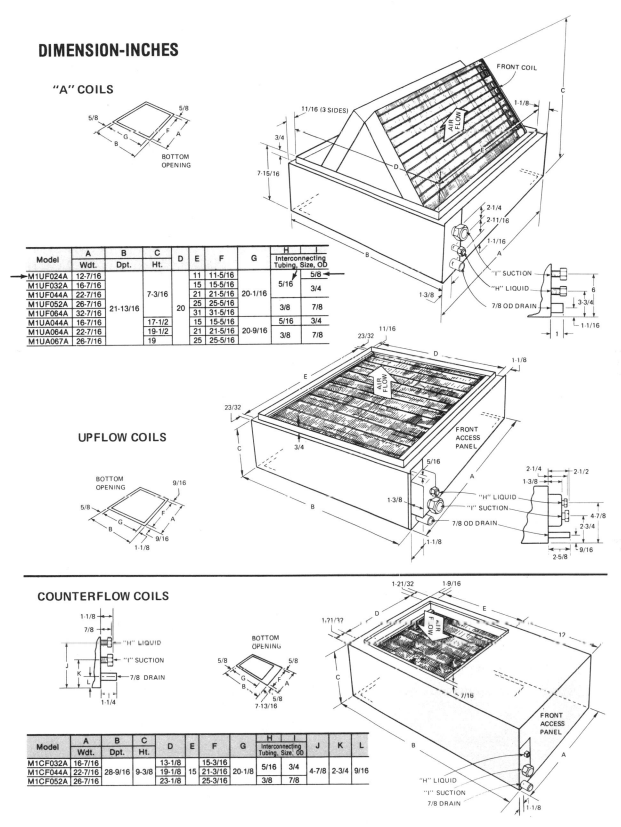

Model	A Wdt.	B Dpt.	C Ht.	D	E	F	G	H Interconnecting Tubing, Size, OD	
M1UF024A	12-7/16				11	11-5/16		5/16	5/8
M1UF032A	16-7/16				15	15-5/16			3/4
M1UF044A	22-7/16		7-3/16		21	21-5/16	20-1/16		
M1UF052A	26-7/16	21-13/16		20	25	25-5/16		3/8	7/8
M1UF064A	32-7/16				31	31-5/16			
M1UA044A	16-7/16		17-1/2		15	15-5/16		5/16	3/4
M1UA064A	22-7/16		19-1/2		21	21-5/16	20-9/16	3/8	7/8
M1UA067A	26-7/16		19		25	25-5/16			

COUNTERFLOW COILS

Model	A Wdt.	B Dpt.	C Ht.	D	E	F	G	H Interconnecting Tubing, Size, OD	I	J	K	L
M1CF032A	16-7/16			13-1/8		15-3/16		5/16	3/4			
M1CF044A	22-7/16	28-9/16	9-3/8	19-1/8	15	21-3/16	20-1/8			4-7/8	2-3/4	9/16
M1CF052A	26-7/16			23-1/8		25-3/16		3/8	7/8			

Evaporator coil data
Figure 6-5 (continued)

PHYSICAL DATA

UPFLOW COILS

	Model	M1UF024A	M1UF032A	M1UF044A	M1UF052A
Coil	Rows Deep	4-1/2			
	Total Face Area, Sq. Ft.	1.54	2.05	2.82	3.34
Weight	Operating, Lbs.	20	26	35	41
	Model	M1UF064A	M1UA44A	M1UA64A	M1UA67A
Coil	Rows Deep	4-1/2	4		3
	Total Face Area, Sq. Ft.	4.11	2.61	4.44	5.44
Weight	Operating, Lbs.	50	35	50	52

COUNTERFLOW COILS

	Model	M1CF032A	M1CF044A	M1CF052A	
Coil	Rows Deep	4-1/2			
	Total Face Area, Sq. Ft.	2.05	2.82	3.34	
Weight	Operating, Lbs.	30	40	45	

TYPICAL WIRING AND PIPING Champion Split System Air Conditioners

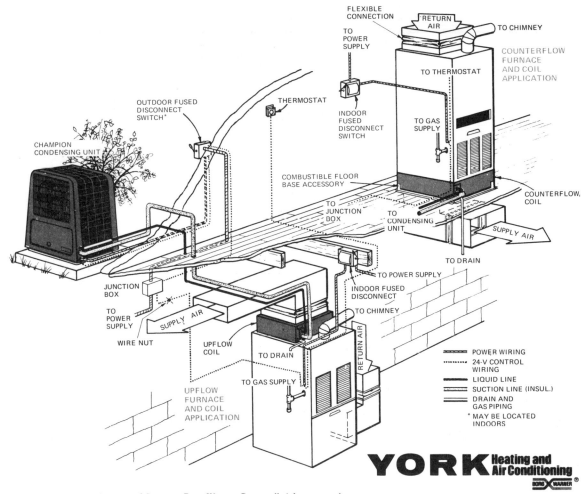

Evaporator coil data
Figure 6-5 (continued)

YORK®

CHAMPION IV

SPLIT SYSTEM CONDENSING UNITS

Nominal Cooling Capacity 12.2 thru 46.0 Mbh with matched evaporators

DESCRIPTION

The Champion IV condensing unit is the outdoor part of a versatile system of air conditioning. It is designed to be custom-matched with one of YORK's complete line of evaporator sections, each designed to serve a specific function. This enables a Champion IV system to cool your home no matter what type of heating system you may presently have. There are various types of evaporator coils to be used with your present warm-air furnace, for up-flow, counterflow or horizontal applications, and evaporator blowers if you do not have a forced air heating system. The evaporator blower line even offers electric heat if you require heating.

The CHAMPION IV unit has an air-cooled, "U"-shaped condenser coil. Within this coil is enough Refrigerant-22 for an entire cooling system, including the evaporator and up to 25 feet of interconnecting refrigerant lines.

FEATURES

SELF-CLEANING CONSTRUCTION — The height and position of the fans minimize the possibility of drawing in debris from the surrounding area. The coil shape and air flow pattern combine to give a reversed washing effect by either rain or hose water.

LOWER INSTALLATION COST — Installation time and cost are reduced by the use of Quik-Lok compression couplings on both condenser and matching evaporators. No brazing is required.

The upright cubical configuration of the condenser means smaller foundations requiring less base material, reducing the cost of the installation.

Courtesy: York Environmental Systems, Borg-Warner Corp., all rights reserved.

LOW OPERATING SOUND LEVEL — The condenser fan motor runs at a low speed because of the twin fan concept (one motor - two fan blades). Low speeds produce less noise. Isolator mounted compressor and the rippled fins of the condenser coil muffle the normal fan motor and compressor operating sounds.

QUALITY CONDENSER COIL — The coil is constructed of copper tube and hardened aluminum fins for durability and long lasting efficient operation.

PROTECTED COMPRESSOR — The compressor is internally protected against high pressure and temperature. This is accomplished by the simultaneous operation of high pressure relief valve and the temperature sensors which stop the compressor if operating temperatures get too high.

Crankcase Heaters further protect the compressor on model H2CC042A, H2CC048A.

EASY ACCESS — All condensing sections are provided with an easily removed access door. To gain access to the electrical controls and the refrigerant valves simply remove the screw.

LONGER FAN MOTOR LIFE — Condenser fan motor life is extended because the equal and opposing thrust of the fan blades reduce the bearing load. With bearing load reduced, less starting torque is required. This, combined with the fact that there is less oil lost from the bearing systems of horizontal shaft motors, leads to longer motor life.

DURABLE SURFACE — 9-step processing guarantees high adhesive, baked-on finish which protects front and rear panels.

Condensing unit data
Figure 6-6

PHYSICAL DATA—CONDENSING SECTION

Model			H2CC012A06	H2CC018A06	H2CC024A06
Unit Power Supply			208/230-1-60		
Comp.	Rated Load Amps		6.2	9.7	14.3
	Locked Rotor Amps		33.5	48.0	60.0
Condenser	Fan Motor	Diameter	12		
		HP	1/8		
		Power Supply	208/230-1-60		
		Rated Load Amps	0.6		
		Locked Rotor Amps	1.4		
	Coil	Rows Deep	1		
		Face Area Sq Ft	6.5		6.6
Operating Weight Lbs			117	128	147

Model			H2CC030A06	H2CC036A06	H2CC042A06*	H2CC048A06*
Unit Power Supply			208/230-1-60			
Comp.	Rated Load Amps		16.4	19.9	22.5	26.5
	Locked Rotor Amps		70.0	83.5	93.0	114
Condenser	Fan Motor	Diameter	14		16	
		HP	1/5		1/2	
		Power Supply	208/230-1-60			
		Rated Load Amps	1.1		2.4	
		Locked Rotor Amps	2.6		6.7	
	Coil	Rows Deep	1			
		Face Area Sq Ft	7.0		11.7	11.8
Operating Weight Lbs			155	163	177	212

*Compressor has built-in insertion type crankcase heater.

COOLING CAPACITY—MBH—CONDENSING SECTION ONLY

Model	Suction Temp. And Press. At Compressor		Air Temperature on Condenser, F DB					
			75		95		115	
	°F	PSIG	MBH	KW	MBH	KW	MBH	KW
H2CC012A	30	55.2	10	1.2	7	1.3	4	1.4
	35	61.9	12	1.2	9	1.3	6	1.5
	40	69.0	14	1.3	11	1.4	8	1.5
	45	76.6	16	1.3	13	1.4	10	1.6
	50	84.7	18	1.4	15	1.5	12	1.6
H2CC018A	30	55.2	15	1.6	11	1.7	6	1.9
	35	61.9	18	1.7	13	1.8	9	2.0
	40	69.0	20	1.7	15	1.9	11	2.1
	45	76.6	22	1.8	17	2.0	12	2.2
	50	84.7	24	1.9	18	2.1	14	2.3
H2CC024A	30	55.2	19	2.0	16	2.1	9	2.3
	35	61.9	22	2.1	18	2.2	11	2.5
	40	69.0	25	2.2	21	2.3	15	2.7
	45	76.6	28	2.4	23	2.6	18	2.8
	50	84.7	30	2.5	26	2.8	21	3.0
H2CC030A	30	55.2	23	2.5	18	2.6	11	2.9
	35	61.9	26	2.6	21	2.8	13	3.1
	40	69.0	30	2.8	25	3.0	17	3.3
	45	76.6	33	2.9	27	3.2	21	3.5
	50	84.7	35	3.0	30	3.4	25	3.7
H2CC036A	30	55.2	26	2.9	21	3.0	13	3.3
	35	61.9	30	3.1	25	3.2	15	3.5
	40	69.0	34	3.2	28	3.3	20	3.8
	45	76.6	38	3.4	31	3.7	24	4.0
	50	84.7	40	3.6	34	4.0	29	4.3
H2CC042A	30	55.2	30	3.3	24	3.4	15	3.8
	35	61.9	35	3.5	29	3.7	18	4.0
	40	69.0	40	3.7	33	3.9	23	4.3
	45	76.7	44	3.9	36	4.2	28	4.5
	50	84.7	47	4.0	40	4.5	33	4.8
H2CC048A	30	55.2	34	3.8	28	4.0	16	4.4
	35	61.9	40	4.1	32	4.3	20	4.7
	40	69.0	45	4.3	37	4.5	26	5.0
	45	76.7	50	4.5	41	4.9	32	5.2
	50	84.7	54	4.7	46	5.2	38	5.6

Condensing unit data
Figure 6-6 (continued)

Sizing refrigerant lines— To size the refrigerant lines between the evaporator and condenser, we refer to the Evaporator Coil Specifications of the manufacturer. For model M1UF024A, Figure 6-5, the suction line will be a 5/8'' outside diameter (O.D.). The liquid line is 5/16'' O.D. The condensate drain line is 7/8'' O.D.

Typically these lines are copper. Common diameters are available from HVAC suppliers, with quick-connect fittings, in lengths of 10', 15', 25' and on up. The package will include 3/4'' flexible foam insulation for the suction line. Since evaporators, condensers and lines are available from the manufacturer pre-charged, these line kits simplify installation.

Line diameters in data tables are good up to a distance of 50 feet between the evaporator and condenser. If the condenser is located between 50 and 100 feet from the evaporator, increase line sizes by one size. Avoid distances over 100 feet. Run the lines as directly as possible, without traps or loops in the suction line.

Air pressure drop— Another item to check when selecting cooling equipment is the air pressure drop over the evaporator coil.

We'll discuss air pressure drops in other sections, also. Basically, warm air systems need a minimum amount of air pressure to insure that the air reaches the furthest outlet at the rated cfm. Pressure reduction, or drops, will take place at outlets, return inlets, filters, ducts and the cooling coil.

Air pressure drops over the cooling coil depend on the size of the coil and the amount of the air passing through it. For a typical residential system, and similar commercial systems, try to use a pressure drop of about 0.20 inches of water across the cooling coil. This pressure drop, along with others for the system shown in Figure 6-7, will let you calculate a fan size that can push the air through the system without excessive noise. The 0.20 cooling coil pressure drop also provides more efficient operation, since the air velocity across the coil is low enough that it doesn't blow the moisture off the coil.

Model M1UF024A has a static pressure drop of 0.20 at 700 cfm, according to the data in Figure 6-5. Our design needs a minimum of 530 cfm, and

we'll see later that we'll need more than that for efficient operation. So the 0.20 drop at 700 cfm will work out well for our residence.

Return air inlet	0.05
Filter when dirty	0.10
Cooling coil	0.20
Supply air outlet	0.05
Total pressure without ductwork	0.40

**Typical warm air system pressure drops
Figure 6-7**

Ease of Maintenance
The third factor to consider when choosing the furnace is its ease of maintenance. Keep in mind that any well-designed unit will probably provide easy maintenance. A good quality furnace will reduce the amount of maintenance needed because it's better built. But any warm air unit will need routine work such as filter replacement, belt adjustments and bearing lubrication. These components should be easy to get at inside the cabinet. Quality manufacturers are aware of such considerations. They'll outline maintenance advantages in their literature.

If you have any doubts about a certain furnace, contact the wholesaler or manufacturer's representative for information.

You may run into a job where space is at a premium in the mechanical room. First of all, check the codes. Most codes require a minimum clearance around the furnace. Next, check your choice of unit. The York furnace we've chosen, for example, has a cabinet 54'' high. The supply duct is top-mounted and the return can be located at the bottom on the right, left or rear. Although floor space is no problem in the sample residence, this type of versatility in a unit can help you to make a proper installation in good time. It's also important for routine maintenance.

Reasonable Cost
The cost of the furnace deserves some attention. For bid work, you'll be restricted by the specs, which tell you what type of unit is acceptable to the

owner. You'll then have to shop around to get the best price for yourself. You'll find more about this in Chapter 10.

For custom work, you'll want a furnace that provides the heating and cooling you need. But an overpriced furnace will cut into your profit margin, or perhaps raise the cost of the job to a point where the owner objects. Fortunately, there are many reliable furnaces to choose from. Improvements in design and technology have enabled manufacturers to produce quality units at reasonable prices.

Most manufacturers have a basic design for each size they build. Additional refinements can be added to suit a wide range of needs and budgets.

Furnace Location
The furnace should be located in the center of the building so the ductwork runs reach each supply and intake register efficiently. But some plans will restrict you to the architect's mechanical room placement. A good floor plan locates the mechanical room in the center, if possible. But other floor space considerations may put the heating unit off in one corner. This will mean longer duct runs, more pressure drops and more heat loss. It may mean an increase in overall size of the system, increasing the cost as well. This can mean a little or a lot to the owners, depending on how badly they need the open floor space. If it's not apparent on the plan why the mechanical room isn't in the center, you may want to give the architect or owner a call and check it out.

After finding the location, draw the heating unit on the floor plan (Figure 6-2). You may need to add a section detail later, after you've sized and located the ductwork.

Step 3: Size and Lay Out the Ductwork
Duct size is based on the amount of air flowing in the duct, the velocity, and the pressure drop. If duct velocities are too high, you'll have noise and excessive pressure drop. The fan may not have the capacity to push the air through the system.

From the definitions at the start of this section, we know that duct velocity is the speed of the air stream measured in feet per minute. The amount of air traveling through the duct is measured in cubic feet per minute. Static pressure is the outward force of the air. Velocity pressure is the forward moving force of the air. The total pressure is the static pressure plus the velocity pressure.

All of the above items are related, both mathematically and in actual practice. Here's what it means for design work: One, if we keep our pressure drops low enough for the air flows needed in the system, we'll have an air velocity that's low enough for a quiet and efficient system. Two, we'll be able to size our ductwork based on these air flows and pressure drops.

To have a quiet system, you need to keep the air velocity to less than 2,000 fpm. As long as the pressure drop remains at 0.1 inches of water or less per 100 feet of duct run or less, you'll stay below 2,000.

Figure 6-8 is a chart that relates pressure drops, cfm's, and duct sizes. It's used for round ducts, which are the least costly. But you may not have the space in the building to run round ducts. If square or rectangular ducts are necessary, use Figure 6-9 to convert round duct sizes to rectangles and squares. For a 6'' diameter duct, the square equivalent is 5.5'' by 5.5''.

Begin by noting on your plan the cfm for each room or area and the approximate route for the ductwork. Start at each outlet and work back.

For example, in Bedroom 1, we need 100 cfm at 0.1 pressure drop. Referring to Figure 6-8, follow the thin vertical line above 0.1 to where it intersects the thin horizontal cfm line of 100. Follow the heavy diagonal line upwards to the duct size scale. We find we need a 5.8'' duct diameter, so we'll use the next larger size, 6''.

Bedroom 2:	6''
Living Room:	7''
Entry:	6''
Bathroom:	5''
Storage:	5''

Size each section of duct to fit the available building space at a pressure drop of 0.1 or less per 100 feet. If a duct size falls between whole inches, use the next larger size. For example, 5.8'' would round up to 6''. Note that the larger the duct size, the smaller the pressure drop from friction. (See Figure 6-8.) But larger ducts will cost more. This can be significant if the job has a lot of ductwork. Also, you may not always have room for the larger duct. So try to size it according to cost, room and pressure drop, as long as you stay at 0.1 inches of water or less per 100 feet.

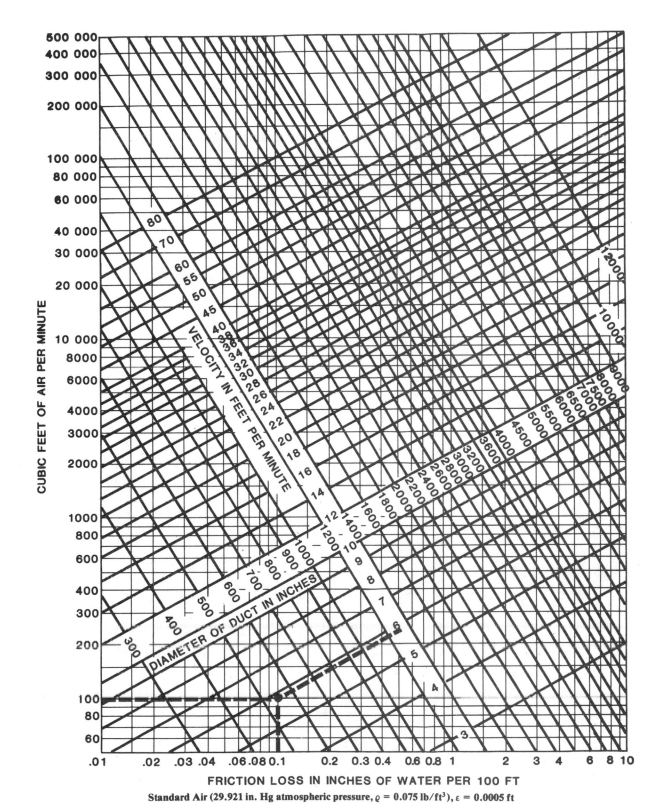

Standard Air (29.921 in. Hg atmospheric pressure, ϱ = 0.075 lb/ft^3), ε = 0.0005 ft

Round duct sizing chart
Figure 6-8

Dimensions in Inches

Side Rectangular Duct	4.0	4.5	5.0	5.5	6.0	6.5	7.0	7.5	8.0	9.0	10.0	11.0	12.0	13.0	14.0	15.0	16.0
3.0	3.8	4.0	4.2	4.4	4.6	4.7	4.9	5.1	5.2	5.5	5.7	6.0	6.2	6.4	6.6	6.8	7.0
3.5	4.1	4.3	4.6	4.8	5.0	5.2	5.3	5.5	5.7	6.0	6.3	6.5	6.8	7.0	7.2	7.5	7.7
4.0	4.4	4.6	4.9	5.1	5.3	5.5	5.7	5.9	6.1	6.4	6.7	7.0	7.3	7.6	7.8	8.0	8.3
4.5	4.6	4.9	5.2	5.4	5.7	5.9	6.1	6.3	6.5	6.9	7.2	7.5	7.8	8.1	8.4	8.6	8.8
5.0	4.9	5.2	5.5	5.7	6.0	6.2	6.4	6.7	6.9	7.3	7.6	8.0	8.3	8.6	8.9	9.1	9.4
5.5	5.1	5.4	5.7	6.0	6.3	6.5	6.8	7.0	7.2	7.6	8.0	8.4	8.7	9.0	9.3	9.6	9.9

Duct diameter

Side Rectangular Duct	6	7	8	9	10	11	12	13	14	15	16	17	18	19	20	22	24	26	28	30	Side Rectangular Duct
6	6.6																				6
7	7.1	7.7																			7
8	7.6	8.2	8.7																		8
9	8.0	8.7	9.3	9.8																	9
10	8.4	9.1	9.8	10.4	10.9																10
11	8.8	9.5	10.2	10.9	11.5	12.0															11
12	9.1	9.9	10.7	11.3	12.0	12.6	13.1														12
13	9.5	10.3	11.1	11.8	12.4	13.1	13.7	14.2													13
14	9.8	10.7	11.5	12.2	12.9	13.5	14.2	14.7	15.3												14
15	10.1	11.0	11.8	12.6	13.3	14.0	14.6	15.3	15.8	16.4											15
16	10.4	11.3	12.2	13.0	13.7	14.4	15.1	15.7	16.4	16.9	17.5										16
17	10.7	11.6	12.5	13.4	14.1	14.9	15.6	16.2	16.8	17.4	18.0	18.6									17
18	11.0	11.9	12.9	13.7	14.5	15.3	16.0	16.7	17.3	17.9	18.5	19.1	19.7								18
19	11.2	12.2	13.2	14.1	14.9	15.7	16.4	17.1	17.8	18.4	19.0	19.6	20.2	20.8							19
20	11.5	12.5	13.5	14.4	15.2	16.0	16.8	17.5	18.2	18.9	19.5	20.1	20.7	21.3	21.9						20
22	12.0	13.0	14.1	15.0	15.9	16.8	17.6	18.3	19.1	19.8	20.4	21.1	21.7	22.3	22.9	24.0					22
24	12.4	13.5	14.6	15.6	16.5	17.4	18.3	19.1	19.9	20.6	21.3	22.0	22.7	23.3	23.9	25.1	26.2				24
26	12.8	14.0	15.1	16.2	17.1	18.1	19.0	19.8	20.6	21.4	22.1	22.9	23.5	24.2	24.9	26.1	27.3	28.4			26
28	13.2	14.5	15.6	16.7	17.7	18.7	19.6	20.5	21.3	22.1	22.9	23.7	24.4	25.1	25.8	27.1	28.3	29.5	30.6		28
30	13.6	14.9	16.1	17.2	18.3	19.3	20.2	21.1	22.0	22.9	23.7	24.4	25.2	25.9	26.6	28.0	29.3	30.5	31.7	32.8	30
32	14.0	15.3	16.5	17.7	18.8	19.8	20.8	21.8	22.7	23.5	24.4	25.2	26.0	26.7	27.5	28.9	30.2	31.5	32.7	33.9	32
34	14.4	15.7	17.0	18.2	19.3	20.4	21.4	22.4	23.3	24.2	25.1	25.9	26.7	27.5	28.3	29.7	31.0	32.4	33.7	34.9	34
36	14.7	16.1	17.4	18.6	19.8	20.9	21.9	22.9	23.9	24.8	25.7	26.6	27.4	28.2	29.0	30.5	32.0	33.3	34.6	35.9	36
38	15.0	16.5	17.8	19.0	20.2	21.4	22.4	23.5	24.5	25.4	26.4	27.2	28.1	28.9	29.8	31.3	32.8	34.2	35.6	36.8	38
40	15.3	16.8	18.2	19.5	20.7	21.8	22.9	24.0	25.0	26.0	27.0	27.9	28.8	29.6	30.5	32.1	33.6	35.1	36.4	37.8	40
42	15.6	17.1	18.5	19.9	21.1	22.3	23.4	24.5	25.6	26.6	27.6	28.5	29.4	30.3	31.2	32.8	34.4	35.9	37.3	38.7	42
44	15.9	17.5	18.9	20.3	21.5	22.7	23.9	25.0	26.1	27.1	28.1	29.1	30.0	30.9	31.8	33.5	35.1	36.7	38.1	39.5	44
46	16.2	17.8	19.3	20.6	21.9	23.2	24.4	25.5	26.6	27.7	28.7	29.7	30.6	31.6	32.5	34.2	35.9	37.4	38.9	40.4	46
48	16.5	18.1	19.6	21.0	22.3	23.6	24.8	26.0	27.1	28.2	29.2	30.2	31.2	32.2	33.1	34.9	36.6	38.2	39.7	41.2	48
50	16.8	18.4	19.9	21.4	22.7	24.0	25.2	26.4	27.6	28.7	29.8	30.8	31.8	32.8	33.7	35.5	37.2	38.9	40.5	42.0	50
52	17.1	18.7	20.2	21.7	23.1	24.4	25.7	26.9	28.0	29.2	30.3	31.3	32.3	33.3	34.3	36.2	37.9	39.6	41.2	42.8	52
54	17.3	19.0	20.6	22.0	23.5	24.8	26.1	27.3	28.5	29.7	30.8	31.8	32.9	33.9	34.9	36.8	38.6	40.3	41.9	43.5	54
56	17.6	19.3	20.9	22.4	23.8	25.2	26.5	27.7	28.9	30.1	31.2	32.3	33.4	34.4	35.4	37.4	39.2	41.0	42.7	44.3	56
58	17.8	19.5	21.2	22.7	24.2	25.5	26.9	28.2	29.4	30.6	31.7	32.8	33.9	35.0	36.0	38.0	39.8	41.6	43.3	45.0	58
60	18.1	19.8	21.5	23.0	24.5	25.9	27.3	28.6	29.8	31.0	32.2	33.3	34.4	35.5	36.5	38.5	40.4	42.3	44.0	45.7	60
62		20.1	21.7	23.3	24.8	26.3	27.6	28.9	30.2	31.5	32.6	33.8	34.9	36.0	37.1	39.1	41.0	42.9	44.7	46.4	62
64		20.3	22.0	23.6	25.1	26.6	28.0	29.3	30.6	31.9	33.1	34.3	35.4	36.5	37.6	39.6	41.6	43.5	45.3	47.1	64
66		20.6	22.3	23.9	25.5	26.9	28.4	29.7	31.0	32.3	33.5	34.7	35.9	37.0	38.1	40.2	42.2	44.1	46.0	47.7	66
68		20.8	22.6	24.2	25.8	27.3	28.7	30.1	31.4	32.7	33.9	35.2	36.3	37.5	38.6	40.7	42.8	44.7	46.6	48.4	68
70		21.0	22.8	24.5	26.1	27.6	29.1	30.4	31.8	33.1	34.4	35.6	36.8	37.9	39.1	41.2	43.3	45.3	47.2	49.0	70
72			23.1	24.8	26.4	27.9	29.4	30.8	32.2	33.5	34.8	36.0	37.2	38.4	39.5	41.7	43.8	45.8	47.8	49.6	72
74			23.3	25.1	26.7	28.2	29.7	31.2	32.5	33.9	35.2	36.4	37.7	38.8	40.0	42.2	44.4	46.4	48.4	50.3	74
76			23.6	25.3	27.0	28.5	30.0	31.5	32.9	34.3	35.6	36.8	38.1	39.3	40.5	42.7	44.9	47.0	48.9	50.9	76
78			23.8	25.6	27.3	28.8	30.4	31.8	33.3	34.6	36.0	37.2	38.5	39.7	40.9	43.2	45.4	47.5	49.5	51.4	78
80			24.1	25.8	27.5	29.1	30.7	32.2	33.6	35.0	36.3	37.6	38.9	40.2	41.4	43.7	45.9	48.0	50.1	52.0	80
82				26.1	27.8	29.4	31.0	32.5	34.0	35.4	36.1	38.0	39.3	40.6	41.8	44.1	46.4	48.5	50.6	52.6	82
84				26.4	28.1	29.7	31.3	32.8	34.3	35.7	37.1	38.4	39.7	41.0	42.2	44.6	46.9	49.0	51.1	53.2	84
86				26.6	28.3	30.0	31.6	33.1	34.6	36.1	37.4	38.8	40.1	41.4	42.6	45.0	47.3	49.6	51.6	53.7	86
88				26.9	28.6	30.3	31.9	33.4	34.9	36.4	37.8	39.2	40.5	41.8	43.1	45.5	47.8	50.0	52.2	54.3	88
90				27.1	28.9	30.6	32.2	33.8	35.3	36.7	38.2	39.5	40.9	42.2	43.5	45.9	48.3	50.5	52.7	54.8	90
92					29.1	30.8	32.5	34.1	35.6	37.1	38.5	39.9	41.3	42.6	43.9	46.4	48.7	51.0	53.2	55.3	92
96					29.6	31.4	33.0	34.7	36.2	37.7	39.2	40.6	42.0	43.3	44.7	47.2	49.6	52.0	54.2	56.4	96

Rectangular duct equivalents
Figure 6-9

Return ducts are the same size as the supply ducts for a given room or area. They should never be smaller than the supply ducts. Sizing them larger is necessary only when the return is serving more than one supply. Then total the supply duct area. The return should be at least as large or slightly larger.

If the heating unit is located in the center of the building, it may be most practical to run a separate duct to each outlet and return. But you may have a building where it's more economical to run a larger duct and then tap off the supplies or returns for two or more rooms, as in our residence. Again, start at each outlet or return. Compute the duct size based on the cfm and the pressure drop of 0.1 per 100 feet or less. The duct for each room or area will be that size until it joins into a larger supply or return duct. That duct will increase in size as you work your way back to the heating unit, depending on how many outlet and return ducts are added.

For example, in our residence, we have two 6'' round ducts for returns in our bedrooms. Duct area equals:

$$\text{pi} \times \frac{d^2}{4}$$
$$3.14 \times \frac{36}{4} = 28 \text{ sq. in. per duct}$$
$$28 \times 2 = 56 \text{ sq. in. for both ducts}$$

Our main return duct will need 56 square inches to supply the two bedroom returns, so we'll use a 6 x 10 rectangular duct. Note that after we add our 14 x 10 return register in the hall, the main duct size increases to 16 x 6 to accommodate the second addition. Actually, the main duct could be even larger for the second increase, but the building construction makes it impractical.

You'll have to lay out duct runs by trial and error until you find the best locations. The sizes of the ducts and the building design will influence where you locate them. Some HVAC equipment manufacturers have easy-to-use adjustable tables that coordinate duct sizes for various cfm, pressures, and lengths, enabling you to quickly determine which sizes are most practical.

The path of single-size ductwork is typically drawn on the tracing using a single line. The size and location is noted where needed. This is called a *one-line layout*. Larger supply and return feeder ducts are drawn to scale and labeled as to size and cfm. Drawing them to scale insures that they'll fit where you locate them in the building. If it's not practical to run ducts between ceiling or floor joists, you can often run them in soffits that run in hallways or along the walls.

Sometimes you may have to call the architect and ask him to include soffits or chases in the building plan to make your HVAC installation easier and therefore cheaper to the owner.

Pressure Drops

As mentioned in Step 2, the static pressure of the fan must equal or exceed the pressure drops throughout the system. The pressure at the fan depends on the losses in the supply outlets, ducts, cooling coil if used, filters, return air ducts, and inlets.

Figure 6-7 shows the pressure drops we'll use for our system. By checking the manufacturer's data used in the previous sections, we see that the pressure drops for the inlet, return, and cooling coil we've chosen for our residence are close to the drops in Figure 6-7. If the equipment drops were too large, we'd have to use different equipment.

Note that so far we don't have any value for duct pressure drop. Here's why. Although we figured our duct sizes within a 0.1 pressure drop per 100 feet or less, it's possible we've exceeded the static pressure limits imposed by the fan size by having ducts that are too long. Since the pressure drops listed above are relatively fixed, the only variable that can be adjusted would be duct length.

Note also that we're using only the pressure drops for one supply outlet and one return inlet. For design purposes, we select those that are the farthest from the furnace. To calculate duct pressure drop, we only need to be concerned with the outlet and return farthest from the fan. The following example will help make all this clear.

After sizing and laying out the ductwork, measure the length of the ductwork from the farthest air return to the fan. Then measure the length of the ductwork from the farthest supply to the fan. Add the two distances and multiply by 1.5. This 1.5 factor takes into account extra losses due to elbows, transitions and take-off ducts. Use this length and multiply it by 0.1 per 100 feet to find the actual pressure drop in the ductwork.

For example, the longest run in our residence

(supply plus return) is 71 feet. The ductwork pressure drop is:

$$71' \times 1.5 \times \frac{.1}{100} = 0.11 \text{ inches of water}$$

Adding this to the other pressure drops listed above gives us a final overall pressure drop of 0.11 plus 0.40, or 0.51 inches for our 530 cfm system.

The furnace we've selected has a fan that will provide 930 cfm of air at 0.60 inches external static pressure, running at a low speed (Figure 6-4). Our design cfm is a minimum of 530. Theoretically, if we have a system that provides only the minimum pressure computed at the system cfm, the air wouldn't move through the ducts, or at best, would be too weak to work efficiently. There have been a few jobs where this has actually happened. However, 930 cfm is too much. This would cause noise and excessive drafts in the system. By placing a damper in the supply duct, we can reduce the 930 cfm to 700. We choose 700 because it gives us enough extra cfm above the design need, and also because the evaporator coil was figured at 700 cfm at our preferred 0.20'' pressure drop.

After the damper is installed, we can adjust it to our desired cfm by first finding the external static pressure using the Fan Law (from Figure 5-11 in Chapter 5):

$$\frac{cfm_1}{cfm_2} = \sqrt{\frac{pressure_1}{pressure_2}}$$

$$cfm_1 = cfm_2 \times \sqrt{\frac{pressure_1}{pressure_2}}$$

Using our known data:

$$930 = 700 \times \sqrt{\frac{.60}{P_2}}$$

$$\sqrt{(P_2)} = .58$$
$$P_2 = .76'' \text{ external static pressure}$$

After initial system start-up, you can put a manometer or similar gauge in the duct on the supply side of the blower, past the damper. Then adjust the damper until the reading is 0.76'', resulting in 700 cfm to supply the system.

In the above calculations, we've used a pressure drop of 0.1 inches per 100 feet or less of duct length as best for most systems. You may find yourself designing a system where duct runs are over 100 feet. In this case, use a pressure drop of 0.05 inches of water per 100 feet for your duct runs. For example, if our residence's longest run was 200 feet, we'd have 200 times 1.5 times 0.0005, or 0.15 inches of water pressure. Adding this to our other pressure drops would bring us very close to the maximum capacity of the fan. However, since the longer duct runs would indicate a larger building, it's likely our heat loss would be larger because of more exterior building area. This means we'd need more cfm per room, requiring a larger overall system cfm. This would mean larger ducts, fan and heating unit; in short, a step up in system size. But the calculation procedure would be the same as the steps we used above for our sample residence.

You may come across a job where the owner wants to wait before installing his air conditioning. Impress upon him that it's cheaper to have the work done while you're installing the system in the first place. But if he still wants to wait, you'll design and install the system sized for the cooling equipment. If the evaporator coil isn't installed right away, you'll have less pressure drop and the fan will move more air. This can make it noisy or drafty. If this happens, install a manually operated damper in the main supply duct, as described above, to reduce the air flow.

Step 4: Thermostat Location

Many types of thermostats are available. Your choice will depend on cost, the type of building where it's installed, and the kind of system.

A standard unit will have a temperature setting, a system selector switch with *Cool, Off* and *Heat* settings, and a fan control switch with *On* and *Auto* settings. The Auto setting starts the fan automatically when heat is needed.

The unit may operate at 24 volts or 115 volts. The lower voltage is preferred because the wiring doesn't have to be run in conduit.

Thermostat wiring is color-coded to match the heating unit terminals. Wiring diagrams are furnished with the thermostat and on the unit control box. On some jobs you may find it cheaper and easier to have an electrician install the thermostat wiring.

A general thermostat location is usually shown on the plan. For work designed by others, be sure to check the specs for precise location instructions. Otherwise, locate it on an inside wall in the stream of the air returning to the furnace. Obviously, don't put the thermostat near heat-producing appliances or directly under light fixtures or lamps.

Step 5: Fuel Piping

We'll consider fuel piping requirements for both gas and oil systems.

Gas Systems

Natural gas and LP gas systems are similar. But you can't use LP gas in a natural gas furnace, or vice versa, without changes in the furnace. Gas furnace manufacturers' data will tell what units can be converted to LP gas and how to do it.

Gas systems require one line from the piping system in the residence to the furnace. If you have no other gas appliances, one line from the meter or LP tank will be suitable. The National Fire Protection Association provides a code, NFPA No. 54, *The National Fuel Gas Code*, for sizing gas piping. This code also shows you how to install a safe system. You can order copies of the code from:

The National Fire Protection Association
Batterymarch Park
Quincy, MA 02269

Be sure to follow local codes. Check with the natural gas utility or LP supplier and follow their recommendations, also.

You may want to have a plumbing contractor do the gas piping. If you do, advise him of the amount and kind of gas used by your equipment. And even if you're subcontracting the work, you'll need to know the code and check the sub's installation to avoid problems later.

Oil Systems

Oil piping requires a line from the oil tank to the furnace or boiler. Most residential installations can be served with 3/8" or 1/2" oil lines, if the following two requirements are met:

1) the distance to the tank is less than 100 feet

2) the tank is on the same level as the furnace, or just one floor level above or below

Sometimes a return line is also required to pump excess oil back to the tank. The manufacturer's catalogs will show this.

The NFPA Code No. 31, *Standard for the Installation of Oil Burning Equipment,* shows how to properly install the oil piping system. Following the code and using common sense will usually result in a good installation. Be sure to check local codes for any special requirements.

You can order the oil burning equipment code from the NFPA at the address just given. It and the gas code are part of a 16-volume set put out by the National Fire Protection Association. Since these volumes are used by many state and local code agencies as the mainstay of their codes, we recommend you order all 16 volumes. Otherwise, you'll surely find that the code you need to review is in a volume you don't have.

Draw fuel piping on the plan as part of the heating unit diagram. Make a note which indicates its size and if it's to be installed by the plumbing contractor. Also show the gas lines sized on the floor plan. Oil line locations are generally left up to the installer. If line locations conflict with the code when the installation is done, the code must be followed. Plans should be changed to indicate any new locations.

Step 6: Electrical Requirements

To determine the electrical requirements for a warm air system, you'll need to check system equipment requirements as shown in the manufacturer's data.

Include in the specifications the size of the furnace fan motor (horsepower), its voltage, and the frequency. Or put it on the plan in a schedule so the electrician can select the proper requirements.

You'll also have to give the electrical information for the air conditioner condensing unit, thermostat, exhaust fans, humidifiers and other equipment. The manufacturer will list the electrical information on the condensing unit. A hot water system will need electricity for the boiler pump and thermostat. If you don't have this information in your plans and specifications, the electrician won't be able to properly size and price the electrical work.

The work must conform with the national and local electrical codes. The electrician should also check the size of the main service to the residence to determine if it's adequate.

Step 7: Chimney, Breeching and Combustion Air

We're getting close to the end, now. The seventh

step includes selecting the chimney, sizing the breeching and providing adequate combustion air.

Chimney and Breeching

Conventional boilers and furnaces require a chimney to remove combustion gases. Locate the boiler or furnace near the chimney, so it can work through natural draft. If the chimney is more than 25 feet away, you'll need to add a draft fan to get the combustion gases to the chimney. Check your local codes.

The selection of a chimney depends on the fuel used. Check the NFPA No. 31 or 54 code. Chimneys may be included in your HVAC contract. You can use prefabricated metal chimneys selected from manufacturers' catalogs.

If you aren't required to furnish the chimney, it'll be part of the general contractor's work. He'll need to know the fuel you're using and the chimney size required.

The chimney cross-section area must be at least equal to the cross-section area of the furnace flue size shown in the manufacturer's catalog. Flues for several gas appliances may be combined into one chimney. The chimney area must equal the total flue area of all the connected equipment. Don't combine flues of gas equipment with oil or a fireplace. Each fuel type needs a separate chimney. Again, check your codes.

The *breeching,* the pipe from the furnace to the chimney, must be sized so its cross section is equal to the sum of the cross sections of the equipment connected to it. This means the size of the breeching will increase as pieces of equipment are added to it on its way to the chimney. The breeching's clearance from combustible material must be sufficient to prevent fire. (See NFPA 90B.) Oil furnaces may have a forced draft burner. Take special care when more than one piece of forced draft equipment is connected to a breeching. Be sure the draft is properly adjusted on each unit to prevent the escape of combustion gases inside the residence. (See NFPA No. 31.)

Some newer high-efficiency units have draft fans. They won't require natural draft chimneys. The manufacturer's catalogs tell you how to select the proper exhaust system.

Chimneys and breeching must be installed with care. Combustion gases can be deadly to building occupants. Don't forget to check chimneys and breeching when you're doing routine maintenance and repair as well.

Locate and label breeching and chimneys both on the floor plan and the equipment detail. Draw them to scale and note sizes and special requirements.

Combustion Air

The amount of air required for combustion depends on the fuel and the burning rate. This air may be supplied from the space around the equipment or brought from the outside by a duct. The NFPA Codes No. 31 and 54 and local codes specify when a combustion air intake is required and its size, depending on the Btu's of the unit.

It's important to provide adequate combustion air to insure safe operation. A system that takes air from a closed space without replacing it will pull enough oxygen from the air to endanger the occupants. Good combustion air also insures that the equipment operates at its rated capacity and efficiency.

If you are using ductwork for combustion air in cold climates and it runs through interior areas, you may have to insulate the duct to prevent condensation. A 1'' cover of fiberglass with vapor barrier will do the job. Be sure all insulation joints are sealed to maintain the vapor barrier.

Draw, size and label the combustion air intakes on the floor plan as they relate to the heating equipment. You may also need to add details and diagrams to prevent confusion in the installation.

Step 8: Design Checklist

So far we've covered a lot of material on HVAC design. None of it has been too difficult. But by now you can see that you need to work carefully.

The only way you'll ever be able to do design work on your own is to dig out all the information you need for a job. In time it'll become second nature to you. But even then, always remember that each job will be different. So the bottom line on design work is to *check and recheck your work.*

The Design Checklist, Figure 6-10, gathers together all the various steps in the design process and summarizes them. To work through the steps, you'll need your planning sheets, worksheets, plan sets, tracings, code books and this manual — in

Design Checklist

1) Calculate heat loss and gain in Btu/hr for each area or room.

2) Summarize heat loss and gain data.

3) Determine heating, cooling and ventilation requirements.

4) Determine volume of supply, return, and exhaust air.

5) Select furnace, cooling and other equipment.

6) Size and locate on the plan: ductwork, intakes, outlets, exhausts, thermostats, piping, venting, electrical connections, and equipment.

7) Label all of the above. Use notes, diagrams, sections, details and schedules to insure clarity.

8) Check for size, location, installation procedures and code compliance of:

   Fuel piping
   Chimney and breeching
   Combustion air intake

9) Is insulation necessary for exhaust air and combustion air ducts?

10) Is distance between fresh air intakes, exhausts and vents 10' or more from plumbing vents? (Or as required by local code?)

11) Is the AC condenser located on the plan?

12) Will all equipment have enough room to be serviced?

13) Is the noise level of equipment satisfactory?

Design checklist
Figure 6-10

Plan and Spec Checklist

1) Plan shows north arrow; job site location; symbol schedule; plan scale; and name, address and phone number of designer.

2) Title blocks and page numbers are completed and correct.

3) Plan shows location and size of all equipment, registers, grilles, thermostats, switches, and all ducts or piping for the building.

4) Plan shows any concealed or underground piping or duct.

5) Plan details, sections and diagrams are correctly coded to floor plan.

6) Plan labels indicate the intended use of all rooms.

7) Plan notes and details indicate size and location of combustion air intake.

8) Plan notes and details indicate insulation protection to be applied to any exhaust or intake ducts.

9) Plan shows distance between fresh air intakes and exhaust or plumbing vents.

10) Plan shows fuel piping location and size.

11) Plan shows location of air conditioner condenser and piping.

12) Specifications describe general job conditions, job payment, and the quality, type and manufacturer of equipment, piping, ductwork, insulation and controls.

13) Arrows from notes on plan are curved — not straight lines which can be confused with building lines.

14) Notes on drawings are printed in clear spaces, not across building lines.

15) All lettering is done so it can be read without turning the plan.

16) Plans and specifications have approval of owner and inspectors.

17) The plans and specifications have all the information needed so other trades can perform HVAC-related work.

Plan and spec checklist
Figure 6-11

short, everything you've used so far for design work. We won't tell you where to look to find the information for each step. Or where it goes after you've found it. From now on, that's up to you.

Step 9: Final Plan and Specification Check

The plans and specifications are the language that describes the job. They're part of the contract. If you're trying to get a custom job, they'll help you show the owner exactly what he's getting, how it'll be installed, how he'll have to pay for it, and other details.

The plans and specs should complement each other. There shouldn't be any conflicts. The specifications should describe the type and quality of equipment and material, how to install it, and how the system will work. They should also cover any special installation procedures or job site details the owner wants. The plans should show the size and extent of the work. You may want to review Chapter 4 on plans and specs.

If you're working with a general contractor on a project, the specifications and plans drawn up by someone else will be part of your contract. They should clearly delegate the work to each contractor. If you, the HVAC contractor, have electrical, plumbing or general work related to your job, be sure that it's clear who will pay for and do that work. Chapter 11 explains how plans and specs drawn up by others are used for HVAC bidding and estimating.

The final check for the plans and specs you've drawn up breaks down into two parts. First, check to make sure you haven't left out a piece of equipment or made a mistake in labeling the work. Second, you'll want to check the plan and specs for spelling, neatness, clarity, missing notes and similar minor, but potentially expensive, mistakes.

To check your plans, run a final set of prints.

Start by taking an imaginary walk through and around your building. Mark up the plans with a red pencil as you discover the errors. Recheck the plans by starting over again in the upper left-hand corner. Work clockwise until you've covered each item on the sheet. Cross-check information in details and schedules to make sure it's correctly coded to the right location on the building plans. Make sure this information is complete and accurate.

Now check your specifications. Make sure they include all work not shown in the plans. Review the specs in Chapter 4 to see how a typical job is covered. Check yours for misspelled words and incomplete sentences. Make all corrections and check again to be sure you haven't missed anything. Use the checklist in Figure 6-11.

If possible, have another person do the final check of your plans and specs. It's always easier for someone else to find your mistakes.

One final word on your plan set. During the design work, you ran off several sets of plans from your tracings as the work progressed. Save those working sets where you developed major design steps. They'll come in handy if you ever have to do modifications as the job progresses. They'll also be very helpful when you design similar jobs.

If you're working with an architect, he may want the HVAC tracings as part of his set. Note as part of your job record that the architect has them. That way, if anyone contacts you for a plan set, you'll know who to refer him to. Tracings which are your property should be stored in a dry protected place, laid flat, with cardboard sheets between them. Drawing desks have large drawers for this purpose. But you may want to save this space for working sets of tracings. You can buy or build cabinets designed for tracing and plan set storage.

In the next chapter, we'll do a residential design with hot water and electric heat.

Designing Hot Water and Electric Heating Systems

One reason that HVAC contracting is one of the toughest of all trades is that there are so many systems that'll do the job. An HVAC contractor has to wear many hats — designer, plumber, sheet metal fabricator, electrician — as well as business owner and salesperson. If he's an expert in all these different areas, he can talk knowledgeably to his customers, bid jobs accurately and stay ahead in today's tough market.

In this chapter, we'll build on the system design steps you've already learned. We'll cover hot water systems and electric heat. Since you're now familiar with warm air design, these new jobs will be easier. Many of the steps are alike. We won't repeat those. Instead, review Chapter 6 as needed.

Residential Hot Water System
We'll start our forced hot water — also called *hydronic* — design using the same plans we used to design the warm air system. Then we'll install a hot water system in an apartment building.

Definitions
Before we begin, review the following definitions:

Air cushion tank— Commonly called an expansion tank, this is a closed tank in a hydronic system located above the boiler to provide for the expansion of the water. Typically, when the system is filled with water, air is trapped in the tank. As the water is heated, it expands and compresses the air in the tank, allowing room for the additional water without excessive increase in system pressure. A *diaphragm* tank uses a factory-installed diaphragm to separate the air and water, preventing any air from being drawn into the system when the water cools.

Air vent— A manual or automatic valve, usually found on the heat distributing units and the system's high points, that allows air to be bled from the system.

Altitude gauge— Often found with a temperature gauge, the altitude gauge registers water pressure within the boiler in both feet of water and psi.

Balancing ell— A type of elbow fitting that controls the amount of water flowing through a heat distribution unit.

Balancing valve— A valve used to control the

water flow, usually found in the return piping of a multiple circuit system. Unlike the balancing ell, these valves are adjustable and are used to balance the heating units in the system for overall heat distribution.

Branch piping— The supply and return piping used to connect heating units to main supply and return piping.

Circuit— A complete path of water flow from the boiler supply to the boiler return.

Dip tube— A downward extension of supply piping that's connected into the vertical supply tapping of the boiler. Water entering the supply pipe enters the low end of the dip tube below the level of any air accumulated at the top of the boiler chamber.

A dip tube requires that the air cushion tank be connected to the boiler chamber rather than the supply line.

Drain cock— A valve installed at the boiler and system low points to allow water drainage from the system.

Expansion tank— Common trade name for an air cushion tank.

Flow control valve— A check valve installed in the supply line to prevent gravity circulation of hot water when the pump is shut off.

Heat distributing units— These units can be grouped into four broad types; baseboard, convector, finned tube radiation, and radiator.

1) Baseboard units come in two types. First is the finned tube type: a finned metal pipe or tube enclosed in a slotted metal cover that allows air to circulate by gravity. It's usually located along the walls at the floor. Second is the cast iron type. These units have front and rear surface areas which are heated by water. Then the air passes by gravity through openings in the enclosure.

2) Convector units are enclosed on all sides, with air inlets at the bottom and outlets at the top or upper front. Typically they are freestanding units found in offices, schools and other commercial buildings.

3) Finned tube radiation units are basically elements made up of a finned tube with or without an enclosure. They're often used in commercial jobs where appearance isn't important and where they can be installed so there's little chance of accidental damage.

4) Radiators are the original, freestanding cast iron heating units made up of connected tubular sections. Many tradespeople incorrectly use "radiator" as a generic term for any type of hot water heat distribution unit.

Horizontal supply tapping— A boiler-supply line connection in a horizontal position.

Main— Used to refer to the main supply and return pipes in the system. Also called a trunk line.

Make-up water line— The water pipe line to the boiler used to fill the system and add water when needed.

Pressure reducing valve— A valve used in the make-up water line to coordinate city water pressures with the lower working pressure of the boiler. It also allows make-up water to enter the system as needed.

Pump— Usually a centrifugal type, the pump circulates the water through the entire system. Zone systems will sometimes have a pump for each zone, separately controlled.

Radiator valve— A manual control valve to adjust the flow of water through the heat distributing unit.

Return branch— Piping that carries water from the heating unit to the return main.

Return main— The main piping from the heating units to the boiler.

Reverse flow— A method where the water in the boiler flows from top to bottom rather than the conventional bottom to top.

Riser— A term used to describe any vertical piping in the system.

Safety relief valve— A valve that protects the boiler from excessive pressure by opening when a pre-set pressure is reached by the boiler. According to code, all hot water and steam boilers must be equipped with a properly rated and installed safety valve.

Single circuit system— There are two basic types of single circuit systems. The *series loop* has heating units in series with a single pipe that both supplies and returns the water. The units act as portions of this main pipe. A *one pipe* system uses a single main to both supply and return the water. Heating units are connected to the main by branch piping using special one-pipe tee fittings that force part of the water into the unit and back to the main.

Supply branch— Piping that carries water from the supply main to the heating unit.

Supply main— The main piping from the boiler to the heating units or supply branches.

Supply tapping— Boiler opening, either vertical or horizontal, where the supply main is connected.

Zone— A thermostatically controlled area made up of one or more circuits.

Design Steps

In designing a hot water system, we'll follow these steps:

1) Select design temperature and radiation units. Size and locate units on the floor plan.

2) Size, select and locate the boiler.

3) Size and locate ventilating fans.

4) Select, size and locate a one-line layout of the piping system. Size and locate the circulating pump or pumps.

5) Select and locate thermostat and boiler controls.

6) Select fuel piping.

7) Determine electrical requirements.

8) Consider chimney, breeching and combustion air.

9) Make the final plan and specification check.

Figures 7-1 and 7-2 have the manufacturer's data for the baseboards and hot water boiler we'll be using. Figure 7-3 gives dimensions and properties of copper tubing. Pump data is found in Figure 7-4. The hot water system is shown on the sample residence floor plan in Figure 7-5. Figure 7-6 is a detail of the boiler and related piping.

Step 1: Design Temperature and Radiation Supply Units

Hot water radiation units, commonly called radiators or baseboards in the trade, are economical and easy to fit into rooms. Like the outlets in a warm air system, radiators are located along the outside walls, under the windows if possible, to offset the heat loss.

Figure 7-1 shows a catalog sheet of baseboard radiation data. But before we can size each baseboard, we must decide on the average temperature of the water as it passes through the system. This is based on the design water temperature drop — the temperature difference between the water leaving and returning to the boiler.

Temperature drops of 20°F, 30°F, 40°F and 50°F are commonly used. The drop is important since it's the basis for finding how many gallons per minute (gpm) of water must circulate to give off the needed Btu's. This in turn affects pipe, pump and heating unit size and ultimately, equipment and materials cost. Here's how to figure it.

If one pound of water drops 1°F as it circulates through the system, then one Btu is given off. One gallon of water weighs about 8.33 pounds, so it will give off 8.33 Btu if it drops 1°F as it circulates. If we have one gallon per minute, the calculation is:

$$8.33 \text{ Btu} \times 60 = 499.8 \text{ Btu/hr/gal}$$

$$\text{Rounded off} = 500 \text{ Btu/hr/gal}$$

For various design temperature drops at 1 gpm:

20°F drop = 20 x 500 = 10,000 Btu/hr

30°F drop = 30 x 500 = 15,000 Btu/hr

40°F drop = 40 x 500 = 20,000 Btu/hr

50°F drop = 50 x 500 = 25,000 Btu/hr

MODEL 740 & 750 I=B=R APPROVED HOT WATER RATINGS
(Aluminum Fins: 2¼" x 2¼" x .008", 54.2 per foot)

Model	Flow Rate		Pressure Drop (Mil. Ins.)	BTU/Hr/Linear Foot with Damper Fully Open									
				Average Water Temperature °F									
	GPM	Lbs/Hr		160°	170°	180°	190°	200°	210°	215°	220°	230°	240°
740	1	500	47	450	510	580	640 ◄—	700	760	800	830	890	950
	4	2000	525	480	540	610	680	740	800	850	880	940	1006
750	1	500	47	430	490	560	620	680	740	780	810	870	930
	4	2000	525	450	520	590	660	720	780	820	860	920	980

Ratings are based on active finned length (3" less than overall length) and include 15% heating effect factor. The ratings based on 4-GPM should be used for all normal loop systems where the flow rate is known to be equal to or greater than a 4-GPM flow rate. Where water flow rates are known, or in piping arrangements resulting in lower flow rates, the 1-GPM ratings should be used.

SPECIFICATIONS—HEATING ELEMENT: Furnish and install FLOORLEVEL MODEL 740 (MODEL 750) HEATING ELEMENT as manufactured by Hydrotherm, Inc., consisting of 3/4" nominal seamless copper tubing with 2¼" wide x 2¼" high x .008" aluminum fins, spaced 54.2 fins per foot to allow for proper air flow. A vertical flange shall be formed on front and on rear edges of each fin. Flange shall be of such width, maximum of 3/32", as not to allow fin-to-fin contact. One end of each tube shall be swagged to receive standard tube or fitting. Continuous plastic rails shall be factory installed on entire length of element, only at the cool bottom corners of the fins.

SPECIFICATIONS—ENCLOSURE ASSEMBLY:
Furnish and install FLOORLEVEL MODEL 740 (MODEL 750) ENCLOSURE ASSEMBLY as manufactured by Hydrotherm, Inc., consisting of one-piece, 22-gauge precoated back panel and top moulding, which shall have no holes or die-formed protrusions which would affect air flow; one-piece, 22-gauge and rigidized front panel cold rolled formed of precoated steel and one-piece precoated cold rolled formed damper. Brackets shall be die-formed and of rigidized, one-piece, (16-gauge for Model 740) (18-gauge for Model 750) construction so that they shall snap in at any point within the entire enclosure, reinforcing the top moulding, and shall have a die-cut pivot that measures no less than 1/2" in diameter to accommodate the damper. The damper shall pivot to any position from full open to full closed by finger-tip control, and shall be fully operative after initial installation regardless of how many additional coats of paint are applied. Provision within the cover assembly shall be made to accommodate the element and assure free expansion and eliminate metal-to-metal contact. Back panel, front panel and damper shall be precoated before forming with white or beige appliance finish, or shall have woodgrain finish, on the outside and shadow black finish on the inside. All accessories are to be die-formed and snapped into position without the use of screws, and shall be precoated with white or beige appliance finish or have a woodgrain finish.

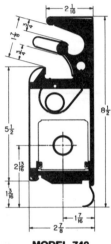

MODEL 750   MODEL 740

Courtesy: Hydrotherm, Inc.

Hot water baseboard data
Figure 7-1

You can see that the higher temperature drops deliver more Btu's, so less water is needed. So, at least in theory, the higher your temperature drop, the smaller the pipe and the less it will cost you for the system. But if you try to choose the highest temperature simply to cut costs, you'll be in trouble. There are other variables to consider.

A higher average temperature means a higher water temperature in the boiler. There's a danger with higher temperatures. If water boils in the system, steam will form, causing severe problems.

RATINGS NATURAL AND LP* GASES

Boiler Number	AGA & CGA Input MBH(1)	DOE Heating Capacity MBH	I=B=R Net Rating MBH(2)	Net Rating Water Sq. Ft.(3)	CGA Gross Output (MBH)	Recommended Chimney Size Round Dia. In. x FT.(4)	DOE Annual Efficiency, %			
							Standing Pilot	Standing Pilot and Vent Damper	EI	EI and Vent Damper
►203	62	51	►44.3	295	49.6	4 x 15	69.1	74.7	72.1	80.7
204	96	79	68.7	458	76.8	5 x 15	69.4	75.4	72.1	80.7
►205	130	107	►93.0	620	104.0	6 x 15	69.6	76.1	72.1	80.7
206	164	135	117.4	783	131.2	6 x 15	69.9	76.8	72.1	80.8
207	198	162	140.9	939	158.4	7 x 15	70.2	77.4	72.1	80.8
208	232	190	165.2	1101	185.6	7 x 15	70.4	78.1	72.1	80.8
209	266	218	189.6	1264	212.8	8 x 15	70.7	78.8	72.1	80.8
210	299	245	213.0	1420	239.2	8 x 15	70.9	79.5	72.1	80.9

DOE heating capacity and annual efficiency are based on U.S. Government standard tests.
*Propane available for standing pilot only.

(1) A.G.A. ratings shown are for installations at sea level and elevations up to 2,000 ft. For elevations above 2,000 ft. A.G.A. ratings should be reduced at the rate of four percent (4%) for each 1,000 ft. above sea level.

(2) Net I=B=R ratings shown are based on normal I=B=R piping and pick-up allowance of 1.15. Consult the manufacturer for installations having unusual piping and pick-up requirements such as intermittent system of operation, extensive piping systems, etc.

(3) Based on 170° F average water temperature in radiators (heat emission rate of 150 BTU/HR/Sq. Ft.). For higher water temperatures, select boiler on basis of net ratings in BTU/HR.

(4) Recommended chimney for rectangular application is 8 in. X 8 in. X 15 ft. for sizes 203 through 209, and 8 in. X 12 in. X 15 ft. for size 210. (15 ft. chimney height is from bottom of drafthood opening to top of chimney.)

DIMENSIONS (In Inches)

Boiler Number	Dim A	Dim B	Dim C	(Flue Outlet) Dim D	Dim E	Dim F	Gas Connection for Gas Valve	
							24 Volt	EI & MV
203	27½	14½	8	4	44⅞	8½	¾	½
204	30¾	17¾	9⅝	5	45 13/16	9⅛	¾	½
205	34	21	11¼	6	46¾	9¾	¾	½
206	37¼	24¼	12⅞	6	46¾	9¾	¾	½
207	40½	27½	14½	7	48 3/16	10⅜	¾	½
208	43¾	30¾	16⅛	7	48 3/16	10⅜	¾	½
209	47	34	17¾	8	49½	11	¾	¾
210	50½	37¼	19⅜	8	49½	11	¾	¾

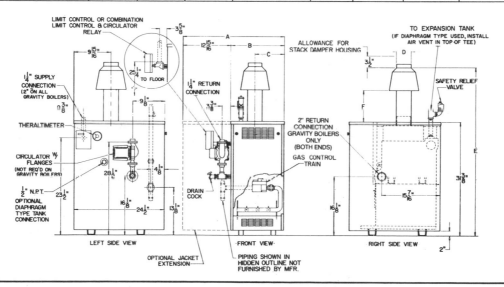

Courtesy: Burnham Corp.

Gas boiler data
Figure 7-2

Dimensions and Properties of Copper Tube
(All Types Except DWV are based on ASTM B-88)

Nominal Size, in. [a] (×25.4=mm)	ASTM[b] Schedule	Diameter, in. (×25.4=mm)		Wall Thickness, in. (×25.4=mm)	Surface Area, ft²/lin. ft (×0.3048=m²/lin. m)		Section Area, in.² (×645=mm²)		Area of Metal, in.² (×645=mm²)	Volume, gal/lin. ft (×12.43= l/lin. m)	Weight,[c] lb/lin. ft (×1.49= kg/lin. m)	Working Pressure, psia (×6.89=kPa)
		OD	ID		OD	ID	OD	ID				
1/4	K[c]	0.375	0.305	0.035	0.0982	0.0798	0.110	0.0730	0.0374	0.00379	0.145	918
	L[c]	0.375	0.315	0.030	0.0982	0.0825	0.110	0.0779	0.0324	0.00404	0.126	764
3/8	K	0.500	0.402	0.049	0.131	0.105	0.196	0.127	0.0695	0.00660	0.269	988
	L	0.500	0.430	0.035	0.131	0.113	0.196	0.145	0.0512	0.00753	0.198	677
1/2	K	0.625	0.527	0.049	0.164	0.138	0.306	0.218	0.0887	0.0113	0.344	779
	L	0.625	0.545	0.040	0.164	0.143	0.306	0.233	0.0735	0.0121	0.285	625
5/8	K	0.750	0.652	0.049	0.193	0.171	0.441	0.334	0.108	0.0174	0.418	643
	L	0.750	0.666	0.042	0.193	0.174	0.441	0.348	0.0934	0.0181	0.362	547
3/4	K	0.875	0.745	0.065	0.229	0.195	0.601	0.436	0.165	0.0227	0.641	747
	L	0.875	0.785	0.045	0.229	0.206	0.601	0.484	0.117	0.0250	0.455	497
1	K	1.125	0.995	0.065	0.295	0.260	0.994	0.778	0.216	0.0405	0.839	574
	L	1.125	1.025	0.050	0.295	0.268	0.994	0.825	0.169	0.0442	0.655	432
1 1/4	K	1.375	1.245	0.065	0.360	0.326	1.48	1.22	0.268	0.0634	1.04	466
	L	1.375	1.265	0.055	0.360	0.331	1.48	1.26	0.228	0.0655	0.884	387
	M	1.375	1.291	0.042	0.360	0.338	1.48	1.31	0.176	0.0681	0.682	293
	DWV	1.375	1.295	0.040	0.360	0.339	1.48	1.32	0.163	0.0684	0.650	—
1 1/2	K	1.625	1.481	0.072	0.425	0.388	2.07	1.72	0.351	0.0894	1.36	421
	L	1.625	1.505	0.060	0.425	0.394	2.07	1.78	0.295	0.0925	1.14	359
	M	1.625	1.527	0.049	0.425	0.400	2.07	1.83	0.243	0.0950	0.940	289
	DWV	1.625	1.541	0.042	0.425	0.403	2.07	1.86	0.205	0.0969	0.809	—
2	K	2.125	1.959	0.083	0.556	0.513	3.56	3.01	0.532	0.157	2.06	376
	L	2.125	1.985	0.070	0.556	0.520	3.56	3.10	0.452	0.161	1.75	316
	M	2.125	2.009	0.058	0.556	0.526	3.56	3.17	0.377	0.164	1.46	255
	DWV	2.125	2.041	0.042	0.556	0.534	3.56	3.27	0.288	0.142	1.07	—
2 1/2	K	2.625	2.435	0.095	0.687	0.638	5.41	4.66	0.755	0.242	2.93	352
	L	2.625	2.465	0.080	0.687	0.645	5.41	4.77	0.640	0.247	2.48	295
	M	2.625	2.495	0.065	0.687	0.653	5.41	4.89	0.523	0.254	2.03	234
3	K	3.125	2.907	0.109	0.818	0.761	7.67	6.64	1.03	0.345	4.00	343
	L	3.125	2.945	0.090	0.818	0.771	7.67	6.81	0.858	0.354	3.33	278
	M	3.125	2.981	0.072	0.818	0.780	7.67	6.98	0.691	0.362	2.68	220
	DWV	3.125	3.035	0.045	0.818	0.796	7.67	7.23	0.435	0.376	1.69	—
3 1/2	K	3.625	3.385	0.120	0.949	0.886	10.3	9.00	1.32	0.468	5.12	324
	L	3.625	3.425	0.100	0.949	0.897	10.3	9.21	1.11	0.478	4.29	268
	M	3.625	3.459	0.083	0.949	0.906	10.3	9.40	0.924	0.489	3.58	218
4	K	4.125	3.857	0.134	1.08	1.01	13.3	11.7	1.68	0.607	6.51	315
	L	4.125	3.905	0.110	1.08	1.02	13.3	12.0	1.39	0.623	5.38	256
	M	4.125	3.935	0.095	1.08	1.03	13.3	12.2	1.20	0.634	4.66	217
	DWV	4.125	4.009	0.058	1.08	1.05	13.3	12.6	0.67	0.656	2.87	—
5	K	5.125	4.805	0.160	1.34	1.26	20.7	18.1	2.50	0.940	9.67	307
	L	5.125	4.875	0.125	1.34	1.28	20.7	18.7	1.96	0.971	7.61	234
	M	5.125	4.907	0.109	1.34	1.29	20.7	18.9	1.72	0.981	6.66	203
6	K	6.225	5.741	0.192	1.60	1.50	29.4	25.9	3.50	1.35	13.9	308
	L	6.125	5.845	0.140	1.60	1.53	29.4	26.8	2.63	1.39	10.2	221
	M	6.125	5.881	0.122	1.60	1.54	29.4	27.2	2.30	1.42	8.92	190
	DWV	6.125	5.959	0.083	1.60	1.56	29.4	27.9	1.58	1.55	6.10	—
8	K	8.125	7.583	0.271	2.13	1.99	51.8	45.2	6.69	2.34	25.9	330
	L	8.125	7.725	0.200	2.13	2.02	51.8	46.9	4.98	2.43	19.3	239
	M	8.125	7.785	0.170	2.13	2.04	51.8	47.6	4.25	2.47	16.5	200
10	K	10.125	9.449	0.338	2.65	2.47	70.5	70.1	10.4	3.65	40.3	332
	L	10.125	9.625	0.250	2.65	2.52	80.5	72.8	7.76	3.79	30.1	241
	M	10.125	9.701	0.212	2.65	2.54	80.5	73.9	6.60	3.84	25.6	202
12	K	12.125	11.315	0.405	3.17	2.96	115.	101.	14.9	5.24	57.8	334
	L	12.125	11.565	0.280	3.17	3.03	115.	105.	10.4	5.45	40.4	225
	M	12.125	11.617	0.254	3.17	3.04	115.	106.	9.47	5.50	36.7	204

Copper pipe diameters
Figure 7-3

Closed Systems

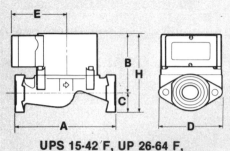

**UPS 15-42 F, UP 26-64 F,
UP 26-96 F and UP 43-75 F**

Electrical:

Model	Speed	Hp	Watts	Volts	Amps	RPM	Capacitor
UPS 15-42 F	3	1/25	85	115	0.76	2475	10MF/160V
3-Speed	2	1/40	63	115	0.55	2190	
	1	1/75	45	115	0.39	1685	
UP 26-64 F	1	1/12	185	115	1.65	3100	8MF/180V
1-Speed	1	1/12	175	230	0.76	2700	2.5MF/380V
UP 26-96 F	1	1/12	205	115	1.85	2515	10MF/160V
High Head	1	1/12	205	230	0.90	2515	2.5MF/380V
UP 43-75 F High Capacity	1	1/6	225	115	2.15	3250	10MF/160V

Dimensions and Weights: Dimensions in Inches

Model	A	B	C	D	E	H	Shipping Carton L x W x H-in	Pack Vol. F³	Net Weight lbs	Ship Weight lbs
UPS 15-42 F 3 Speed	6-1/2	4	1-5/16	4-3/16	3-1/4	5-3/8	7-1/2 x 5-3/4 x 6	1/7	7	7-1/4
UP 26-64 F UP 26-96 F 1-Speed	6-1/2	5-1/16	1-5/16	4-3/16	3-9/16	6-3/8	8 x 6 x 7-1/4	1/5	10-3/4	11-1/4
UP 43-75 F High Capacity	8-9/16	5-3/16	1-7/16	4-3/4	3-5/8	6-5/8	9-1/2 x 6-1/2 x 7-1/2	1/3	12-3/4	13-1/2

Performance:

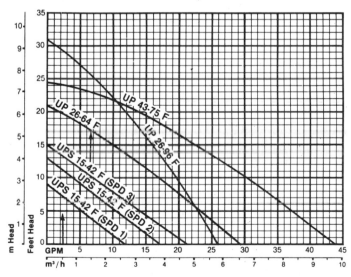

Construction Materials:

Metal Impregnated Carbon: Thrust bearing.
Cast Iron: Pump housing (volute).
EP (ethylene propylene rubber): O-ring, gasket, flange gaskets.

Stainless-Steel: Impeller, inlet cone, bearing plate and bearing retainers, rotor can, rotor cladding, shaft retainer.
Aluminum: Stator housing.
Aluminum Oxide Ceramic: Shaft, upper and lower radial bearings.

Courtesy: Grundfos Corp.

**Circulating pump data
Figure 7-4**

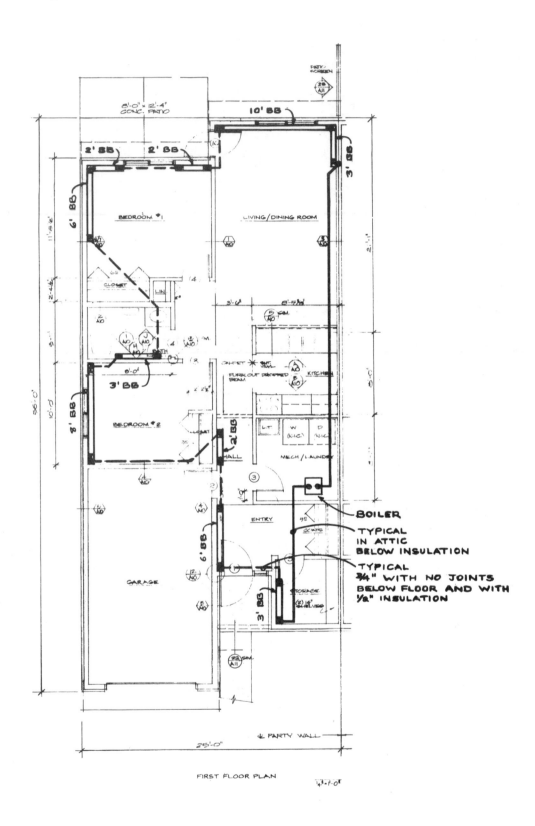

Sample residence, hot water system
Figure 7-5

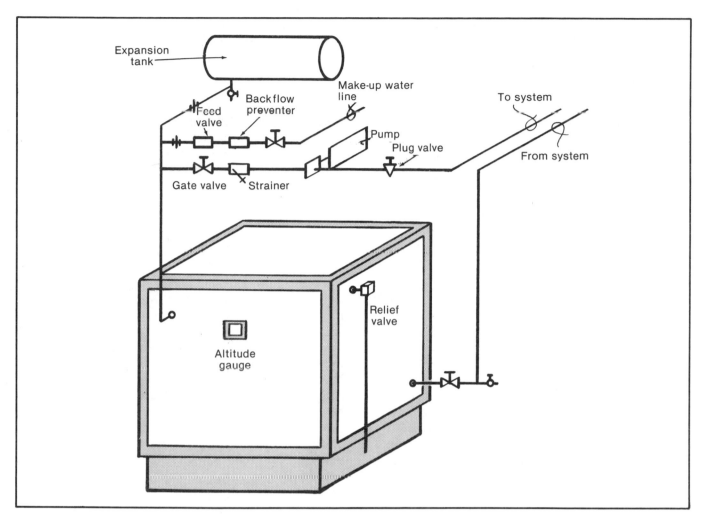

Typical boiler piping
Figure 7-6

Normally water boils at 212°F. But a typical hydronic system will have a minimum operating pressure of 15 pounds per square inch (psi) pressure. Since increased pressure raises the boiling point of water, 15 psi of pressure would mean a boiling point of about 250°F. So, for example, a system could use a boiler water temperature of 230°F and have a 50°F temperature drop and still be effective. But your boiler would be operating close to the system's boiling point.

Research conducted by the Hydronics Institute (formerly the Institute of Boiler and Radiator Manufacturers) has shown that higher system temperatures have very little effect on yearly fuel consumption. Higher temperatures will mean smaller piping and heat distributing units as noted above, so installation costs can be reduced somewhat. But some installers feel there's a safety consideration in having higher temperature water running through a residence or business, especially if there's a sudden drop in pressure without a reduction in boiler temperature. There may also be a more rapid deterioration of equipment operating over time at the higher temperature. Smaller piping means greater head loss due to friction, and you'll need a larger pump. You may end up with a noisy system. You'll have to weigh these factors and make your own decision if you feel a large design temperature drop would be to your advantage. Your choice may also be limited by built-in features of the heat distributing units. Be sure to check the manufacturer's recommendations for the units required.

Overall, research has indicated that a 20°F temperature drop usually is the most economical for sizing the pump and piping. During actual operation, the temperature drop is usually considerably less than the computed drop, so there's not much need to exceed 20°F to maintain heating unit ratings and efficiency. We feel it's the best choice for most residential and many commercial systems. Consult with an engineer if you feel your job may benefit by a larger temperature drop.

For our residence we'll use a 20°F temperature drop, with the water leaving the boiler at 200°F. The 20°F drop will mean it returns at 180°F, giving us an average or design temperature of 190°F. This is, in our opinion, the best temperature for most residential systems installed in northern climates. It insures a quiet system at ample capacity and a boiler temperature that will allow upward adjustment if needed.

A lower average temperature will take a longer time to heat up the building. It will also mean more feet of radiation and larger pipe diameters to meet the Btu needs, increasing the system cost. It will, however, have a lower operating cost. This should be considered when the building is located in a climate where fewer Btu's will permit a smaller system.

We'll see in a later section how the temperature drop in the system is held to 20 degrees by the rate at which the water flows through the system. For design purposes, we assume that this drop will be accurate. But in actual systems, it stands to reason that the water entering the first baseboard in the system will be at a higher temperature than the water entering the last. For floor areas up to 2,000 square feet, you can generally discount this difference. However, in larger systems, some HVAC designers increase the length of the last one or two baseboards in the run to compensate for the lower water temperature. For the first baseboards, where temperature may be on the high side, you can install manual valves to reduce slightly the amount of water passing through, or manual dampers to reduce the air flow across the baseboard.

Figure 7-1 has two lines of flow rate. Later we'll show you how to calculate the actual flow rate for a system. For now, use the 1 gpm flow rate. You'll find it works for most systems you'll be designing,

since it covers rates *up to* 4 gpm. At flow rates of 4 gpm and higher, you'll have to increase the size of the piping. The Btu gain from a larger system often won't justify the cost unless the system needed is very large.

Also in Figure 7-1, you'll find radiation unit pressure drop, listed in milli-inches. It isn't a determining factor in our selection because we'll figure it in when we do the system pressure drop.

Based on our average temperature of 190°F, and the above considerations, the capacity of the baseboards we'll be using is 640 Btu/hr/linear ft.

Bedroom 1 of our residence has a heat loss of 5,646 Btu/hr. Note that this amount, when figured for the forced air system, included a 20% factor for duct loss. Whether a hydronic system will have a similar loss has been debated by designers for quite some time. Because hot water piping takes up much less space than ducts, you can often run it within the insulated areas of the building. Some designers claim that any heat given off by the piping isn't really "lost," since it will help to heat the building. But others argue that if the heat is given off in a basement or a ceiling, it isn't effectively heating the living space. Our opinion is that the added Btu's in the system due to the 20% loss won't add that much to the system cost, but will add a worthwhile safety factor for this climate. So we'll use the Btu loss that includes a 20% increase for duct or piping loss.

Divide the Btu loss of each room or area by the Btu capacity per foot of baseboard to get the length needed.

Baseboard comes in standard lengths of 2 to 10 feet in 1-foot increments. Three inches on each end don't have fins, reducing the operating capacity by 1/2 foot. Always select a standard length that exceeds the computed length by a minimum of 1/2 foot.

Note that the specs in Figure 7-1 tell us that the outside diameter of the pipe in the baseboard is 3/4''. This will be important later on.

For our residence:

Bedroom 1

$$\frac{5646}{640} = 8.8 = 10 \text{ ft. baseboard}$$

Bedroom 2

$$\frac{4547}{640} = 7.1 = 8 \text{ ft. baseboard}$$

Living room (plus 1/2 hall and kitchen)

$$\frac{8045}{640} = 12.6 = 13 \text{ ft. baseboard}$$

Entry (plus 1/2 hall and mech room)

$$\frac{4447}{640} = 6.9 = 8 \text{ ft. baseboard}$$

Bathroom

$$\frac{1282}{640} = 2.0 = 3 \text{ ft. baseboard}$$

Storage

$$\frac{1387}{640} = 2.2 = 3 \text{ ft. baseboard}$$

Total   45 ft. baseboard

Be sure each baseboard unit has a manual radiator valve so heat output can be adjusted.

Locate the baseboards on the plan where they'll be installed in the building. Draw them to scale and label as to size and identity. See Figure 7-5.

Step 2: Boiler Selection and Location
The boiler's capacity must meet the heat loss of the building. However, it's best to size it larger, if practical, to handle very cold weather and to provide a pick-up allowance. Figure 7-2 shows the catalog capacities of a Burnham boiler, rated in MBH — thousands of Btu/hr. We need a 28,255 Btu/hr net capacity, so we'll select boiler number 203.

Much of the technology used for hot water heat evolved from steam heat design. For example, in Figure 7-2, we have four MBH ratings — input, heating capacity, gross output, and the Hydronics Institute IBR rating. The *input* is the Btu fuel capacity of the boiler. The *gross output* is the Btu's after taking into account stack and other losses. The *DOE heating capacity* is based on government tests and is often cited by local codes as the rating used to comply with their regulations. The *IBR rating* is based on the Btu's available for room heat after taking into account cold piping, radiators and room areas if the system was shut down overnight, a common practice in the past with steam systems. The difference between the IBR rating and the gross output is the *pick-up* allowance. Without this pick-up allowance available to boost the system to

overcome the cold temperatures an overnight shut-down causes, the system theoretically could fail to heat the building on a very cold day.

The *Net Rating Water,* in square feet, was used in the past for sizing steam heat and later adopted for water systems. It relates the heating area of the radiator to the boiler output. It's no longer used for sizing so we don't have to be concerned with it for our system.

Locate the boiler in the center of the building, in the same area that you'd put the furnace. Put it as close to the chimney as practical. This keeps the smoke pipe short and provides for the least number of turns. Check with the architect or owner to see if the location will interfere with other building plans. Follow the code and manufacturer's instructions concerning clearance, insulation, piping, combustion air and the like. Also review manufacturer's data for ease of maintenance, economy and quiet operation. These are important selling points to the owner.

Consult with the builder or architect if a special foundation is needed for the boiler. You have to determine who will install it and how it should be done.

Draw the boiler to scale on both the floor plan and in a detail or section (Figure 7-6). Include labels and notes to clarify equipment, piping and its installation.

Adjust ratings for boilers located more than 2,000 feet above sea level, as indicated by Note (1) in the data table in Figure 7-2.

Step 3: Ventilating Fans
The design and installation of ventilating fans for buildings with hot water systems is the same as in warm air designs. Review Chapter 6 if necessary.

Step 4: Piping System Layout
The piping layout used in our residence is a *series loop* system. The water is pumped through a piping loop in which each radiation unit is installed in series. See Figure 7-5. The flow follows a single route along the outer walls of the building through the pump, boiler, and radiation units. This system is simple and economical for residences and smaller commercial buildings.

Several other piping systems are available for larger buildings. Later we'll detail an apartment using a two-pipe, reverse return setup. For your design work, use either the series loop or the two-pipe, reverse return. These will cover almost all the hot water jobs you'll ever do. We don't recommend any of the other systems. They often have balancing problems and require extra materials and controls which add to the cost. Consult the ASHRAE Systems Volume if you need an explanation of any piping systems not covered here.

Sizing the pipe— The pipe size in our series loop depends upon the amount of water that's needed to do the job at each baseboard along the way. This is called the *flow rate* and is figured in gallons per minute (gpm). For all piping, the rate of flow of the water in gpm is directly related to the pipe size. But with hot water heat, we also have to figure in some other factors.

Recall that in sizing the baseboards, we used water that leaves the boiler at 200°F and returns at 180°F, for an average temperature of 190°F. To keep that temperature at 190°F, we have to move the water through the system quickly enough so that it doesn't lose any more than 20°F. On the other hand, if we move the water too quickly, we'll have excessive noise and friction loss. The system will also lose efficiency, moving too fast to supply the total Btu's needed to overcome the building's heat loss.

We begin by making sure we've calculated enough feet of baseboard to overcome our building's heat loss.

$$\text{Total design footage of baseboard} = 45$$

$$\text{Btu/ft/hr} = 640$$

$$45 \times 640 = 28{,}800 \text{ Btu/hr}$$

Our building's total heat loss is 28,255 Btu/hr. So we have enough baseboard.

We can now figure the flow rate by taking the system's total heat capacity and dividing it by the allowable temperature drop. But we must also figure in the mathematical factors that take into account the different volume, heat and time variables.

$$\text{gpm} = \frac{\text{Heat capacity}}{(\text{System temp drop}) \times 1.0 \text{ Btu/lb/deg} \times 60 \text{ min/hr} \times 8.33 \text{ lbs/gal}}$$

For our residence:

$$\text{gpm} = \frac{28{,}255 \text{ Btu/hr}}{(200°F - 180°F) \times 1.0 \text{ Btu/lb/deg.} \times 60 \text{ min/hr} \times 8.33 \text{ lbs/gal}}$$

$$\text{gpm} = \frac{28{,}255}{9{,}996} = 2.8$$

Each section of pipe in our series loop will need to carry 2.8 gpm.

The rate of flow determines the amount of water that will move through the pipe. The speed at which it travels is the *velocity*. The higher the velocity, the greater the friction loss. If the velocity is too high, the flow will cause noise. Also, there will be too much friction loss for the pump to move the water efficiently through the pipe.

A good rule to follow for hot water piping is to limit the velocity to less than 3 feet per second. Determine the pipe size based on that velocity limit as follows:

$$\text{Pipe cross section area} = \frac{\text{gpm}}{\text{velocity}}$$

Again we have to use math factors to allow for different volumes and rates of our values.

$$\text{Pipe area} = \frac{2.8 \text{ gal/min} \times 0.133 \text{ cu ft/gal} \times 144 \text{ sq in/sq ft}}{60 \text{ sec/min} \times 3 \text{ ft/sec}}$$

$$\text{Area} = 53.63/180 = 0.298 \text{ sq in}$$

Before using Figure 7-3 to choose a pipe diameter based on our area, let's review several items.

First, although commonly called "pipe" in the trade, copper tubing is sized differently than steel pipe. Be careful when using tables and sizing charts. Almost all modern hot water systems will be piped using copper tubing. Make sure you're using a copper tubing table for your figuring.

Next, copper piping comes in two wall thicknesses — K and L. You can use Type L, the thinner wall, for low pressure systems like the one we're using for our residence. It costs less and works for all systems operating at less than 30 psi.

According to Figure 7-3, we estimate a cross sectional area of 0.298 square inches, which indicates a diameter of 0.61 inches. That's between 1/2'' and

5/8'' nominal size OD. Typically we'd use the next larger nominal size, or 5/8''. But, although tubing is available in sizes like 5/8'', 7/8'' and so on, preferred industry practice for hot water heat is to go with 3/8'', 1/2'', 3/4'', 1'', 1¼'', 1½'' and 2'' sizes. So we'll use 3/4''.

Sizing up from 5/8'' to 3/4'' wouldn't be possible if we had to stay exactly at our calculated velocity pipe size. But recall that the 5/8'' pipe is the minimum diameter we can use to make sure the velocity doesn't exceed 3 ft/sec. A larger pipe size will reduce the velocity to less than 3 ft/sec, which is actually to our advantage, as it also means lower head pressure. As we'll explain in the next section, this will mean a smaller pump, using less horsepower. That means cheaper operation. We also benefit since our system pipe will match our baseboard pipe diameter of 3/4'' (from Step 1). No reduction fittings will be needed. You won't find many baseboards with pipe diameters sized smaller than 3/4''. Early on, the manufacturers chose 3/4'' as the minimum size that would cover a wide range of applications efficiently. Also, water system piping in general was always 3/4'' or larger. Elbows, tees and other piping parts were readily available for the installer. And there was a widespread fear of putting in piping that was too small. Even if you have a boiler and baseboards with extra capacity, there's little you can do about customer complaints during a prolonged cold spell if the piping is too small. And by staying with one standard size, installations go faster and easier. For all of these reasons, many installers still never put in a system smaller than 3/4'', even if its size would permit it.

But what if the calculations you do for a job show you can use a pipe with a diameter smaller than 3/4''? The Hydronics Institute has conducted tests showing that labor and material costs could be substantially reduced for some jobs by using diameters such as 3/8'' or 1/2'', without any loss in system comfort or quality. For your jobs, you'll have to make your own decision. Just remember you can't size down below the minimum computed size where the velocity will go above 3 ft/sec. If you do go to a size smaller than 3/4'', check available baseboard pipe sizes to see if you'll need additional fittings.

Draw the piping on the plan using a single line, arrowed as to size. A dashed line indicates that it's passing under a concrete slab or other special in-stallation. Some designers prefer to used dashed lines for the return lines when more than one circuit is used. Vertical pipe running up is shown by using a small circle with the line touching. Vertical pipe running down is indicated by the line inside the circle. Check the plan carefully to make sure your piping can be run where you indicate and that it doesn't pass through stairs, hallways or other undesirable areas. Spell out the actual location in a note on the plan or in the specs, such as: "All piping shall be run between floor joists except where noted otherwise on plan." As a rule, all piping should be installed so it slants slightly in the direction of the boiler. Install drain valves in the system's low points.

For some jobs you may first want to draw a *piping layout*. This is a separate drawing showing the boiler, heat distributing units and the system piping. It doesn't have to be detailed, but it should be drawn to scale to permit you to estimate pipe lengths. There's a piping layout for a two-story apartment building in the next section (Figure 7-7). It's a good idea to do this layout using tracing paper over the plan. You can make several rough layouts to determine which is the best one.

Sizing the pump— Head pressure, commonly called *head*, is the pressure needed in the system to overcome any resistance. Basically, head pressures are determined when manufacturers measure the pump's ability to pump a certain amount of water a specified vertical distance. As the height increases, the gpm drops, since the pump has to work harder to overcome friction and gravity. The relationship of the head and the gpm is shown as a curved line on charts. Manufacturers indicate on those charts the best efficiency range for their pumps. Head is measured in feet of water, often called in the trade *feet of head*.

Our pump will have to develop enough pressure to overcome the loss by friction of water flowing through the piping, boiler and baseboards. This is called *head loss*. We'll start by measuring the length of actual pipe in our loop, including the baseboard lengths. The total length is 110 feet. Multiply this length by 1.5 to account for the extra friction losses in fittings and valves. For our residence:

$$110 \times 1.5 \ = \ 165 \text{ ft. adjusted length}$$

Pipe size, flow, head loss and velocity are all

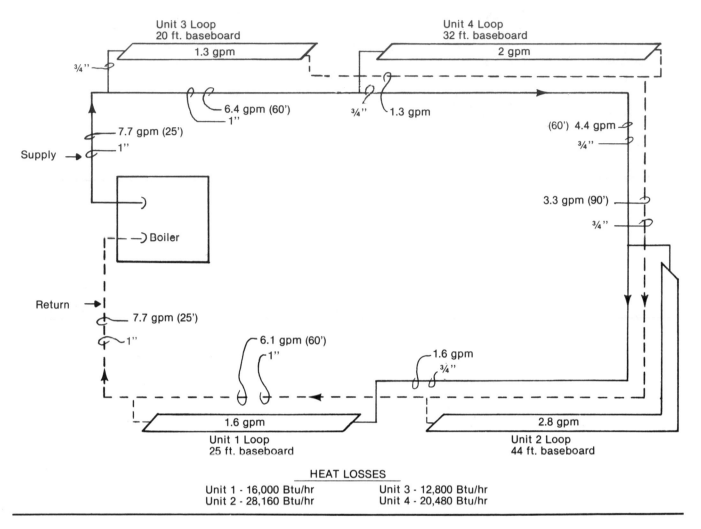

Reverse return hot water system
Figure 7-7

related. Here's the data we know for copper piping:

Size	Flow	Head loss	Velocity
1/2"	3 gpm	10.0 ft/100 ft	3 ft/sec
3/4"	5 gpm	6.6 ft/100 ft	3 ft/sec
1"	8 gpm	5.0 ft/100 ft	3 ft/sec
1¼"	14 gpm	3.2 ft/100 ft	3 ft/sec

Head loss can be found for any given flow, pipe size and velocity using tables in the ASHRAE Fundamentals Handbook. But the process can be simplified by using the data in a formula relating our known values with those we wish to find:

$$\frac{H_2}{H_1} = \left(\frac{W_2}{W_1}\right)^2$$

H_1 is head loss at flow W_1, taken from our known data table, for 3/4" pipe. H_2 is the head loss we need to find, at our design flow of W_2 (the 2.8 gpm we figured in the previous section). So:

$$\frac{H_2}{6.6} = \left(\frac{2.8}{5}\right)^2$$

$$H_2 = \left(\frac{2.8}{5}\right)^2 \times 6.6$$

$$H_2 = .314 \times 6.6 = 2.07 \text{ ft hd/100 ft}$$

This is the head loss at our design flow rate of 2.8 for 100 feet of pipe. We now apply it to our adjusted pipe length for the system.

System piping head loss
= 2.07 x 165 ft/100 = 3.4 feet

Here's a tip that will help you check your work. When you calculate friction loss in piping, it shouldn't exceed 8 to 10 feet of head per 100 feet of adjusted pipe length. If you stay with the guideline limiting your velocity to 3 gpm or less, this should be no problem. If your calculations result in head loss greater than 10 feet per 100 feet of pipe, using the recommended velocity, it's usually a sign that you've figured wrong.

Total head for the entire system must include friction loss in the boiler. Allow a loss of at least 1.0 minimum for the boiler.

Total system head loss = 3.4 + 1.0 = 4.4 feet

Select a pump that will provide the flow at the system head loss. Pump manufacturers' catalog sheets like the one in Figure 7-4 show performance curves. You select the pump from the curve that provides the flow and head for your system. If the curve doesn't match your system exactly, use the next larger capacity pump curve. Don't pick a motor that's less than what's required for any operating point on the performance curve.

For our system, we have a flow of 2.8 gpm and 4.4 feet of head. In Figure 7-4, a performance table for a Grundfos model UPS 15-42 F three-speed pump is shown. Correlating our data on the table places us close to the performance curve shown for speed 1. This is the lowest speed and works out well, since it will mean longer pump life and cheaper operation.

Be sure to check the pump electrical data to determine the motor horsepower and other specs for your equipment schedule. Draw the pump on the boiler diagram of the plan (Figure 7-6). Although some designers install pumps on the return line, we recommend you locate the pump on the supply line of the system. This assures that the pump suction will have an adequate supply of water to prevent cavitation. We'll talk more about cavitation later in the chapter. It also prevents air leakage into the system at topmost points due to low pressure.

You can vary pump capacities by using different diameter impellers in the same pump. Don't select a pump with the maximum size impeller, as it may be noisy.

Expansion— An expansion tank is necessary for our system because it's a closed system. When it's filled with cold water and started up, the water will expand as it heats. Without an expansion tank to relieve the pressure, the piping or equipment would break.

The tank contains both air and water. When the water expands, the air is compressed. We must be sure the tank is of adequate size to handle the expansion. Here's a good rule to use for tank sizing: Provide one gallon of tank capacity for every 5,000 Btu of heat loss. Check the manufacturer's data for the final word on proper tank capacity and installation.

Based on our building heat loss of 28,255 Btu/hr, we'll need a tank holding at least 6 gallons.

Location and installation of the expansion tank, pump, boiler and piping must be done with care. Look at the arrangement in Figure 7-6. Water for the initial fill and make-up water for small amounts of leakage is provided by an automatic, pressure-operated feed valve in the make-up water line. Typically, the water pressure in the plumbing is at about 40 psi. The boiler will operate at about 15 to 25 psi. By sensing the pressure on the downstream side, the feed valve opens to allow water in until the correct boiler pressure is reached. Boiler installation data will list feed valve pressure for the unit.

The make-up water line should have a backflow preventer to prevent boiler water from going back into the domestic water system. Many building codes require a specific type. You can tap the line into the system at one of two places: at the pipe connecting the expansion tank to the boiler or system, or at the supply line near the boiler, provided it's between the pump and the expansion tank.

Expansion of the pipe and baseboard may be a problem. Copper tubing will increase in length when heated. If we assume our piping is installed at 70°F and the boiler and pump circulates water at 200°F, the temperature difference of 130°F will cause expansion. Expansion of copper is 0.0012 in/ft/degree. So every foot of tube will expand 0.015 in. Fifty feet of copper pipe could expand up to 3/4". Provide clearance at each end of the run to allow for this expansion. Don't locate pipe turns and ends tight against building surfaces. Also avoid clamps which hold the pipe too tight. The pipe has to be able to slide a little in the clamp.

If you don't have the clearance, you'll need an expansion joint in the run. Expansion joints are U-shaped sections that allow the legs of the "U" to bend to absorb the expansion. In a 50' run, a "U" with 12" legs and a 6" bottom is adequate. In-line expansion compensators are also available from manufacturers.

Step 5: Thermostat and Boiler Controls
In our hot water system, the thermostat will start the pump and boiler burner control when heat is needed. A high-limit boiler control will turn off the boiler when the water temperature exceeds the setting. An operating control will cycle the burner to maintain water temperature as long as heat is called for in the building. This control is set at the water design temperature of the system: 190°F. The pump will continue to run after the boiler shuts down. When the circulating water temperature drops below 190°F, the operating control will allow the burner to restart. When heat is no longer needed, the thermostat will shut off both the pump and the boiler.

The code will require that the boiler has a relief valve to relieve excessive pressure buildup in the boiler. Most domestic boilers include basic controls as part of the unit. For larger units, you may have to order them separately. Refer to the manufacturer's data.

Local ordinances may require additional boiler controls such as a low water cut-off control or a second high-limit control, set 10°F higher than the first one. These are often available as options from the manufacturers.

Some installations are wired so that the thermostat controls only the burner. The pump runs continuously. Some designers feel this avoids noise and permits more even heat flow as the water circulates. On the other hand, constant pumping will add to operating costs. If you've designed your system at a low enough velocity, noise and heat flow shouldn't be a problem when the pump is restarted. We feel thermostat control of both the burner and the pump is the better installation.

Thermostat wiring and location is the same as for the warm air system design in Chapter 6. Draw and label the boiler controls on the boiler diagram. Write the controls specifications in the job specs or note them on the plan.

The Rest of the Steps
To finish your hot water system design, you'll need to do the following steps. Review these steps in Chapter 6. Common sense will tell you where to adapt them to your hot water system.

Step 6: Fuel piping
Step 7: Electrical requirements
Step 8: Chimney, breeching and combustion air
Step 9: Final plan and spec check

Cooling
A hot water system has the disadvantage of not including air conditioning as part of the system. There are a few exceptions to this, but they're usually only in very large systems. If you need air conditioning for residences and commercial jobs with a hot water system, you'll need to design two systems — one for heating and one for cooling.

The forced-air conditioning system is designed like a warm air system, with a few differences. You won't, or course, have a furnace. Instead, you'll have an evaporator coil, condensing unit, ductwork and a fan. Your design data will be made based on heat gain alone. Otherwise, follow the steps used for a warm air system. Obviously, some steps won't apply.

A simpler option is to install individual AC wall units. Check the plan to make sure the architect has included wall openings and electrical outlets if these units are desired. Also, individual AC units may be too noisy to suit some occupants. Be sure to review this point with the architect or owner if you suggest them as an option.

Apartment Hot Water Systems
Apartment buildings in cold climates often heat with a hot water system. With a central boiler and a single piping network, the system is economical to install and operate. It also adjusts easily to separate controls for each unit in the building.

The single loop system used in our residence isn't adequate for an apartment building. It won't allow for individual apartment control. Instead, you'll need to install a two-pipe, *reverse return* system.

As the name says, in a two-pipe, reverse return system, two pipes are used. One is the supply and the other is the return. These are the main flow pipes. Loops for each heating unit or series of units in each apartment are tapped off these. A control valve for each loop is operated by a thermostat in the apartment.

With a reverse return, flow in the return main is in the same direction as the supply main flow. If you trace the flow of the water from the boiler through any loop and back to the boiler, the total length of the piping will be about the same for both the supply and return line. The pressure drops through the mains and through each loop are about the same. This has two advantages. First, you won't have to install balancing valves. Second, with similar pressure drops, head loss in the system will be easy to figure.

Other two-pipe circuit designs are sometimes used. We don't recommend them. The direct return circuit, for example, is sometimes used because less main pipe length is needed. But circuit balancing valves are needed on each sub-circuit. Pump operating cost is also higher because of the added pressure drops due to the balancing valves. The system can be difficult to balance.

Heat Loss and Gain

Heat loss and gain for our apartment can be calculated using the steps described in Chapter 5. We won't detail the building construction of our apartment and go through the process of pulling the information off the plans to figure our heat loss, since we'd just be repeating steps we've already studied. Instead, our schematic in Figure 7-7 lists the Btu/hr loss for each unit. These are the losses we'll use to size our reverse return system.

We're using a wide range of losses to give our figuring a good workout. Typically, your jobs will be easier to calculate since large buildings usually have more uniform heat losses over their exposed surfaces.

Baseboard Sizing

We'll size the baseboards for the apartment the same way as we sized the hot water system in our residence. Using the same baseboard data from Figure 7-1 (1 gpm, 190°F average water temp, model number 740), we calculate:

Unit 3

$$\frac{12,800}{640} = 20' \text{ of baseboard}$$

Unit 4

$$\frac{20,480}{640} = 32' \text{ of baseboard}$$

Unit 2

$$\frac{28,160}{640} = 44' \text{ of baseboard}$$

Unit 1

$$\frac{16,000}{640} = 25' \text{ of baseboard}$$

Later we'll show how these total baseboard lengths are divided up and placed where they're needed to account for the building's heat loss. But for now, we'll use the total lengths for our calculations.

Boiler Selection

Select the boiler by following the same steps we used for a single residence. The boiler must have enough capacity to equal the total heat loss of our building, with spare capacity for pick up. Using Figure 7-2, we'll choose Burnham model number 205.

Sizing the Pipe

We can't size the pipe for a reverse return system using the same method we used for a series loop. Here the supply and return mains must be a different size at each baseboard loop to maintain the same pressure drops through the system. But it's not as difficult to figure as it might sound.

We'll use the same equation we used for the residence in Figure 7-5:

$$\text{gpm} = \frac{\text{Heat capacity}}{\text{(System temp drop) x 1.0 Btu/lb/deg x 60 min/hr x 8.33 gals/min}}$$

It's important to note that the heat capacity data used here is for each baseboard loop, rather than the entire system. For example:

In Unit 3:

$$\text{gpm} = \frac{12,800 \text{ Btu/hr}}{(200°F - 180°F) \times 1 \times 60 \times 8.33} = 1.3 \text{ gpm}$$

In Unit 4:

$$\text{gpm} = \frac{20,480 \text{ Btu/hr}}{(200°F - 180°F) \times 1 \times 60 \times 8.33} = 2 \text{ gpm}$$

In Unit 2:

$$gpm = \frac{28,160 \, Btu/hr}{(200°F - 180°F) \times 1 \times 60 \times 8.33} = 2.8 \, gpm$$

In Unit 1:

$$gpm = \frac{16,000 \, Btu/hr}{(200°F - 180°F) \times 1 \times 60 \times 8.33} = 1.6 \, gpm$$

$$Total = 7.7 \, gpm$$

Label the gpm for each baseboard loop on the schematic as shown in Figure 7-7.

To find the supply and return main sizes between the baseboard loops, begin by adding the gpm of the baseboards. The total is 7.7 gpm. This is the flow rate needed between the boiler and the two loops closest to the boiler. Keep in mind that the system's supply flow is largest where it leaves the boiler. The return flow is largest where it enters the boiler. The flow is the least for both the supply and return at the loop furthest from the boiler *on their respective runs.*

To find the reduction in flow at each loop, start with the closest loop and work your way to the loop farthest away, subtracting each loop's gpm from the total gpm as you go. Figure the supply main first and then the return. Label the reduction or increase in flow rate for the supply and return between the loops on the schematic. When you're finished, check each loop to make sure you have the same amount of water both entering and leaving. Do your loop amounts add up to your total gpm?

Now get the pipe sizes by taking the different flow rates and applying the pipe sizing formula. Again we'll use a 3 ft/sec velocity rate.

$$Pipe \ area = \frac{flow \ rate \times 0.133 \ cu \ ft/gal \times 144 \ sq \ in/sq \ ft}{60 \ sec/min \times 3 \ ft/sec}$$

For the return main:

$$\frac{1.3 \times 0.133 \times 144}{60 \times 3} = 0.138 \ sq. \ in.$$

$$\frac{3.3 \times 0.133 \times 144}{60 \times 3} = 0.351 \ sq. \ in.$$

$$\frac{6.1 \times 0.133 \times 144}{60 \times 3} = 0.649 \ sq. \ in.$$

$$\frac{7.7 \times 0.133 \times 144}{60 \times 3} = 0.819 \ sq. \ in.$$

For the supply main:

$$\frac{7.7 \times 0.133 \times 144}{60 \times 3} = 0.819 \ sq. \ in.$$

$$\frac{6.4 \times 0.133 \times 144}{60 \times 3} = 0.681 \ sq. \ in.$$

$$\frac{4.4 \times 0.133 \times 144}{60 \times 3} = 0.468 \ sq. \ in.$$

$$\frac{1.6 \times 0.133 \times 144}{60 \times 3} = 0.170 \ sq. \ in.$$

Using Figure 7-3 convert pipe area into tube size:

Flow/gpm	Area ID	Tube size
7.7	.819	1''
6.4	.680	1''
6.1	.649	1''
4.4	.468	3/4''
3.3	.350	3/4''
1.6	.170	1/2''
1.3	.140	1/2''

According to the above data, our system can be made up of three pipe sizes: 1'', 3/4'' and 1/2''. But as we mentioned earlier in this chapter, the baseboards and valves used in hot water systems come in the 3/4'' size. It's also perfectly all right to size *up* one or even two pipe sizes, since this will lower the velocity and head loss in the system. The big advantage to you, the contractor, is a reduced installation time, since reduction fittings don't have to be installed between the baseboard and valves and the 1/2'' tubing. So we're going to go with a 3/4'' size as the smallest size tubing in our system.

Keep in mind that you can't size *down* in tubing or it will increase your velocity to over the acceptable 3 ft/sec. So our 1'' tubing stays at 1'' (Figure 7-7).

Earlier, we mentioned that in actual installations, the total baseboard length in a loop would probably be divided among two or more baseboards to cover outside walls, windows, or other heat loss areas. You can calculate the branch piping size between these baseboards and to and from the mains according to the flow rate needed for that loop.

For example, for Unit 3 we have 20' of baseboard with a 1.3 gpm flow rate. We could divide this up into two 10' baseboards on the two outside walls. The branch piping would be:

$$\frac{1.3 \times 0.133 \times 144}{} = 0.138 \text{ sq in}$$

From Figure 7-3, we could use 3/8'' Type L copper tube for the loop. But we'll again go the 3/4'' size to match our baseboard and valves. Each apartment loop will have its own thermostat and control, and manual control valves.

Pump Sizing

To size the pump, begin by measuring the total length of the *longest* circuit from a scaled floor plan. This circuit will run from the boiler, along the supply main, through the longest baseboard loop furthest from the boiler, and back to the boiler along the return main. Just think of a single drop of water traveling the longest possible route through the system. For our apartment building we'll use the piping to, from, and including Unit 2's baseboard loop.

To find the total head loss, figure the head loss for each of the different sizes of piping in the longest circuit. For our apartment (Figure 7-7), we'll assume we have measured lengths of:

```
25' of   1''    (supply)        @ 7.7 gpm
60' of   1''    (supply)        @ 6.4 gpm
60' of 3/4''    (supply)        @ 4.4 gpm
50' of 3/4'' (Unit 2 loop)      @ 2.8 gpm
60' of   1''    (return)        @ 6.1 gpm
25' of   1''    (return)        @ 7.7 gpm
```

We can combine lengths of the same size and flow rate. Multiply all lengths by 1.5 to account for extra friction loss in fittings and valves.

```
50' x 1.5 = 75' of    1''  @ 7.7 gpm
60' x 1.5 = 90' of    1''  @ 6.4 gpm
60' x 1.5 = 90' of  3/4''  @ 4.4 gpm
50' x 1.5 = 75' of  3/4''  @ 2.8 gpm
60' x 1.5 = 90' of    1''  @ 6.1 gpm
```

Look back to the data table for copper tube on page 134.

We need to substitute our design flow rates into the formula as we did on page 134, in Step 4: Piping System Layout. The formula is:

$$\frac{H_2}{H_1} = \left(\frac{W_2}{W_1}\right)^2$$

For 1'' @ 7.7 gpm:

$$\frac{H_2}{5.0} = \left(\frac{7.7}{8}\right)^2 = 4.6$$

H_2 = 4.6 ft. head loss per 100 ft. @ 7.7 gpm

For 1'' @ 6.4 gpm:

$$\frac{H_2}{5.0} = \left(\frac{6.4}{8}\right)^2 = 3.2$$

H_2 = 3.2 ft. head loss per 100 ft. @ 6.4 gpm

For 3/4'' @ 4.4 gpm:

$$\frac{H_2}{6.6} = \left(\frac{4.4}{5}\right)^2 = 5.1$$

H_2 = 5.1 ft. head loss per 100 ft. @ 4.4 gpm

For 3/4'' @ 2.8 gpm:

$$\frac{H_2}{6.6} = \left(\frac{2.8}{5}\right)^2 = 2.1$$

H_2 = 2.1 ft. head loss per 100 ft. @ 2.8 gpm

For 1'' @ 6.1 gpm:

$$\frac{H_2}{5.0} = \left(\frac{6.1}{8}\right)^2 = 2.9$$

H_2 = 2.9 ft. head loss per 100 ft. @ 6.1 gpm

Totaling:

$$1'' \text{ @ 7.7 gpm} = \frac{75}{100} \times 4.6 = 3.45 \text{ ft.}$$

$$1'' \text{ @ 6.4 gpm} = \frac{90}{100} \times 3.2 = 2.88 \text{ ft.}$$

$$1'' \text{ @ 6.1 gpm} = \frac{90}{100} \times 2.9 = 2.61 \text{ ft.}$$

$$3/4'' \text{ @ 4.4 gpm} = \frac{90}{100} \times 5.1 = 4.59 \text{ ft.}$$

$$3/4'' \text{ @ 2.8 gpm} = \frac{75}{100} \times 2.1 = 1.58 \text{ ft.}$$

15.11 ft. head loss

This is the total head loss for the adjusted pipe length of 420 feet. Adding 1.0 foot for boiler friction loss, we get a total of 16.11 feet. Now we can use the manufacturer's table in Figure 7-4 to choose a pump. Our data for 7.7 gpm and 16.11 head put us above the performance curve for a model UPS 15-42F at speed 3, so we'll select the larger UP 26-64F.

Cavitation

When a control valve on any loop in the reverse return system opens, the total pressure will be enough to force the water through the loop. This holds true even if all the control valves are open, since each loop is designed as part of the overall system.

In the event all the loops are closed, the flow in the system will be blocked, due to the reverse return system design. The pump, however, will continue to run, overloading and burning out. Or the energy put into the water by the pump at the impeller will cause *cavitation* — the water turns to steam, damaging the impeller.

One solution to this problem is to install a bypass valve to allow the water to continue to flow through the mains. For some systems, a small, spring-loaded bypass valve will open at high pump head and allow sufficient flow. With small pumps, up to 1/3 hp, you can eliminate the bypass valve. Instead, install a thermostat to sense the outdoor air temperature and shut down the pump and boiler when outside temperatures reach 60 or 65°F.

Air Venting

Hot water heating systems need to have air vents in the piping. These are small manual or automatic valves installed at the physical high points in the system — where the piping runs reach their highest point in relation to the overall system.

Air enters the system when the piping is filled with water. It's also is caused by the gases that result when the water is heated. Leaks in the system where the pressure is below atmospheric pressure are possible sources of air in the system. A typical situation is a leaking mechanical seal on the pump when the pump is installed on the return piping to the boiler. (This is why we recommend placing the pump on the supply line.) Usually, after the air is vented from the system during start-up, air will only enter the system and cause problems if there's a malfunction in the system.

Cooling

The cooling for this type of job is usually done with through-the-wall or window air conditioning units in each apartment. The AC units are electric and are controlled separately by each apartment's occupants. Since the power for each AC unit is drawn from the apartment's system, the operating cost of the air conditioning is paid by the occupant. You should suggest to apartment owners or managers that they seal off the AC units to prevent heat loss in winter, using plastic film or some other method. Also, AC filters should be cleaned periodically to maintain efficiency and reduce operating costs to the tenants.

Heating Costs

Multiple-unit apartment owners must decide how the cost of heat will be handled. Since individual metering of heat from a central boiler is very expensive, it's usually not done. Instead, average heating costs are generally divided up among the tenants. We suggest you recommend the following book to help the owner in allocating heat costs:

Encouraging Energy Conservation in Multifamily Housing

The RUBS Project
Campus Box 468
The University of Colorado
Boulder, CO 80309   (303) 492-6749

You're an HVAC contractor who's always looking for work. Arrange with the building owner for routine maintenance and emergency service on a contract basis. It'll save the owner money in the long run, and give you guaranteed work. And even though the heating costs will be passed on to the tenants, encourage the owner to follow an energy management plan. In fact, many states now have mandatory energy programs for new and existing rental units. The owner benefits through tax breaks and smaller rent increases, resulting in better relations with the tenants and fewer turnovers. You benefit through increased opportunities for remodeling work. Besides the government agencies, many utilities also offer low- or no-cost energy conservation assistance.

Zoning, Multiple Circuits and Combined Piping Layouts

It's easy to get confused over hot water system terminology. The following definitions may be helpful in keeping things straight. But keep in mind that any one system may be made up of items from all three categories.

Zoning— This is a method of maintaining individual temperature control in various sections of the system or building. Each area of the zone will need its own piping circuit. If you install motorized zone valves in each zone supply line, only one pump is needed. If each zone has a pump, flow control valves are needed in the supply lines. In either case, install manual balancing valves in the return lines.

The zone thermostat can operate in several ways. Typically, when the zone calls for heat, it will start both the boiler and the pump or zone valve, depending on the system. Boiler controls are like the series loop system.

Often zone systems will use two-stage thermostats, wired so the first stage will operate the pump and valve and the second stage will start the boiler as needed.

Multiple circuits— In some systems, a single piping circuit would result in pipe diameters that were too large and runs that were too long so multiple piping circuits are installed.

For multiple circuits in a series loop system, calculate the flow rate for each circuit and size the circuit pipe using the pipe area formula. For the trunk piping — those pipe sections that run between the boiler and the circuit piping — the flow rate is the sum of the circuit flow rates. Figure the size using the pipe area formula.

Compute the head loss for each circuit and include the loss from the boiler. Use the circuit with the highest head for selecting the pump.

The series loop multiple circuit system can operate with one thermostat and pump, or you can install zone controls. Install balancing valves in the return circuits.

Combined piping layouts— These layouts are used to design a system that will suit a variety of needs within the same building. The system will use series loops and reverse returns in combination to get the advantages of each method in a single system.

For example, you may have a job heating a trucking company building. The building has a warehouse area for storing goods prior to transfer, as well as executive offices, reception area, rest rooms, drivers' locker room and lounge, and loading dock. A logical arrangement for the building would be to divide it into zones using a combined piping layout, since heat losses and system controls are different for each area.

The warehouse and drivers' facilities could be handled by a series loop system, using finned tube radiation or baseboard units. The executive offices, reception area and rest rooms could have a two-pipe, reverse-return circuit, to permit separate thermostats in each area. The loading dock would need a series loop with a high-capacity unit heater to handle the high infiltration.

Provide each circuit with its own pump and controls. Add up the circuit values to find trunk line sizes and capacities. A single boiler, sized to handle the entire system, is an efficient source of hot water.

Electrical System Design

Electric heat can be supplied by individual baseboard or wall heaters in each room. Another way to provide electric heat is to install an electric heating coil in a warm air furnace system.

Size electric furnaces and cooling coils based on the job requirements, using manufacturer's data. These installations work best in climates with moderate temperatures. The installation should be done by a qualified electrical contractor after you determine the necessary Btu/hr based on heat loss and gains. He'll know the codes and utility rules for your area. Before you begin designing any system that draws a large electrical load, check with your local building code and utility. Some of them may limit the size.

An all-electric baseboard system is laid out much like the hot water system. If you want cooling, you'll have to install a separate system. Vent fans and a humidifier will also have to be added. Calculate heat losses for the various rooms and areas of the building. Size the baseboards from the manufacturer's data. Locate them on the outer edge of the building, under the windows. Follow the manufacturer's recommendations and local codes. Some locations, such as behind draperies, may not be permitted. Typically each room will have a separate thermostat and be on a separate circuit.

After finding the heat losses and determining the general system equipment and layout, we recommend that you consult with a qualified electrical designer. Electrical circuits and loads must be sized carefully to prevent system failure and fire hazards. You must follow the codes. Coordinate the electric heat system with the building's electrical system for a practical, safe and economical installation.

That brings us to the end of the residential design material. In the next chapter, we'll do some typical commercial designs.

Designing Commercial Installations

We'll build on the knowledge and techniques we've learned so far to design a couple of commercial installations. The first one is a convenience store. These stores, which have become very popular over the last decade, have design conditions that are quite different from residences. But many of these conditions are common in commercial buildings.

Convenience Store

This convenience store is typical, with large window areas, heavy door traffic and long operating hours. These factors all result in high heat losses or gains, depending on the season. Display lighting and equipment loads add to the heat gains in summer. These stores are architecturally simple. Generally the building is a concrete block or steel shell with no bearing walls inside. Interior walls divide the floor space into the shopping area, a storage room, a toilet and a small office/utility room. There's a flat insulated roof above a dropped ceiling. Walls are block or steel with foam insulation covered by a brick veneer, built on a slab floor. This type of building is ideal for rooftop heating and cooling units. We'll use them in our design.

Figure 8-1 is the floor plan and elevation for our store. Because of its simple architecture, it's easy to get the U-values from Figure 5-12 in Chapter 5. We've noted these values on the plan. This works well for these jobs, but make sure you don't overlook anything.

Design Conditions

Our store is located in Minneapolis, Minnesota. Using Figure 5-10, we've chosen these design conditions:

Summer — Outside: 89°Fdb, 73°Fwb

Winter — Outside: -12°F, 15 mph wind or less

Summer — Inside: 78°F, 50% relative humidity

Winter — Inside: 67°F 20% relative humidity

Heat Loss

Our store has heat losses from windows, walls, roof, floor and infiltration. We'll use tabular calculations to figure them, rather than the datasheet we used in Chapter 5.

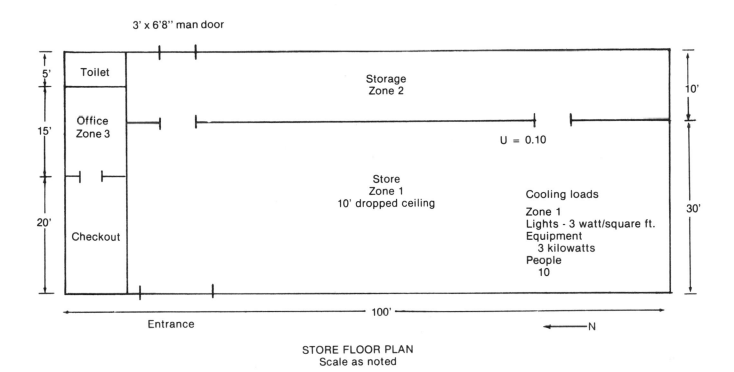

STORE FLOOR PLAN
Scale as noted

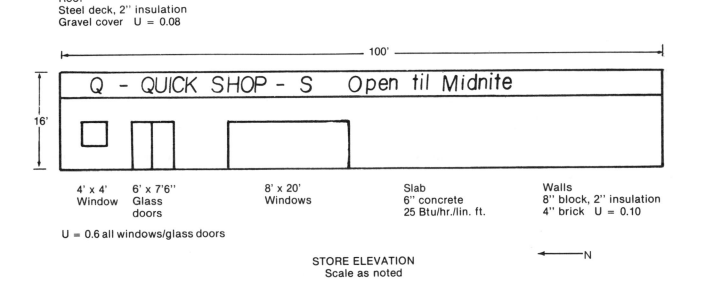

STORE ELEVATION
Scale as noted

Convenience store, floor plan and elevation
Figure 8-1

Our basic heat loss equation for building material is:

Heat loss = area x heat transfer coefficient x temp difference

or

$$Btu/hr = A \times U \times (T_i - T_o)$$

For infiltration:

Infiltration = volume x air chg/hr x heat/lb of air x lbs air/cu ft x temp difference

or

$$Btu/hr = V \times ac \times .24 \times .075 \times (T_i - T_o)$$

We've simplified the heat loss equation from the one we used in Chapter 5. Before we begin calculating, let's review a few facts:

- Infiltration volume is measured in cubic feet.

- Wall, door, window, and roof area are measured in square feet.

- Floor perimeter is measured in linear feet.

Zoning— To simplify the figuring, we'll divide different areas of the building into *zones*. We'll figure the HVAC data separately for each zone. In Figure 8-1, Zone 1 is the store area, Zone 2 is the storage room and Zone 3 includes the office and bathroom.

Let's talk about zoning for a moment. When we figured the heat losses and gains for the residence in Chapter 5, we figured each room separately. This was necessary because of the different room sizes, number of windows, number of outside walls, ceilings and all the other items that contribute to heat loss and gain.

But suppose you were evaluating a single-story medical clinic with six examining rooms located in the center of the clinic. All the rooms have the same construction — interior walls and a heat loss or gain through the ceiling. To figure, you could group all six rooms into a single zone. Calculate heat loss and gain for the entire area. Then divide your total according to the size of each room. Zoning can shorten your figuring of heat loss and gain.

More important, zoning allows separate temperature control where building use and construction require it. Using our medical clinic again as an example, let's assume the floor plan shows a large waiting room and reception area with a row of windows. In summer, this area will have heat gain from sunlight and the considerable volume of occupants. It will also have outside air infiltration, as people go in and out. In the winter, these same conditions will cause heat loss. If the examining rooms were controlled by the same thermostat as the reception area, it would be a disaster. Since they are smaller rooms and will usually only have just a doctor and a patient in them, they would be much too cold in the summer and too hot in the winter. So the waiting room/reception area will be a separate zone with its own thermostat. A central unit would still supply heating and cooling for the entire building. You'd install automatic controls to operate dampers in the ductwork to supply each zone with the right amount of heat or cooling.

To find different zones on your building projects, study your plan set. The key variables are room use and room construction. You need to use common sense in determining the zones. The plans will show the use of each room. The plan set and specs will tell you how they're built. Using this information, outline and label your zones in red on a working plan set (not the tracing). It's also helpful to draw up a sheet placing room names or numbers, and their uses, under the appropriate zone heading. Generally you won't label the zones on your final plan. Instead, the actual HVAC system layout will show how the building is zoned, since thermostats, automatic dampers, ductwork and other details will show if more than one zone is to be installed in the system.

With our convenience store, the zone areas closely follow the actual rooms. Zone 1, the store area, has windows and pedestrian traffic influencing thermal transfer. Zone 2, the storeroom, has little heat loss or gain outside of its walls and ceiling. Its comfort conditions are also less important than the store area. Zone 3 is made up of the office and toilet. After consultation with the owners, we've decided to omit air conditioning from Zones 2 and 3.

Now let's figure all three zones for our model store. Get out your calculator and do them along with us for practice.

Zone 1 ────────────────────────────────────

Outside wall = [(area) - (window + door area)] x U x (Ti-To)
        = [(150 x 16) - (6 x 32 + 6 x 7.5)] x .1 x [67-(-12)]
        = 2,400 - 237 x .1 x 79
        = 17,088 Btu/hr

Infiltration = V x air chg/hr x .24 x .075 x (Ti-To)
        = (2,900 x 10') x 2.5 x .24 x .075 x 79
        = 103,095 Btu/hr

Windows and = A x U x (Ti-To)
doors   = 221 x .6 x 79
        = 10,475 Btu/hr

Roof   = A x U x (Ti-To)
        = 2,900 x .08 x 79
        = 18,328 Btu/hr

Floor   = Perimeter x loss
        = 150' x 25 Btu/hr/lin ft (Chapter 5)
        = 3,750 Btu/hr

	Zone 1 total
	17,088
	10,475
	18,328
	3,750
	103,095
	152,736 Btu/hr

Zone 2 ────────────────────────────────────

Outside wall = 1600 - 20 (approx) x .1 x 79
        = 12,482 Btu/hr

Infiltration = 9,000 x .8 x .24 x .075 x 79
        = 10,238 Btu/hr

Man door = 20 x .59 (Chapter 5) x 79
        = 932 Btu/hr

Roof   = 900 x .08 x 79
        = 5,688 Btu/hr

Floor   = 100 x 25
        = 2,500 Btu/hr

	Zone 2 total
	12,482
	932
	5,688
	2,500
	10,238
	31,840 Btu/hr

Zone 3 ────────────────────────────────────

Outside walls = 480 - (0) x .1 x 79
        = 3,792 Btu/hr

Windows and doors = 0

Roof   = 300 x .08 x 79
        = 1,896 Btu/hr

Infiltration = 3,000 x .8 x .24 x .075 x 79
        = 3,413 Btu/hr

Floor   = 30 x 25
        = 750 Btu/hr

The total heat loss for the building is:

Zone 1 = 152,736
Zone 2 = 31,840
Zone 3 = 9,851
Total = 194,427 Btu/hr

	Zone 3 total
	3,792
	1,896
	750
	3,413
	9,851 Btu/hr

Outside walls	No glass	Windows & hinged doors	Sliding doors
1 15 mph wind or more	0.0 AC ———	0.8 AC 1.0 AC	1.0 AC 1.5 AC
2 15 mph wind or more	0.8 AC 1.0 AC	1.5 AC 2.0 AC	2.0 AC 2.0 AC
3, 4 15 mph wind or more	1.5 AC 2.0 AC	2.0 AC 2.5 AC	2.5 AC 2.5 AC
Entrance lobbies/vestibules	3.0 AC — All wind conditions		

AC = Air changes per hour

Commercial buildings — air change table
Figure 8-2

For infiltration we'll use the 10' dropped ceiling height, rather than the total building height of 16', because all the doors and windows are in the 10' high area. The solid block construction above 10' will allow very little infiltration and can be ignored.

Look at Figure 8-2 to find the number of air changes we need for the infiltration formula. For example, in Zone 1 we have three outside walls, windows and hinged doors, and a wind of 15 mph average — so there are 2.5 air changes per hour.

Heat Gain

We figure the heat gain for the building using the same heat transfer equation used for heat loss. But we won't use the design temperatures. Instead we'll use the equivalent temperature differences (ETD's), as we did in Chapter 5, to take into account solar gain and other cooling loads.

We'll only air condition the store area and enough of Zone 3 to cover a 30' by 100' area. The back one-third of the building won't be air conditioned because the owner doesn't feel it's sary. Instead he chose to insulate the wall between zones 1 and 2.

Roof, wall and glass heat gains all have to be figured. In order to apply our tables, we need to know the orientation of the building. Using an atlas, we determine that Minneapolis, Minnesota is located at about 44 degrees north latitude. The front of the building faces the west. According to the wall construction groups in Figure 8-3, the store has Group B walls.

There's one other thing we need to know — the hour of the day when the heat gain loads are at their maximum. Figures 8-4 and 8-5 give the values of several types of walls and roofs, for each hour, using a 24-hour clock. They also give the hour when the loads are at their maximum or "peak." The problem you'll run into is that the different building areas — walls, glass and roof — peak at different times. In this building, the walls peak at about midnight while the roof peaks at 3 p.m. To find the wall peak time, look at the west-facing Group B wall in Figure 8-4.

The glass peak gain is based on complicated tables that measure shadow lengths at different times during different months. As a general rule, however, there's very little lag in heat gain from glass. You can safely assume that in most areas it will be at its maximum at the hottest time of the day. So the glass will also be peaking about 3 p.m. for our building. By now you're beginning to see the problem. As the roof and glass loads decline during the evening, the wall load begins to climb toward its maximum.

If you're fortunate enough to have a building with no glass, and walls that are equal in U-values and area, you won't have any problem selecting a peak solar time. But you probably won't be that lucky. You'll have a building where the peaks are all different, like ours. So how do you figure your loads?

First, find the peak times for the different building areas. Then take into account what percent each area makes up of the total building. Give it all some thought and make an "educated

Wall Construction Group Description

Group No.	Description of Construction	Weight (lb/ft²)	U-Value Btu/(hr·ft²·F)	Heat Capacity Btu/(ft²·F)
4-in. Face Brick+(*Brick*)				
C	Air Space+4-in. Face Brick	83	0.358	18.3
D	4-in. Common Brick	90	0.415	18.4
C	1-in. Insulation or Air space+4-in. Common Brick	90	0.174-0.301	18.4
B	2-in. Insulation+4-in. Common Brick	88	0.111	18.5
D	8-in. Common Brick	130	0.302	26.4
A	Insulation or Air space+8-in. Common Brick	130	0.154-0.243	26.4
4-in. Face Brick+(*H.W. Concrete*)				
C	Air Space+2-in. Concrete	94	0.350	19.7
B	2-in. Insulation+4-in. concrete	97	0.116	19.8
A	Air Space or Insulation+8-in. or more Concrete	143-190	0.110-0.112	29.1-38.4
4-in. Face Brick+(*L.W. or H.W. Concrete Block*)				
E	4-in. Block	62	0.319	12.9
D	Air Space or Insulation+4-in. Block	62	0.153-0.246	12.9
D	8-in. Block	70	0.274	15.1
C	Air Space or 1-in. Insulation+6-in. or 8-in. Block	73-89	0.221-0.275	15.5-18.5
→B	2-in. Insulation+8-in. Block	89	0.096-0.107←	15.5-18.6
4-in Face Brick+(*Clay Tile*)				
D	4-in. Tile	71	0.381	15.1
D	Air Space+4-in. Tile	71	0.281	15.1
C	Insulation+4-in. Tile	71	0.169	15.1
C	8-in. Tile	96	0.275	19.7
B	Air Space or 1-in. Insulation+8-in. Tile	96	0.142-0.221	19.7
A	2-in. Insulation+8-in. Tile	97	0.097	19.8
H.W. Concrete Wall+(*Finish*)				
E	4-in. Concrete	63	0.585	12.5
D	4-in. Concrete+1-in. or 2-in. Insulation	63	0.119-0.200	12.5
C	2-in. Insulation+4-in. Concrete	63	0.119	12.7
C	8-in. Concrete	109	0.490	21.9
B	8-in. Concrete+1-in. or 2-in. Insulation	110	0.115-0.187	22.0
A	2-in. Insulation+8-in. Concrete	110	0.115	21.9
B	12-in. Concrete	156	0.421	31.2
A	12-in. Concrete+Insulation	156	0.113	31.3
L.W. and H.W. Concrete Block+(*Finish*)				
F	4-in. Block+Air Space/Insulation	29-36	0.161-0.263	5.7-7.2
E	2-in. Insulation+4-in. Block	29-37	0.105-0.114	5.8-7.3
E	8-in. Block	41-57	0.294-0.402	6.3-11.3
D	8-in. Block+Air Space/Insulation	41-57	0.149-0.173	8.3-11.3
Clay *Tile*+(*Finish*)				
F	4-in. Tile	39	0.419	7.8
F	4-in. Tile+Air space	39	0.303	7.8
E	4-in. Tile+1-in. Insulation	39	0.175	7.9
D	2-in. Insulation+4-in. Tile	40	0.110	7.9
D	8-in. Tile	63	0.296	12.5
C	8-in. Tile+Air Space/1-in. Insulation	63	0.151-0.231	12.6
B	2-in. Insulation+8-in. Tile	63	0.099	12.6
Metal Curtain Wall				
G	With/without Air Space+1-in./2-in./3-in. Insulation	5-6	0.091-0.230	0.7
Frame Wall				
G	1-in. to 3 in. Insulation	16	0.081-0.178	2.2

Courtesy: American Society of Heating, Refrigerating and Air-Conditioning Engineers, Inc. (ASHRAE)
1981 Fundamentals Handbook, all rights reserved.

Wall construction groups
Figure 8-3

estimate'' (a guess) as to the best average peak time. Then stay with it for all your figuring. The only other choice is to calculate each hour for each area separately and then plot them on a graph to find some sort of average.

For our building, we've chosen 5 p.m., or the 17th hour of solar time, as our peak value for all areas.

Outside Walls

The walls have 4'' face brick, 2'' of insulation and 8'' concrete block construction. They're Group B walls, according to Figure 8-3. The U-value is from 0.096 to 0.107. We'll settle for 0.10. To find the ETD, use Figure 8-4. Look at the Group B wall section, facing different directions, at 17 hours solar time. The formula is:

Cooling Load Temperature Differences for Calculating Cooling Load from Sunlit Walls

North Latitude Wall Facing	1	2	3	4	5	6	7	8	9	10	11	12	13	14	15	16	17	18	19	20	21	22	23	24	Hr of Maximum CLTD	Minimum CLTD	Maximum CLTD	Difference CLTD
Group A Walls																												
N	14	14	14	13	13	13	12	12	11	11	10	10	10	10	10	10	10	11	11	12	12	13	13	14	2	10	14	4
NE	19	19	19	18	17	17	16	15	15	15	15	15	16	16	17	18	18	18	19	19	20	20	20	20	22	15	20	5
E	24	24	23	23	22	21	20	19	19	18	19	19	20	21	22	23	24	24	25	25	25	25	25	25	22	18	25	7
SE	24	23	23	22	21	20	20	19	18	18	18	18	19	20	21	22	23	24	24	24	24	24	24	24	22	18	24	6
S	20	20	19	19	18	18	17	16	16	15	14	14	14	14	14	15	16	17	18	19	19	20	20	20	23	14	20	6
SW	25	25	25	24	24	23	22	21	20	19	18	18	17	17	17	17	17	18	19	20	22	23	24	25	24	17	25	8
W	27	27	26	26	25	24	24	23	22	21	20	19	19	18	18	18	18	19	20	22	23	25	26	26	1	18	27	9
NW	21	21	21	20	20	19	19	18	17	16	16	15	15	14	14	15	15	16	17	18	19	20	21		1	14	21	7
Group B Walls																												
N	15	14	14	13	12	11	11	10	9	9	9	8	9	9	9	10	11	12	13	14	14	15	15	15	24	8	15	7
NE	19	18	17	16	15	14	13	12	12	13	14	15	16	17	18	19	19	20	20	21	21	20	20	20	21	12	21	9
E	23	22	21	20	18	17	16	15	15	15	17	19	21	22	24	25	26	26	27	27	26	26	25	24	20	15	27	12
SE	23	22	21	20	18	17	16	15	14	14	15	16	18	20	21	23	24	25	26	26	26	25	24		21	14	26	12
S	21	20	19	18	17	15	14	13	12	11	11	11	11	12	14	15	17	19	20	21	22	22	22	21	23	11	22	11
SW	27	26	25	24	22	21	19	18	16	15	14	14	13	13	14	15	17	20	22	25	27	28	28	28	24	13	28	15
W	29	28	27	26	24	23	21	19	18	17	16	15	14	14	15	17	19	22	25	27	29	29	30		24	14	30	16
NW	23	22	21	20	19	18	17	15	14	13	12	12	12	11	12	12	13	15	17	19	21	22	23	23	24	11	23	12
Group C Walls																												
N	15	14	13	12	11	10	9	8	8	7	7	8	9	10	12	13	14	15	16	17	17	17	16		22	7	17	10
NE	19	17	16	14	13	11	10	10	11	13	15	17	19	20	21	22	22	23	23	23	23	22	21	20	20	10	23	13
E	22	21	19	17	15	14	12	12	14	16	19	22	25	27	29	29	30	30	30	29	28	27	26	24	18	12	30	18
SE	22	21	19	17	15	14	12	12	12	13	16	19	22	24	26	28	29	29	29	28	27	26	24		19	12	29	17
S	21	19	18	16	15	13	12	10	9	9	9	10	11	14	17	20	22	24	25	26	25	25	24	22	20	9	26	17
SW	29	27	25	22	20	18	16	15	13	12	11	11	11	13	14	18	22	26	29	32	33	33	32	31	22	11	33	22
W	31	29	27	25	22	20	18	16	14	13	12	12	12	13	14	16	20	24	29	32	35	35	35	33	22	12	35	23
NW	25	23	21	20	18	16	14	13	11	10	10	10	11	12	13	15	18	22	25	27	27	27	26		22	10	27	17
Group D Walls																												
N	15	13	12	10	9	7	6	6	6	6	6	7	8	10	12	13	15	17	18	19	19	19	18	16	21	6	19	13
NE	17	15	13	11	10	8	7	8	10	14	17	20	22	23	24	24	25	25	24	23	22	20	18		19	7	25	18
E	19	17	15	13	11	9	8	9	12	17	22	27	30	32	33	33	32	32	31	30	28	26	24	22	16	8	33	25
SE	20	17	15	13	11	10	8	8	10	13	17	22	26	29	31	32	32	32	31	30	28	26	24	22	17	8	32	24
S	19	17	15	13	11	9	7	6	6	7	9	12	16	20	24	27	29	29	29	27	24	22			19	6	29	23
SW	28	25	22	19	16	14	12	10	9	8	8	8	10	12	16	21	27	32	36	38	38	37	34	31	21	8	38	30
W	31	27	24	21	18	15	13	11	10	9	9	10	11	14	18	24	30	36	40	41	40	38	34		22	9	41	32
NW	25	22	19	17	14	12	10	9	8	7	7	8	9	10	12	14	18	22	27	31	32	32	30	27	22	7	32	25
Group E Walls																												
N	12	10	8	7	5	4	3	4	5	6	7	9	11	13	15	17	19	20	21	23	20	18	16	14	20	3	22	19
NE	13	11	9	7	6	4	5	9	15	20	24	25	25	26	26	26	26	25	24	22	19	17	15		16	2	26	22
E	14	12	10	8	6	5	6	11	18	26	33	36	37	36	34	33	32	30	28	25	22	20	17	13	15	5	38	33
SE	15	12	10	8	7	5	5	8	12	19	25	31	35	37	37	36	34	33	31	28	26	23	20	17	15	5	37	32
S	15	12	10	8	7	5	4	3	4	5	9	13	19	24	29	32	34	33	31	29	26	23	20	17	17	3	34	31
SW	22	18	15	12	10	8	6	5	5	6	7	9	12	18	24	32	38	43	45	44	40	35	30	26	19	5	45	40
W	25	21	17	14	11	9	7	6	6	6	7	9	11	14	20	27	36	43	49	49	45	40	34	29	20	6	49	43
NW	20	17	14	11	9	7	6	5	5	5	6	8	10	13	16	20	26	32	37	38	36	32	28	24	20	5	38	33
Group F Walls																												
N	8	6	5	3	2	1	2	4	6	7	9	11	14	17	19	21	22	23	24	23	20	16	13	11	19	1	24	23
NE	9	7	5	3	2	1	5	14	23	28	30	29	28	27	27	27	26	24	22	19	16	13	11		11	1	30	29
E	10	7	6	4	3	2	6	17	28	38	44	45	43	39	36	34	32	30	27	24	21	17	15	12	12	2	45	43
SE	10	7	6	4	3	2	4	10	19	28	36	41	43	42	39	36	34	31	28	25	21	18	15	12	13	2	43	41
S	10	8	6	4	3	2	1	3	7	13	20	27	34	38	39	38	35	31	26	22	18	15	12		16	1	39	38
SW	15	11	9	6	5	3	2	4	5	8	11	17	26	35	44	50	53	52	45	37	28	23	18		18	2	53	51
W	17	13	10	7	5	4	3	4	6	8	11	14	20	28	39	49	57	60	54	43	34	27	21		19	3	60	57
NW	14	10	8	6	4	3	2	2	3	5	8	10	13	15	21	27	35	42	46	43	35	28	22	18	19	2	46	44
Group G Walls																												
N	3	2	1	0	-1	2	7	8	9	12	15	18	21	23	24	24	25	26	22	15	11	9	7	5	18	-1	26	27
NE	3	2	1	0	-1	9	27	36	39	35	30	26	26	27	27	26	25	22	18	14	11	9	7	5	9	-1	39	40
E	4	2	1	0	-1	11	31	47	54	55	50	40	33	30	29	27	24	19	15	12	10	8	6		10	-1	55	56
SE	4	2	1	0	-1	5	18	32	42	49	51	48	42	36	32	30	27	24	19	15	12	10	8	6	11	-1	51	52
S	4	2	1	0	-1	0	1	5	12	22	31	39	45	46	43	37	31	25	20	15	12	10	8	5	14	-1	46	47
SW	5	4	3	1	0	0	2	5	8	12	16	26	38	50	59	63	61	52	37	24	17	13	10	8	16	0	63	63
W	6	5	3	2	1	1	2	5	8	11	15	19	27	41	56	67	72	67	48	29	20	15	11	8	17	1	72	71
NW	5	3	2	1	0	0	2	5	8	11	15	18	21	27	37	47	55	55	41	25	17	13	10	7	18	0	55	55

(1) *Direct Application of the Table Without Adjustments:*

Values in the table were calculated using the same conditions for walls as outlined for the roof CLTD table, Table 3.8. These values may be used for all normal air conditioning estimates usually without correction (except as noted below) when the load is calculated for the hottest weather.

For totally shaded walls use the North orientation values.

(2) *Adjustments to Table Values:*

The following equation makes adjustment for conditions other than those listed in Note (1).

$$CLTD_{corr} = (CLTD + LM) \times K + (78 - T_R) + (T_o - 85)$$

where CLTD is from Table 3.10 at the wall orientation.

  (a)   LM is the latitude-month correction from Table 3.12.

  (b) K is a color adjustment factor and is applied after first making latitude-month adjustment

        $K = 1.0$ is dark colored or light in an industrial area

        $K = 0.83$ if permanently medium colored (rural area)

        $K = 0.65$ if permanently light colored (rural area)

Wall CLTD table — north latitudes
Figure 8-4

Cooling Load Temperature Differences for Calculating Cooling Load from Flat Roofs

Roof No	Description of Construction	Weight lb/ft²	U-value Btu/(h·ft²·°F)	1	2	3	4	5	6	7	8	9	10	11	12	13	14	15	16	17	18	19	20	21	22	23	24	Hour of Maximum CLTD	Minimum CLTD	Maximum CLTD	Difference CLTD	Heat Capacity Btu/(ft²·F)
										Without Suspended Ceiling																				
1	Steel sheet with 1-in. (or 2-in.) insulation	7 (8)	0.213 (0.124)	1	-2	-3	-3	-5	-3	6	19	34	49	61	71	78	79	77	70	59	45	30	18	12	8	5	3	14	-5	79	84	2.13
2	1-in. wood with 1-in. insulation	8	0.170	6	3	0	-1	-3	-3	-2	4	14	27	39	52	62	70	74	74	70	62	51	38	28	20	14	9	16	-3	74	77	3.73
3	4-in. l.w concrete	18	0.213	9	5	2	0	-2	-3	-3	1	9	20	32	44	55	64	70	73	71	66	57	45	34	25	18	13	16	-3	73	76	4.45
4	2-in. h.w. concrete with 1-in. (or 2-in.) insulation	29	0.206 (0.122)	12	8	5	3	0	-1	-1	3	11	20	30	41	51	59	65	66	66	62	54	45	36	29	22	17	16	-1	67	68	6.57
5	1-in. wood with 2-in. insulation	19	0.109	3	0	-3	-4	-5	-7	-6	-3	5	16	27	39	49	57	63	64	62	57	48	37	26	18	11	7	16	-7	64	71	3.83
6	6-in. l.w. concrete	24	0.158	22	17	13	9	6	3	1	1	3	7	15	23	33	43	51	58	62	64	62	57	50	42	35	28	18	1	64	63	5.79
7	2.5-in. wood with 1-in. insulation	13	0.130	29	24	20	16	13	10	7	6	6	9	13	20	27	34	42	48	53	55	56	54	49	44	39	34	19	6	56	50	6.51
8	8-in. l.w. concrete	31	0.126	35	30	26	22	18	14	11	9	7	7	9	13	19	25	33	39	46	50	53	54	53	49	45	40	20	7	54	47	7.13
9	4-in. h.w. concrete with 1-in. (or 2-in.) insulation	52 (52)	0.200 (0.120)	25	22	18	15	12	9	8	8	10	14	20	26	33	40	46	50	53	53	52	48	43	38	34	30	18	8	53	45	11.21
10	2.5-in. wood with 2-in. insulation	13	0.093	30	26	23	19	16	13	10	9	8	9	13	17	23	29	36	41	46	49	51	50	47	43	39	35	19	8	51	43	6.61
11	Roof terrace system	75	0.106	34	31	28	25	22	19	16	14	13	13	15	18	22	26	31	36	40	44	45	46	45	43	40	37	20	13	46	33	15.98
12	6-in. h.w. concrete with 1-in. (or 2-in.) insulation	75 (75)	0.192 (0.117)	31	28	25	22	20	17	15	14	14	16	18	22	26	31	36	40	43	45	45	44	42	40	37	34	19	14	45	31	15.89
13	4-in. wood with 1-in. (or 2-in) insulation	17 (18)	0.106 (0.078)	38	36	33	30	28	25	22	20	18	17	16	17	18	21	24	28	32	36	39	41	43	43	42	40	22	16	43	27	9.27
										With Suspended Ceiling																				
1	Steel Sheet with 1-in. (or 2-in.) insulation	9 (10)	0.134 (0.092)	2	0	-2	-3	-4	-4	-1	9	23	37	50	62	71	77	78	74	67	56	42	28	18	12	8	5	15	-4	78	82	2.50
2	1-in. wood with 1-in. insulation	10	0.115	20	15	11	8	5	3	2	3	7	13	21	30	40	48	55	60	62	61	58	51	44	37	30	25	17	2	62	60	4.11
3	4-in. l.w. concrete	20	0.134	19	14	10	7	4	2	0	0	4	10	19	29	39	48	56	62	65	64	61	54	46	38	30	24	17	0	65	65	4.83
4	2-in. h.w. concrete with 1-in. insulation	30	0.131	28	25	23	20	17	15	13	13	14	16	20	25	30	35	39	43	46	47	46	44	41	38	35	32	18	13	47	34	6.94
5	1-in. wood with 2-in. insulation	10	0.083	25	20	16	13	10	7	5	5	7	12	18	25	33	41	48	53	57	57	56	52	46	40	34	29	18	5	57	52	4.21
6	6-in. l.w. concrete	26	0.109	32	28	23	19	16	13	10	8	8	11	16	22	29	36	42	48	52	54	54	51	47	42	37	32	20	7	54	47	6.17
7	2.5-in. wood with 1-in. insulation	15	0.096	34	31	29	26	23	21	18	16	15	15	16	18	21	25	30	34	38	41	43	44	42	40	37	32	21	15	44	29	6.89
8	8-in. l.w. concrete	33	0.093	39	36	33	29	26	23	20	18	15	14	14	15	17	20	25	29	34	38	42	45	46	45	44	42	21	14	46	32	7.51
9	4-in. h.w. concrete with 1-in. (or 2-in.) insulation	53 (54)	0.128 (0.090)	30	29	27	26	24	22	21	20	20	21	22	24	27	29	32	34	36	38	38	38	37	36	34	33	19	20	38	18	11.58
10	2.5-in. wood with 2-in. insulation	15	0.072	35	33	30	28	26	24	22	20	18	18	18	20	22	25	28	32	35	38	40	41	41	40	39	37	21	18	41	23	6.98
11	Roof terrace system	77	0.082	30	29	28	27	26	25	24	23	22	22	22	23	23	25	26	28	29	31	32	33	33	33	33	32	22	22	33	11	16.36
12	6-in. h.w. concrete with 1-in. (or 2-in) insulation	77 (77)	0.125 (0.088)	29	28	27	26	25	24	23	22	21	21	22	23	25	26	28	30	32	33	34	34	34	33	32	31	20	21	34	13	16.26
13	4-in. wood with 1-in (or 2-in.) insulation	19 (20)	0.082 (0.064)	35	34	33	32	31	29	27	26	24	23	22	21	22	22	24	25	27	30	32	34	35	36	37	36	23	21	37	16	9.64

(1) *Direct Application of Table 3.8 Without Adjustments:*

Values in Table 3.8 were calculated using the following conditions:

- Dark flat surface roof ("dark" for solar radiation absorption)
- Indoor temperature of 78 F
- Outdoor maximum temperature of 95 F with outdoor mean temperature of 85 F and an outdoor daily range of 21 deg F
- Solar radiation typical of 40 deg North latitude on July 21
- Outside surface resistance, $R_o = 0.333$ (hr·ft²·F)/Btu
- Without and with suspended ceiling, but no attic fans or return air ducts in suspended ceiling space
- Inside surface resistance, $R_i = 0.685$ (hr·ft²·F)/Btu

(2) *Adjustments to Table 3.8 Values:*

The following equation makes adjustments for deviations of design and solar conditions from those listed in (1) above.

$$CLTD_{corr} = [(CLTD + LM) \times K + (78 - T_R) + (T_o - 85)] \times f$$

where CLTD is from this table

(a) LM is latitude-month correction from Table 3.12 for a horizontal surface.

(b) K is a color adjustment factor and is applied after first making month-latitude adjustments. Credit should not be taken for light-colored roof except where permanence of light color is established by experience, as in rural areas or where there is little smoke.
    K = 1.0 if dark colored or light in an industrial area
    K = 0.5 if permanently light-colored (rural area)

(c) (78 - T_R) is indoor design temperature correction.
    Table 3.13 can be used when indoor design is other than 78 F.

(d) (T_o - 85) is outdoor design temperature correction, where T_o is the average outside temperature on design day.
    Table 3.13 is based on the local design outside dry-bulb temperature and daily range, as given in Column 6 (2 1/2%) and 7, Table 2.1, Climatic Conditions.

(e) f is a factor for attic fan and/or ducts above ceiling and is applied after all other adjustments have been made.
    f = 1.0 no attic or ducts
    f = 0.75, positive ventilation
    Values in Table 3.8 were calculated without and with a suspended ceiling, but made no allowance for positive ventilation or return ducts thru the space. If ceiling is insulated and a fan is used between ceiling and roof, CLTD may be reduced by 25% (f = .75). Use of the suspended ceiling space for a return air plenum or with return air ducts should be analyzed separately.

(3) *Roof Constructions Not Listed In Table:*

The U-values listed are to be used only as guides. The actual value of U as obtained from tables such as Table 3.2 or as calculated for the actual roof construction should be used.

An actual roof construction not in this table would be thermally similar to a roof in the table, if it has similar mass, m lb/ft², and similar heat capacity Btu/(ft²·F). In such a case, use the CLTD from this table as corrected by Note (2) above.

Example: A flat roof without a suspended ceiling has properties m = 18.0 lb/ft², U = 0.20 Btu/(hr·ft²·F), and heat capacity = 9.5 Btu/(ft²·F). Use $CLTD_{uncorr}$ from Roof No. 13, to obtain $CLTD_{corr}$ and use the actual U value to calculate q/A = U $(CLTD)_{corr}$ = 0.20 $(CLTD_{corr})$.

(4) *Additional Insulation*

For each R-7 increase in R-value due to insulation added to the roof structure (Table 3.7), use a CLTD for a roof whose weight and heat capacity are approximately the same, but whose CLTD has a maximum value 2 hr later. If this is not possible, because a roof with the longest time lag has already been selected, use an effective CLTD in the cooling load calculation equal to 29 deg F.

Flat roof CLTD table — north latitudes
Figure 8-5

Heat gain = area x U x ETD

West wall	= (100 x 16) - (160 + 45) x 0.10 x 17	= 2,372 Btu/hr
South wall	= (30 x 16) x 0.10 x 17	= 816
North wall	= (30 x 16) x 0.10 x 11	= 528
Wall between Zone 1 & Zones 2 & 3	= (110 x 16) x 0.10 x (89-78)	= 1,936
	Total	5,652 Btu/hr

Roof

The equivalent temperature of the roof depends on the type of construction and the time of day. Using Figure 8-5, we find Roof No. 1 with a suspended ceiling fits our building. At 17 solar hours, we find an ETD of 67 and a U-value of 0.134.

Roof = (30 x 100) x 0.134 x 67 = 26,934 Btu/hr

Glass

For our glass area, we use a Solar Heat Gain Factor (SHGF) and a Cooling Load Factor (CLF). This is because we have two types of gain through glass, one due to conduction and the other through fenestration (the arrangement of the windows). Use Figure 8-6 to find the SHGF, which depends on the latitude, month and orientation of our location. You'll find the CLF, which depends on orientation, solar time and room construction, in Figure 8-7. The formula is:

Heat gain = A x SHGF x CLF

For our building:

West wall glass = 221 x 215 x 0.64 = 30,410 Btu/hr

Lights, People and Equipment Loads

We'll ignore floor gain and gain due to infiltration, since these will be slight for our type of building. We should, however, figure the heat load gained with lights, people and equipment.

Lights— As a general rule, one watt of lighting will give off 3.4 Btu/hr in heat. In our store, the lights will be on continuously from 6 a.m. until midnight. From the plan drawn up by the electrical designer, we find he's averaged two watts per square foot for lighting.

Lights =
100 x 30 x 2 watts x 3.4 Btu/wt = 20,400 Btu/hr

This formula will be suitable for most jobs you run into. If the lighting load is very large, consult with the electrical designer, since you'll need a load factor to account for the increase in the load over time. Other items to consider are the amount of ventilation the light fixtures receive, the type of fixtures, and how long the lights remain on.

People— Many of the convenience stores have a very high traffic rate. It's difficult to estimate the number of occupants at any one time, since customers come and go quickly. For our store, however, we're going to assume a very good business. Based on its location and his prior business experience, the owner expects an average of eight customers in the store at all times. He'll need two full-time clerks.

Figure 8-8 gives the rates of heat gain from people in occupied spaces. Two types of heat are given off, sensible and latent, and both have to be considered. The formula used for both is:

Heat load =
heat gain per person x number of people

For our job, Figure 8-8 shows 315 Btu/hr sensible and 325 Btu/hr latent for a retail store typical application, using the adjusted group category for men, women and children (see Note B).

Heat load — sensible =	315 x 10 =	3,150 Btu/hr
Heat load — latent =	325 x 10 =	3,250 Btu/hr
	Total =	6,400 Btu/hr

Maximum Solar Heat Gain Factor, Btu/(hr·ft^2) for Sunlit Glass, North Latitudes

0 Deg

	N	NNE/ NNW	NE/ NW	ENE/ WNW	E/ W	ESE/ WSW	SE/ SW	SSE/ SSW	S	HOR
Jan.	34	34	88	177	234	254	235	182	118	296
Feb.	36	39	132	205	245	247	210	141	67	306
Mar.	38	87	170	223	242	223	170	87	38	303
Apr.	71	134	193	224	221	184	118	38	37	284
May	113	164	203	218	201	154	80	37	37	265
June	129	173	206	212	191	140	66	37	37	255
July	115	164	201	213	195	149	77	38	38	260
Aug.	75	134	187	216	212	175	112	39	38	276
Sep.	40	84	163	213	231	213	163	84	40	293
Oct.	37	40	129	199	236	238	202	135	66	299
Nov.	35	35	88	175	230	250	230	179	117	293
Dec.	34	34	71	164	226	253	240	196	138	288

16 Deg

	N	NNE/ NNW	NE/ NW	ENE/ WNW	E/ W	ESE/ WSW	SE/ SW	SSE/ SSW	S	HOR
Jan.	30	30	55	147	210	244	251	223	199	248
Feb.	33	33	96	180	231	247	233	188	154	275
Mar.	35	53	140	205	239	235	197	138	93	291
Apr.	39	99	172	216	227	204	150	77	45	289
May	52	132	189	218	215	179	115	45	41	282
June	66	142	194	217	207	167	99	41	41	277
July	55	132	187	214	210	174	111	44	42	277
Aug.	41	100	168	209	219	196	143	74	46	282
Sep.	36	50	134	196	227	224	191	134	93	282
Oct.	33	33	95	174	223	237	225	183	150	270
Nov.	30	30	55	145	206	241	247	220	196	246
Dec.	29	29	41	132	198	241	254	233	212	234

4 Deg

	N	NNE/ NNW	NE/ NW	ENE/ WNW	E/ W	ESE/ WSW	SE/ SW	SSE/ SSW	S	HOR
Jan.	33	33	79	170	229	252	237	193	141	286
Feb.	35	35	123	199	242	248	215	152	88	301
Mar.	38	77	163	219	242	227	177	96	43	302
Apr.	55	125	189	223	223	190	126	43	38	287
May	93	154	200	220	206	161	89	38	38	272
June	110	164	202	215	196	147	73	38	38	263
July	96	154	197	215	200	156	85	39	38	267
Aug.	59	124	184	215	214	181	120	42	40	279
Sep.	39	75	156	209	231	216	170	93	44	293
Oct.	36	36	120	193	234	239	207	148	86	294
Nov.	34	34	79	168	226	248	232	190	139	284
Dec.	33	33	62	157	221	250	242	206	160	277

20 Deg

	N	NNE/ NNW	NE/ NW	ENE/ WNW	E/ W	ESE/ WSW	SE/ SW	SSE/ SSW	S	HOR
Jan.	29	29	48	138	201	243	253	233	214	232
Feb.	31	31	88	173	226	244	238	201	174	263
Mar.	34	49	132	200	237	236	206	152	115	284
Apr.	38	92	166	213	228	208	158	91	58	287
May	47	123	184	217	217	184	124	54	42	283
June	59	135	189	216	210	173	108	45	42	279
July	48	124	182	213	212	179	119	53	43	278
Aug.	40	91	162	206	220	200	152	88	57	280
Sep.	36	46	127	191	225	225	199	148	114	275
Oct.	32	32	87	167	217	236	231	196	170	258
Nov.	29	29	48	136	197	239	249	229	211	230
Dec.	27	27	35	122	187	238	254	241	226	217

8 Deg

	N	NNE/ NNW	NE/ NW	ENE/ WNW	E/ W	ESE/ WSW	SE/ SW	SSE/ SSW	S	HOR
Jan.	32	32	71	163	224	250	242	203	162	275
Feb.	34	34	114	193	239	248	219	165	110	294
Mar.	37	67	156	215	241	230	184	110	55	300
Apr.	44	117	184	221	225	195	134	53	39	289
May	74	146	198	220	209	167	97	39	38	277
June	90	155	200	217	200	141	82	39	39	269
July	77	145	195	215	204	162	93	40	39	272
Aug.	47	117	179	214	216	186	128	51	41	282
Sep.	38	66	149	205	230	219	176	107	56	290
Oct.	35	35	112	187	231	239	211	160	108	288
Nov.	33	33	71	161	220	245	233	200	160	273
Dec.	31	31	55	149	215	246	247	215	179	265

24 Deg

	N	NNE/ NNW	NE/ NW	ENE/ WNW	E/ W	ESE/ WSW	SE/ SW	SSE/ SSW	S	HOR
Jan.	27	27	41	128	190	240	253	241	227	214
Feb.	30	30	80	165	220	244	243	213	192	249
Mar.	34	45	124	195	234	237	214	168	137	275
Apr.	37	88	159	209	228	212	169	107	75	283
May	43	117	178	214	218	190	132	67	46	282
June	55	127	184	214	212	179	117	55	43	279
July	45	116	176	210	213	185	129	65	46	278
Aug.	38	87	156	203	220	204	162	103	72	277
Sep.	35	42	119	185	222	225	206	163	134	266
Oct.	31	31	79	159	211	237	235	207	187	244
Nov.	27	27	42	126	187	236	249	237	224	213
Dec.	26	26	29	112	180	234	247	247	237	199

12 Deg

	N	NNE/ NNW	NE/ NW	ENE/ WNW	E/ W	ESE/ WSW	SE/ SW	SSE/ SSW	S	HOR
Jan.	31	31	63	155	217	246	247	212	182	262
Feb.	34	34	105	186	235	248	226	177	133	286
Mar.	36	58	148	210	240	233	190	124	73	297
Apr.	40	108	178	219	227	200	142	64	40	290
May	60	139	194	220	212	173	106	40	40	280
June	75	149	198	217	204	161	90	40	40	274
July	63	139	191	215	207	168	102	41	41	275
Aug.	42	109	174	212	218	191	135	62	142	282
Sep.	37	57	142	201	229	222	182	121	73	287
Oct.	34	34	103	180	227	238	219	172	130	280
Nov.	32	32	63	153	214	241	243	209	179	260
Dec.	30	30	47	141	207	242	251	223	197	250

28 Deg

	N (Shade)	NNE/ NNW	NE/ NW	ENE/ WNW	E/ W	ESE/ WSW	SE/ SW	SSE/ SSW	S	HOR
Jan.	25	25	35	117	183	235	251	247	238	196
Feb.	29	29	72	157	213	244	246	224	207	234
Mar.	33	41	116	189	231	237	221	182	157	265
Apr.	36	84	151	205	228	216	178	124	94	278
May	40	115	172	211	219	195	144	83	58	280
June	51	125	178	211	213	184	128	68	49	278
July	41	114	170	208	215	190	140	80	57	276
Aug.	38	83	149	199	220	207	172	120	91	272
Sep.	34	38	111	179	219	226	213	177	154	256
Oct.	30	30	71	151	204	236	238	217	202	229
Nov.	26	26	35	115	181	232	247	243	235	195
Dec.	24	24	24	99	172	227	248	251	246	179

Solar heat gain factors for glass
Figure 8-6

Maximum Solar Heat Gain Factor, Btu/(hr·ft^2) for Sunlit Glass, North Latitudes

32 Deg

	N (Shade)	NNE/ NNW	NE/ NW	ENE/ WNW	E/ W	ESE/ WSW	SE/ SW	SSE/ SSW	S	HOR
Jan.	24	24	29	105	175	229	249	250	246	176
Feb.	27	27	65	149	205	242	248	232	221	217
Mar.	32	37	107	183	227	237	227	195	176	252
Apr.	36	80	146	200	227	219	187	141	115	271
May	38	111	170	208	220	199	155	99	74	277
June	44	122	176	208	214	189	139	83	60	276
July	40	111	167	204	215	194	150	96	72	273
Aug.	37	79	141	195	219	210	181	136	111	265
Sep.	33	35	103	173	215	227	218	189	171	244
Oct.	28	28	63	143	195	234	239	225	215	213
Nov.	24	24	29	103	173	225	245	246	243	175
Dec.	22	22	22	84	162	218	246	252	252	158

48 Deg

	N (Shade)	NNE/ NNW	NE/ NW	ENE/ WNW	E/ W	ESE/ WSW	SE/ SW	SSE/ SSW	S	HOR
Jan.	15	15	15	53	118	175	216	239	245	85
Feb.	20	20	36	103	168	216	242	249	250	138
Mar.	26	26	80	154	204	234	239	232	228	188
Apr.	31	61	132	180	219	225	215	194	186	226
May	35	97	158	200	218	214	192	163	150	247
June	46	110	165	204	215	206	180	148	134	252
July	37	96	156	196	214	209	187	158	146	244
Aug.	33	61	128	174	211	216	208	188	180	223
Sep.	27	27	72	144	191	223	228	223	220	182
Oct.	21	21	35	96	161	207	233	241	242	136
Nov.	15	15	15	52	115	172	212	234	240	85
Dec.	13	13	13	36	91	156	195	225	233	65

36 Deg

	N (Shade)	NNE/ NNW	NE/ NW	ENE/ WNW	E/ W	ESE/ WSW	SE/ SW	SSE/ SSW	S	HOR
Jan.	22	22	24	90	166	219	247	252	252	155
Feb.	26	26	57	139	195	239	248	239	232	199
Mar.	30	33	99	176	223	238	232	206	192	238
Apr.	35	76	144	196	225	221	196	156	135	262
May	38	107	168	204	220	204	165	116	93	272
June	47	118	175	205	215	194	150	99	77	273
July	39	107	165	201	216	199	161	113	90	268
Aug.	36	75	138	190	218	212	189	151	131	257
Sep.	31	31	95	167	210	228	223	200	187	230
Oct.	27	27	56	133	187	230	239	231	225	195
Nov.	22	22	24	87	163	215	243	248	248	154
Dec.	20	20	20	69	151	204	241	253	254	136

52 Deg

	N (Shade)	NNE/ NNW	NE/ NW	ENE/ WNW	E/ W	ESE/ WSW	SE/ SW	SSE/ SSW	S	HOR
Jan.	13	13	13	39	92	155	193	222	230	62
Feb.	18	18	29	85	156	202	235	247	250	115
Mar.	24	24	73	145	196	230	239	238	236	169
Apr.	30	56	128	197	215	224	220	204	199	211
May	34	98	154	198	217	217	199	175	167	235
June	45	111	161	202	214	210	188	162	152	242
July	36	97	152	194	213	212	195	171	163	233
Aug.	32	56	124	169	208	216	212	197	193	208
Sep.	25	25	65	136	182	218	228	228	227	163
Oct.	19	19	28	80	148	192	225	238	240	114
Nov.	13	13	13	39	90	152	189	217	225	62
Dec.	10	10	10	19	73	127	172	199	209	42

40 Deg

	N (Shade)	NNE/ NNW	NE/ NW	ENE/ WNW	E/ W	ESE/ WSW	SE/ SW	SSE/ SSW	S	HOR
Jan.	20	20	20	74	154	205	241	252	254	133
Feb.	24	24	50	129	186	234	246	244	244	180
Mar.	29	29	93	169	218	238	236	216	206	223
Apr.	34	71	140	190	224	223	203	170	154	252
May	37	102	165	202	220	208	175	133	113	265
June	48	113	172	205	216	199	166	116	95	267
July	38	102	163	198	216	203	170	129	109	262
Aug.	35	71	135	185	216	214	196	165	149	247
Sep.	30	30	87	160	203	227	226	209	200	215
Oct.	25	25	49	123	180	225	238	236	234	177
Nov.	20	20	20	73	151	201	237	248	250	132
Dec.	18	18	18	60	135	188	232	249	253	113

56 Deg

	N (Shade)	NNE/ NNW	NE/ NW	ENE/ WNW	E/ W	ESE/ WSW	SE/ SW	SSE/ SSW	S	HOR
Jan.	10	10	10	21	74	126	169	194	205	40
Feb.	16	16	21	71	139	184	223	239	244	91
Mar.	22	22	65	136	185	224	238	241	241	149
Apr.	28	58	123	173	211	223	223	213	210	195
May	36	99	149	195	215	218	206	187	181	222
June	53	111	160	199	213	213	196	174	168	231
July	37	98	147	192	211	214	201	183	177	221
Aug.	30	56	119	165	203	216	215	206	203	193
Sep.	23	23	58	126	171	211	227	230	231	144
Oct.	16	16	20	68	132	176	213	229	234	91
Nov.	10	10	10	21	72	122	165	190	200	40
Dec.	7	7	7	7	47	92	135	159	171	23

44 Deg

	N (Shade)	NNE/ NNW	NE/ NW	ENE/ WNW	E/ W	ESE/ WSW	SE/ SW	SSE/ SSW	S	HOR
Jan.	17	17	17	64	138	189	232	248	252	109
Feb.	22	22	43	117	178	227	246	248	247	160
Mar.	27	27	87	162	211	236	238	224	218	206
Apr.	33	66	136	185	221	224	210	183	171	240
May	36	96	162	201	219	211	183	148	132	257
June	47	108	169	205	215	203	171	132	115	261
July	37	96	159	198	215	206	179	144	128	254
Aug.	34	66	132	180	214	215	202	177	165	236
Sep.	28	28	80	152	198	226	227	216	211	199
Oct.	23	23	42	111	171	217	237	240	239	157
Nov.	18	18	18	64	135	186	227	244	248	109
Dec.	15	15	15	49	115	175	217	240	246	89

60 Deg

	N (Shade)	NNE/ NNW	NE/ NW	ENE/ WNW	E/ W	ESE/ WSW	SE/ SW	SSE/ SSW	S	HOR
Jan.	7	7	7	7	46	88	130	152	164	21
Feb.	13	13	13	58	118	168	204	225	231	68
Mar.	20	20	56	125	173	215	234	241	242	128
Apr.	27	59	118	168	206	222	225	220	218	178
May	43	98	149	192	212	220	211	198	194	208
June	58	110	162	197	213	215	202	186	181	217
July	44	97	147	189	208	215	206	193	190	207
Aug.	28	57	114	169	199	214	217	213	211	176
Sep.	21	21	50	115	160	202	222	229	231	123
Oct.	14	14	14	56	111	159	193	215	221	67
Nov.	7	7	7	7	45	86	127	148	160	22
Dec.	4	4	4	4	16	51	76	100	107	9

Solar heat gain factors for glass
Figure 8-6 (continued)

Cooling Load Factors for Glass without Interior Shading, North Latitudes

Fenestration Facing	Room Construction	1	2	3	4	5	6	7	8	9	10	11	12	13	14	15	16	17	18	19	20	21	22	23	24
N (Shaded)	L	0.17	0.14	0.11	0.09	0.08	0.33	0.42	0.48	0.56	0.63	0.71	0.76	0.80	0.82	0.82	0.79	0.79	0.84	0.61	0.48	0.38	0.31	0.25	0.20
	M	0.23	0.20	0.18	0.16	0.14	0.34	0.41	0.46	0.53	0.59	0.65	0.70	0.74	0.75	0.76	0.74	0.75	0.79	0.61	0.50	0.42	0.36	0.31	0.27
	H	0.25	0.23	0.21	0.20	0.19	0.38	0.45	0.49	0.55	0.60	0.65	0.69	0.72	0.72	0.72	0.70	0.70	0.75	0.57	0.46	0.39	0.34	0.31	0.28
NNE	L	0.06	0.05	0.04	0.03	0.03	0.26	0.43	0.47	0.44	0.41	0.40	0.39	0.39	0.38	0.36	0.33	0.30	0.26	0.20	0.16	0.13	0.10	0.08	0.07
	M	0.09	0.08	0.07	0.06	0.06	0.24	0.38	0.42	0.39	0.37	0.37	0.36	0.36	0.36	0.34	0.33	0.30	0.27	0.22	0.18	0.16	0.14	0.12	0.10
	H	0.11	0.10	0.09	0.09	0.08	0.26	0.39	0.42	0.39	0.36	0.35	0.34	0.34	0.33	0.32	0.31	0.28	0.25	0.21	0.18	0.16	0.14	0.13	0.12
NE	L	0.04	0.04	0.03	0.02	0.02	0.23	0.41	0.51	0.51	0.45	0.39	0.36	0.33	0.31	0.28	0.26	0.23	0.19	0.15	0.12	0.10	0.08	0.06	0.05
	M	0.07	0.06	0.06	0.05	0.04	0.21	0.36	0.44	0.45	0.40	0.36	0.33	0.31	0.30	0.28	0.26	0.24	0.21	0.17	0.15	0.13	0.11	0.09	0.08
	H	0.09	0.08	0.08	0.07	0.07	0.23	0.37	0.44	0.44	0.39	0.34	0.31	0.29	0.27	0.26	0.24	0.22	0.20	0.17	0.14	0.13	0.12	0.11	0.10
ENE	L	0.04	0.03	0.03	0.02	0.02	0.21	0.40	0.52	0.57	0.53	0.45	0.39	0.34	0.31	0.28	0.25	0.22	0.18	0.14	0.12	0.09	0.08	0.06	0.05
	M	0.07	0.06	0.05	0.05	0.04	0.20	0.35	0.45	0.49	0.47	0.41	0.36	0.33	0.30	0.28	0.26	0.23	0.20	0.17	0.14	0.12	0.11	0.09	0.08
	H	0.09	0.09	0.08	0.07	0.07	0.22	0.36	0.46	0.49	0.45	0.38	0.33	0.30	0.27	0.25	0.23	0.21	0.19	0.16	0.14	0.13	0.12	0.11	0.10
E	L	0.04	0.03	0.03	0.02	0.02	0.19	0.37	0.51	0.57	0.57	0.50	0.42	0.37	0.32	0.29	0.25	0.22	0.19	0.15	0.12	0.10	0.08	0.06	0.05
	M	0.07	0.06	0.06	0.05	0.05	0.18	0.33	0.44	0.50	0.51	0.46	0.39	0.35	0.31	0.29	0.26	0.23	0.21	0.17	0.15	0.13	0.11	0.10	0.08
	H	0.09	0.09	0.08	0.08	0.07	0.20	0.34	0.45	0.49	0.49	0.43	0.36	0.32	0.29	0.26	0.24	0.22	0.19	0.17	0.15	0.13	0.12	0.11	0.10
ESE	L	0.05	0.04	0.03	0.03	0.02	0.17	0.34	0.49	0.58	0.61	0.57	0.48	0.41	0.36	0.32	0.28	0.24	0.20	0.16	0.13	0.10	0.09	0.07	0.06
	M	0.08	0.07	0.06	0.05	0.05	0.16	0.31	0.43	0.51	0.54	0.51	0.44	0.39	0.35	0.32	0.29	0.26	0.22	0.19	0.16	0.14	0.12	0.11	0.09
	H	0.10	0.09	0.09	0.08	0.08	0.19	0.32	0.43	0.50	0.52	0.49	0.41	0.36	0.32	0.29	0.26	0.24	0.21	0.18	0.16	0.14	0.13	0.12	0.11
SE	L	0.05	0.04	0.04	0.03	0.03	0.13	0.28	0.43	0.55	0.62	0.63	0.57	0.48	0.42	0.37	0.33	0.28	0.24	0.19	0.15	0.12	0.10	0.08	0.07
	M	0.09	0.08	0.07	0.06	0.05	0.14	0.26	0.38	0.48	0.54	0.56	0.51	0.45	0.40	0.36	0.33	0.29	0.25	0.21	0.18	0.16	0.14	0.12	0.10
	H	0.11	0.10	0.10	0.09	0.08	0.17	0.28	0.40	0.49	0.53	0.53	0.48	0.41	0.36	0.33	0.30	0.27	0.24	0.20	0.18	0.16	0.14	0.13	0.12
SSE	L	0.07	0.05	0.04	0.04	0.03	0.06	0.15	0.29	0.43	0.55	0.63	0.64	0.60	0.52	0.45	0.40	0.35	0.29	0.23	0.18	0.15	0.12	0.10	0.08
	M	0.11	0.09	0.08	0.07	0.06	0.08	0.16	0.26	0.38	0.48	0.55	0.57	0.54	0.48	0.43	0.39	0.35	0.30	0.25	0.21	0.18	0.16	0.14	0.12
	H	0.12	0.11	0.11	0.10	0.09	0.12	0.19	0.29	0.40	0.49	0.54	0.55	0.51	0.44	0.39	0.35	0.31	0.27	0.23	0.20	0.18	0.16	0.15	0.13
S	L	0.08	0.07	0.05	0.04	0.04	0.06	0.09	0.14	0.22	0.34	0.48	0.59	0.65	0.65	0.59	0.50	0.43	0.36	0.28	0.22	0.18	0.15	0.12	0.10
	M	0.12	0.11	0.09	0.08	0.07	0.08	0.11	0.14	0.21	0.31	0.42	0.52	0.57	0.58	0.53	0.47	0.41	0.35	0.29	0.25	0.21	0.18	0.16	0.14
	H	0.13	0.12	0.12	0.11	0.10	0.11	0.14	0.17	0.24	0.33	0.43	0.51	0.56	0.55	0.50	0.43	0.37	0.32	0.26	0.22	0.20	0.18	0.16	0.15
SSW	L	0.10	0.08	0.07	0.06	0.05	0.06	0.09	0.11	0.15	0.19	0.27	0.39	0.52	0.62	0.67	0.65	0.58	0.46	0.36	0.28	0.23	0.19	0.15	0.12
	M	0.14	0.12	0.11	0.09	0.08	0.09	0.11	0.13	0.15	0.18	0.25	0.35	0.46	0.55	0.59	0.59	0.53	0.44	0.35	0.30	0.25	0.22	0.19	0.16
	H	0.15	0.14	0.13	0.12	0.11	0.12	0.14	0.16	0.18	0.21	0.27	0.37	0.46	0.53	0.57	0.55	0.49	0.40	0.32	0.26	0.23	0.20	0.18	0.16
SW	L	0.12	0.10	0.08	0.06	0.05	0.06	0.08	0.10	0.12	0.14	0.16	0.24	0.36	0.49	0.60	0.66	0.66	0.58	0.43	0.33	0.27	0.22	0.18	0.14
	M	0.15	0.14	0.12	0.10	0.09	0.09	0.10	0.12	0.13	0.15	0.17	0.23	0.33	0.44	0.53	0.58	0.59	0.53	0.41	0.33	0.28	0.24	0.21	0.18
	H	0.15	0.14	0.13	0.12	0.11	0.12	0.13	0.14	0.16	0.17	0.19	0.25	0.34	0.44	0.52	0.56	0.56	0.49	0.37	0.30	0.25	0.21	0.19	0.17
WSW	L	0.12	0.10	0.08	0.07	0.05	0.06	0.07	0.09	0.10	0.12	0.13	0.17	0.26	0.40	0.52	0.62	0.66	0.61	0.44	0.34	0.27	0.22	0.18	0.15
	M	0.15	0.13	0.12	0.10	0.09	0.09	0.10	0.11	0.12	0.13	0.14	0.17	0.24	0.35	0.46	0.54	0.58	0.55	0.42	0.34	0.28	0.24	0.21	0.18
	H	0.15	0.14	0.13	0.12	0.11	0.11	0.12	0.13	0.14	0.15	0.16	0.19	0.26	0.36	0.46	0.53	0.56	0.51	0.38	0.30	0.25	0.21	0.19	0.17
W	L	0.12	0.10	0.08	0.06	0.05	0.06	0.07	0.08	0.10	0.11	0.12	0.14	0.20	0.32	0.45	0.57	0.64	0.61	0.44	0.34	0.27	0.22	0.18	0.14
	M	0.15	0.13	0.11	0.10	0.09	0.09	0.09	0.10	0.11	0.12	0.13	0.14	0.19	0.29	0.40	0.50	0.56	0.55	0.41	0.33	0.27	0.23	0.20	0.17
	H	0.14	0.13	0.12	0.11	0.10	0.11	0.12	0.12	0.14	0.14	0.15	0.16	0.21	0.30	0.40	0.49	0.54	0.52	0.38	0.30	0.24	0.21	0.19	0.16
WNW	L	0.12	0.10	0.08	0.06	0.05	0.06	0.07	0.09	0.10	0.12	0.13	0.15	0.17	0.26	0.40	0.53	0.63	0.62	0.44	0.34	0.27	0.22	0.18	0.14
	M	0.15	0.13	0.11	0.10	0.09	0.09	0.10	0.11	0.12	0.13	0.14	0.15	0.17	0.24	0.35	0.47	0.55	0.55	0.41	0.33	0.27	0.23	0.20	0.17
	H	0.14	0.13	0.12	0.11	0.10	0.11	0.12	0.13	0.14	0.15	0.16	0.17	0.18	0.25	0.36	0.46	0.53	0.52	0.38	0.30	0.24	0.20	0.18	0.16
NW	L	0.11	0.09	0.08	0.06	0.05	0.06	0.08	0.10	0.12	0.14	0.16	0.17	0.19	0.23	0.33	0.47	0.59	0.60	0.42	0.33	0.26	0.21	0.17	0.14
	M	0.14	0.12	0.11	0.09	0.08	0.09	0.10	0.11	0.13	0.15	0.16	0.17	0.18	0.21	0.30	0.42	0.51	0.54	0.39	0.32	0.26	0.22	0.19	0.16
	H	0.14	0.12	0.11	0.10	0.10	0.10	0.12	0.13	0.15	0.16	0.18	0.18	0.19	0.22	0.30	0.41	0.50	0.51	0.36	0.29	0.23	0.20	0.17	0.15
NNW	L	0.12	0.09	0.08	0.06	0.05	0.07	0.11	0.14	0.18	0.22	0.25	0.27	0.29	0.30	0.33	0.44	0.57	0.62	0.44	0.33	0.26	0.21	0.17	0.14
	M	0.15	0.13	0.11	0.10	0.09	0.10	0.12	0.15	0.18	0.21	0.23	0.26	0.27	0.28	0.31	0.39	0.51	0.56	0.41	0.33	0.27	0.23	0.20	0.17
	H	0.14	0.13	0.12	0.11	0.10	0.12	0.15	0.17	0.20	0.23	0.25	0.26	0.28	0.28	0.31	0.38	0.49	0.53	0.38	0.30	0.25	0.21	0.18	0.16
HOR.	L	0.11	0.09	0.07	0.06	0.05	0.07	0.14	0.24	0.36	0.48	0.58	0.66	0.72	0.74	0.73	0.67	0.59	0.47	0.37	0.29	0.24	0.19	0.16	0.13
	M	0.16	0.14	0.12	0.11	0.09	0.11	0.16	0.24	0.33	0.43	0.52	0.59	0.64	0.67	0.66	0.62	0.56	0.47	0.38	0.32	0.28	0.24	0.21	0.18
	H	0.17	0.16	0.15	0.14	0.13	0.15	0.20	0.28	0.36	0.45	0.52	0.59	0.62	0.64	0.62	0.58	0.51	0.42	0.35	0.29	0.26	0.23	0.21	0.19

Cooling load factors for glass
Figure 8-7

Cooling Load Factors for Glass without Interior Shading, North Latitudes
(All Room Constructions)

Fenestration Facing	Solar Time, hr																							
	1	2	3	4	5	6	7	8	9	10	11	12	13	14	15	16	17	18	19	20	21	22	23	24
N	0.08	0.07	0.06	0.06	0.07	0.73	0.66	0.65	0.73	0.80	0.86	0.89	0.89	0.86	0.82	0.75	0.78	0.91	0.24	0.18	0.15	0.13	0.11	0.10
NNE	0.03	0.03	0.02	0.02	0.03	0.64	0.77	0.62	0.42	0.37	0.37	0.37	0.36	0.35	0.32	0.28	0.23	0.17	0.08	0.07	0.06	0.05	0.04	0.04
NE	0.03	0.02	0.02	0.02	0.02	0.56	0.76	0.74	0.58	0.37	0.29	0.27	0.26	0.24	0.22	0.20	0.16	0.12	0.06	0.05	0.04	0.04	0.03	0.03
ENE	0.03	0.02	0.02	0.02	0.02	0.52	0.76	0.80	0.71	0.52	0.31	0.26	0.24	0.22	0.20	0.18	0.15	0.11	0.06	0.05	0.04	0.04	0.03	0.03
E	0.03	0.02	0.02	0.02	0.02	0.47	0.72	0.80	0.76	0.62	0.41	0.27	0.24	0.22	0.20	0.17	0.14	0.11	0.06	0.05	0.05	0.04	0.03	0.03
ESE	0.03	0.03	0.02	0.02	0.02	0.41	0.67	0.79	0.80	0.72	0.54	0.34	0.27	0.24	0.21	0.19	0.15	0.12	0.07	0.06	0.05	0.04	0.04	0.03
SE	0.03	0.03	0.02	0.02	0.02	0.30	0.57	0.74	0.81	0.79	0.68	0.49	0.33	0.28	0.25	0.22	0.18	0.13	0.08	0.07	0.06	0.05	0.04	0.04
SSE	0.04	0.03	0.03	0.03	0.02	0.12	0.31	0.54	0.72	0.81	0.81	0.71	0.54	0.38	0.32	0.27	0.22	0.16	0.09	0.08	0.07	0.06	0.05	0.04
S	0.04	0.04	0.03	0.03	0.03	0.09	0.16	0.23	0.38	0.58	0.75	0.83	0.80	0.68	0.50	0.35	0.27	0.19	0.11	0.09	0.08	0.07	0.06	0.05
SSW	0.05	0.04	0.04	0.03	0.03	0.09	0.14	0.18	0.22	0.27	0.43	0.63	0.78	0.84	0.80	0.66	0.46	0.25	0.13	0.11	0.09	0.08	0.07	0.06
SW	0.05	0.05	0.04	0.04	0.03	0.07	0.11	0.14	0.16	0.19	0.22	0.38	0.59	0.75	0.83	0.81	0.69	0.45	0.16	0.12	0.10	0.09	0.07	0.06
WSW	0.05	0.05	0.04	0.04	0.03	0.07	0.10	0.12	0.14	0.16	0.17	0.23	0.44	0.64	0.78	0.84	0.78	0.55	0.16	0.12	0.10	0.09	0.07	0.06
W	0.05	0.05	0.04	0.04	0.03	0.06	0.09	0.11	0.13	0.15	0.16	0.17	0.31	0.53	0.72	0.82	0.81	0.61	0.16	0.12	0.10	0.08	0.07	0.06
WNW	0.05	0.05	0.04	0.03	0.03	0.07	0.10	0.12	0.14	0.16	0.17	0.18	0.22	0.43	0.65	0.80	0.84	0.66	0.16	0.12	0.10	0.08	0.07	0.06
NW	0.05	0.04	0.04	0.03	0.03	0.07	0.11	0.14	0.17	0.19	0.20	0.21	0.22	0.30	0.52	0.73	0.82	0.69	0.16	0.12	0.10	0.08	0.07	0.06
NNW	0.05	0.05	0.04	0.03	0.03	0.11	0.17	0.22	0.26	0.30	0.32	0.33	0.34	0.34	0.39	0.61	0.82	0.76	0.17	0.12	0.10	0.08	0.07	0.06
HOR.	0.06	0.05	0.04	0.04	0.03	0.12	0.27	0.44	0.59	0.72	0.81	0.85	0.85	0.81	0.71	0.58	0.42	0.25	0.14	0.12	0.10	0.08	0.07	0.06

Cooling load factors for glass
Figure 8-7 (continued)

You can calculate most heat loads from building occupants by using this formula and common sense to determine how long the building or room is occupied. For example, if you have an office with 50 employees who are in the office for almost the entire eight-hour work day, you'll surely have to compute the heat load from the occupants. But for a small office with one or two clerical people and a sales staff that's usually out on the road, you can ignore the heat gain due to occupants. It'll be very small compared to the building heat gain. Consult a heating engineer if you feel your job is out of the ordinary and needs a detailed analysis of conditions.

Equipment— Like people, equipment gives off heat. And like occupant heat gain, you only have to figure equipment heat gain if it's a significant amount.

A general rule of thumb is to allow 3.4 Btu for each watt used by any equipment operating for a significant length of time. For example, we'll have a coffee machine, video games, warm fast-food machines, counter equipment, and neon signs in the convenience store.

This 3.4 Btu allowance doesn't include any heat given off by motors. We can ignore the heat gain from small motors used to run fans inside equipment, for example. And most new HVAC installations have the compressor and other large motors outside the air conditioned space, if possible, to avoid added heat gain. (Don't overlook this in your design work!) But if you do have a job where there are motors in the air conditioned space, you can compute their heat gain using Figure 8-9.

Rates of Heat Gain from Occupants of Conditioned Spaces[a]

Degree of Activity	Typical Application	ADULT MALE Total (q_s/person + q_l/person)		ADJUSTED GROUP[b] Total (q_s/person + q_l/person)		ADJUSTED GROUP[b] Sensible q_s/person		ADJUSTED GROUP[b] Latent q_l/person	
		Watts	Btu/hr	Watts	Btu/hr	Watts	Btu/hr	Watts	Btu/hr
Seated at rest	Theater, movie	115	400	100	350	60	210	40	140
Seated, very light work writing	Offices, hotels, apts	140	480	120	420	65	230	55	190
Seated, eating	Restaurant[c]	150	520	170	580[c]	75	255	95	325
Seated, light work, typing	Offices, hotels, apts	185	640	150	510	75	255	75	255
Standing, light work or walking slowly	Retail Store, bank	235	800	185	640	90	315	95	325
Light bench work	Factory	255	880	230	780	100	345	130	435
Walking, 3 mph, light machine work	Factory	305	1040	305	1040	100	345	205	695
Bowling[d]	Bowling alley	350	1200	280	960	100	345	180	615
Moderate dancing	Dance hall	400	1360	375	1280	120	405	255	875
Heavy work, heavy machine work, lifting	Factory	470	1600	470	1600	165	565	300	1035
Heavy work, athletics	Gymnasium	585	2000	525	1800	185	635	340	1165

[a]Note: Tabulated values are based on 78 F room dry-bulb temperature. For 80 F room dry-bulb, the total heat remains the same, but the sensible heat value should be decreased by approximately 8% and the latent heat values increased accordingly.

[b]Adjusted total heat gain is based on normal percentage of men, women, and children for the application listed, with the postulate that the gain from an adult female is 85% of that for an adult male, and that the gain from a child is 75% of that for an adult male.

[c]Adjusted total heat value for eating in a restaurant, includes 60 Btu/hr for food per individual (30 Btu sensible and 30 Btu latent).

[d]For bowling figure one person per alley actually bowling, and all others as sitting (400 Btu/hr) or standing and walking slowly (790 Btu/hr).

Also refer to Tables 4 and 5, Chapter 8, 1977 ASHRAE Handbook of Fundamentals.

Courtesy: American Society of Heating, Refrigerating and Air-Conditioning Engineers, Inc. (ASHRAE)
1981 Fundamentals Handbook, all rights reserved.

Heat gain from people
Figure 8-8

Our store has a large wall cooler for milk, soda and beer and a large wall freezer. We'll assume the compressors and motors are located in the storage area, Zone 2, which won't be air conditioned. The equipment we do have in the store gives off 3,000 watts.

3,000 x 3.4 Btu/hr = 10,200 Btu/hr

Zone 1: Heat gain summary

Outside walls	=	5,652 Btu/hr
Glass at 5 p.m.	=	30,410
Roof at 5 p.m.	=	26,934
People	=	6,400
Lights	=	20,400
Equipment	=	10,200
Total		99,996 Btu/hr

Outside Air Loads
So far we have our Zone 1 heating load figured at 152,736 Btu/hr. The Zone 1 cooling load is 98,060

Btu/hr. But these aren't the only loads we need to figure. The codes require that our building have a minimum of 7 cfm of outside air per person per hour. (See Figure 9-1 in the next chapter). Our heating and cooling loads will increase accordingly.

Heating outside air— To figure the increase for heating, we'll use our math factor of 1.1 (the amount of Btu's needed to heat one cubic foot of air one degree F) to value the amount of heat needed for the outside air. We've determined our store will have an average of 10 occupants.

Outside air heating load = cfm x number of people x 1.1 x (Ti-To)
= 7 x 10 x 1.1 x (67 - [-12])
= 6,083 Btu/hr

The total heating load for Zone 1 is 152,736 plus 6,083, or 158,819 Btu/hour.

Heat Gain From Typical Electric Motors

Motor Nameplate or Rated Horsepower	Motor Type	Nominal rpm	Full Load Motor Efficiency In Percent	Location of Motor and Driven Equipment with Respect to Conditioned Space or Air Stream		
				A Motor in, Driven Equipment in Btu/hr	**B** Motor out, Driven Equipment in Btu/hr	**C** Motor in, Driven Equipment Out Btu/hr
0.05	Shaded Pole	1500	35	360	130	240
0.08	Shaded Pole	1500	35	580	200	380
0.125	Shaded Pole	1500	35	900	320	590
0.16	Shaded Pole	1500	35	1160	400	760
0.25	Split Phase	1750	54	1180	640	540
0.33	Split Phase	1750	56	1500	840	660
0.50	Split Phase	1750	60	2120	1270	850
0.75	3-Phase	1750	72	2650	1900	740
1	3-Phase	1750	75	3390	2550	850
1.5	3-Phase	1750	77	4960	3820	1140
2	3-Phase	1750	79	6440	5090	1350
3	3-Phase	1750	81	9430	7640	1790
5	3-Phase	1750	82	15500	12700	2790
7.5	3-Phase	1750	84	22700	19100	3640
10	3-Phase	1750	85	29900	24500	4490
15	3-Phase	1750	86	44400	38200	6210
20	3-Phase	1750	87	58500	50900	7610
25	3-Phase	1750	88	72300	63600	8680
30	3-Phase	1750	89	85700	76350	9440
40	3-Phase	1750	89	114000	102000	12600
50	3-Phase	1750	89	143000	127000	15700
60	3-Phase	1750	89	172000	153000	18900
75	3-Phase	1750	90	212000	191000	21200
100	3-Phase	1750	90	283000	255000	28300
125	3-Phase	1750	90	353000	318000	35300
150	3-Phase	1750	91	420000	382000	37800
200	3-Phase	1750	91	559000	509000	50300
250	3-Phase	1750	91	699000	636000	62900

Courtesy: American Society of Heating, Refrigerating and Air-Conditioning Engineers, Inc. (ASHRAE) 1981 Fundamentals Handbook, all rights reserved.

Electric motor heat gain
Figure 8-9

Cooling outside air— For the increase in cooling load, we'll use a psychrometric chart. The chart shows the relationship between properties of air, such as humidity, dry bulb temperature, wet bulb temperature and *enthalpy (h)* — the amount of Btu's per pound of dry air at a given temperature and humidity. If we know any two factors, we can find the unknowns.

We'll determine the temperature of the air entering the cooling unit using the temperature of both the return air and the 70 cfm of outside air. This air mix will have to be cooled to the supply temperature (55°F is standard) before it's returned to the building.

We'll also use both the return air and the outside

supply to determine the volume of air mix. Both the volume and temperature can be figured using the same steps, and applying the psychrometric chart. The following example will make this clearer.

The total volume of air (cfm) that the rooftop units need to circulate depends on the sensible heat gain of the building and the temperature difference between the supply air (Ts) leaving the cooling unit and the building design dry bulb temperature (Ti). The supply air temperature should be between 52°F and 60°F. We'll use 55°F, typical for most commercial systems. Our building data gives us a cooling design temperature of 78°F. The sensible heat gain is the total heat gain minus the latent heat load figured earlier for ten occupants.

$$\text{Total cfm} = \frac{\text{sensible heat gain}}{(Ti\text{-}Ts) \times 1.1}$$

$$= \frac{(99{,}996 - 3{,}250)}{(78\text{-}55)\, 1.1}$$

$$= 3{,}824$$

But 70 cfm of this air will be displaced by outside air. So we'll subtract it: 3,824 minus 70 equals 3,754 cfm return air.

The conditions of the outside air are 89°Fdb and 74°Fwb (from Figure 5-10, Chapter 5). The room design conditions are 78°Fdb and 50% rh. We plotted this data on the psychrometric chart as points 1 and 2, then drew a slanted line between them (Figure 8-10).

To find point 3, we first have to find the dry bulb temperature of the air mixture. The formula is:

$$Ti + \frac{(To\text{-}Ti) \times \text{cfm outside air}}{\text{cfm total}}$$

$$78 + \frac{(89\text{-}78) \times 70}{3824} = 78.2$$

On the chart, we draw a vertical line from 78.17°F until it intersects the line from points 1 and 2. This is point 3. This point is very close to point 2 on the chart because the outside air replacement is relatively small compared to the total air supply. At point 3, we find the wet bulb temp is 65.5°Fwb.

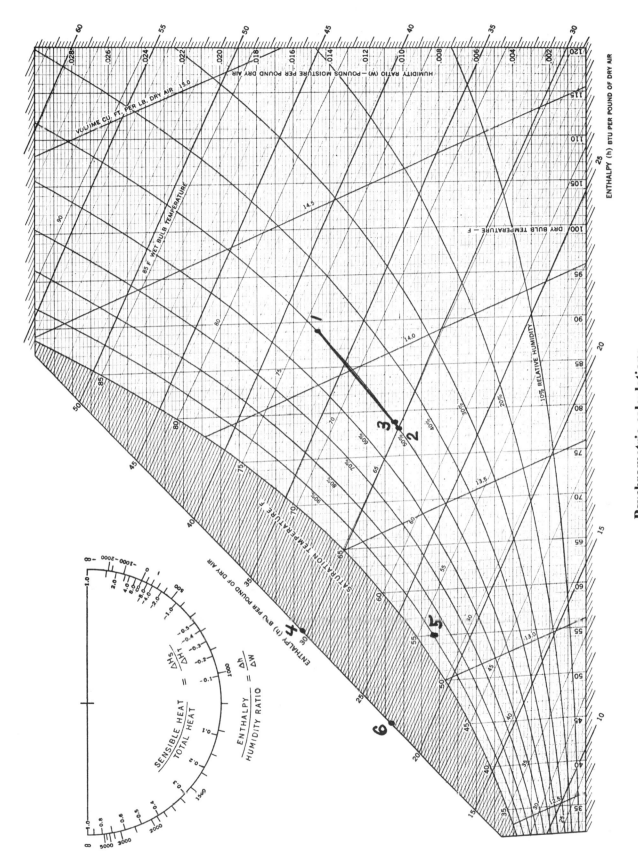

**Psychrometric calculations
Figure 8-10**

The enthalpy is 30.4 Btu/lb of dry air. (See the enthalpy scale on the diagonal line of the chart, point 4.) This is our *inlet enthalpy* (hi).

The rooftop unit must cool 3,824 cfm from 78.2°Fdb and 66°Fwb to 55°Fdb and 54°Fwb, our supply air temperature. The air supply from the coil typically has a relative humidity of 90%. We can plot this supply air data (55°Fdb and 90% rh) as point 5 on the chart. The enthalpy of the supply air, reading to the left at point 6, is 22.5 Btu/lb of dry air. We'll use the inlet enthalpy, outlet enthalpy, total cfm, and math factors to give us the total load of both building and outside air.

Total load

= cfm x 60 min/hr x 0.075 cu ft/lb x (hi-ho)

Total load

= 3,824 x 4.5 x (30.4 - 22.5) = 135,943 Btu/hr

This is our total cooling load, including latent heat. To select the right size unit, we also have to know the sensible heat load.

Sensible heat load

= cfm x 1.1 x (Tmx - Ts)
= 3,824 x 1.1 x (78.2 - 55) = 97,588 Btu/hr

Sensible Heat Ratio for the Rooftop Unit
Using the sensible heat load and the total load, we can find the *sensible heat ratio*. This ratio is necessary to make sure the cooling unit we select will handle both latent and sensible loads.

$$\frac{97,588}{135,943} = .72 = 72\%$$

Cooling and Heating Unit Selection
We can now choose a unit based on the manufacturer's data. Here's the data we have for cooling:

cfm	=	3,824
Outside air	=	89° Fdb, 74° Fwb
Air entering coil	=	78.2° Fdb, 66° Fwb
Air leaving coil	=	55° Fdb, 54° Fwb
Sensible heat gain	=	97,588 Btu/hr
Total heat gain	=	135,943 Btu/hr
Sensible heat ratio	=	.71

If we pick a unit with a sensible heat capacity high enough to meet our cooling needs, the unit's total capacity may exceed that amount. This is all right as long as we meet the sensible capacity and meet or exceed the total.

Also, most manufacturer's data indicates that a heat loss must be included for the motor in the unit. As a rule, figure 3,400 Btu/hr per one horsepower of motor. If you're unsure of how much is needed, check with the manufacturer.

The heating capacity of our unit must meet or equal the building heat loss plus the Btu's needed to heat the outside air (page 155).

Heat loss	152,736
Outside load	+ 6,083
Total	158,819 Btu/hr

Using all of our data, we can pick the best unit for the job from the manufacturer's tables. Look at Figure 8-11. Carrier model number 48DP014, at 95°F outside, 67°Fwb and 3,750 cfm will produce 147,000 Btu/hr total cooling and 103,000 Btu/hr sensible cooling. The heating capacity is 183,000 Btu/hr. It will work well for our job.

We also have to evaluate some other items. For this job, for example, we'll choose a rooftop unit with gas heating and electric cooling, since this reflects the most economical fuel source according to utility data. The best location for the unit is the center of the roof. This will allow the shortest duct runs to all parts of Zone 1.

Zone 2 needs only heating, and can be heated with a heating-only rooftop unit, or unit heater. You'll learn how to select these units in our next example, the auto repair shop.

Zone 3 can be heated economically with small electric wall heaters in the office and toilet or by extending ductwork from the rooftop unit serving Zone 1. Of course if you use ductwork from Zone 1, you'll have to calculate the cooling loads for Zone 3 and add to the rooftop unit load.

To size each register and return grille, use the same method of figuring air volume as we used in the residential section. Then size the ductwork, also using the residential method. Note that for this particular building, we don't need to locate the outlets near the floor. People won't be sitting near the windows here, as they would be in their homes.

Performance data

ARI CAPACITY RATINGS

UNIT	TOTAL KW	COOLING CAP. NET BTU/HR.	STANDARD CFM	EER
48DP,DR012	14.5	120,000	4000	8.3
48DP,DR014	18.6	149,000	5000	8.0
48DP,DR016	21.7	180,000	6000	8.3

NOTE: Capacities for 48DP,DR014,016 are not within scope limitations of ARI Certification Program.

HEATING CAPACITIES

UNIT MODEL	HEATING INPUT (Btuh)	BONNET CAPACITY (Btuh)
48DP012	154,000	123,000
48DR012	231,000	183,000
48DP014	154,000	123,000
➤ 48DR014	231,000	183,000 ◄
48DP016	231,000	183,000
48DR016	270,000	216,000

NOTES:

1. Ratings are approved for altitudes up to 2,000 feet. At altitudes over 2,000 ft, ratings are 4% less for each 1,000 ft above sea level.

2. At altitudes up to 2,000 ft, the following formula may be used to calculate air temperature rise:

$$\Delta t = \frac{\text{bonnet capacity}}{1.09 \times \text{air quantity}}$$

3. At altitudes above 2,000 ft, the following formula may be used:

$$\Delta t = \frac{\text{bonnet capacity}}{(.24 \times \text{specific weight of air} \times 60)(\text{air quantity})}$$

4. Maximum allowable gas pressure is 14.0 in. wg. Minimum allowable gas pressure for full rated input is 5.0 in. wg.

5. Unit may be converted for 100% LP gas, using Carrier factory parts. Unit will maintain AGA certification if conversion is performed, as directed, and conversion nameplate is attached to unit adjacent to existing unit nameplate.

6. Minimum allowable temperature of mixed air entering the heat exchanger during half-rate (first-stage) operation is 35 F. There is no minimum mixture temperature limitation during full-rate operation.

7. Bonnet capacity rated in accordance with ANSIZ 21.47 — 1973.

GROSS COOLING CAPACITIES
48DP,DR012

Temp (F) Air Ent Cond (Edb)		Evap Air — Cfm/BF								
		3000/.153			4000/.187			5000/.216		
		Evap Air — Ewb (F)								
		72	67	62	72	67	62	72	67	62
85	TC	132	123	110	136	128	118	138	131	124
	SHC	65	81	95	70	91	109	75	99	124
	Kw	12.2	11.7	11.1	12.4	11.8	11.4	12.5	12.1	11.7
95	TC	128	116	100	133	123	110	135	126	117
	SHC	63	79	91	70	90	106	75	99	117
	Kw	13.0	12.3	11.9	13.2	12.3	11.9	13.4	12.9	12.3
105	TC	122	107	90	127	114	100	130	119	109
	SHC	61	75	85	68	87	100	73	97	109
	Kw	13.6	13.1	12.6	13.9	13.1	13.1	14.1	13.5	13.1
115	TC	113	96	81	120	103	92	123	109	101
	SHC	58	71	80	66	83	92	72	93	101
	Kw	14.4	13.8	13.3	14.6	14.1	13.8	14.8	14.2	13.9

GROSS COOLING CAPACITIES
48DP,DR014

Temp (F) Air Ent Cond (Edb)		Evap Air — Cfm/BF								
		3750/.058			5000/.090			6250/.125		
		Evap Air — Ewb (F)								
		72	67	62	72	67	62	72	67	62
85	TC	169	156	140	174	163	149	180	169	163
	SHC	85	106	125	92	121	145	105	143	163
	Kw	15.0	14.3	13.5	15.4	14.8	14.1	15.6	15.1	14.8
95	TC	162	147	128	168	154	140	175	162	156
	SHC	82	103	120	92	117	140	106	142	156
	Kw	16.0	15.1	14.4	16.4	15.6	14.9	16.7	16.0	15.7
105	TC	153	135	115	159	144	130	167	152	148
	SHC	79	98	113	88	114	130	105	140	148
	Kw	16.8	16.0	15.3	17.3	16.4	15.9	17.7	16.8	16.6
115	TC	143	121	103	149	129	119	156	140	138
	SHC	76	98	103	85	109	119	102	135	138
	Kw	17.6	16.9	16.4	18.1	17.3	16.9	18.5	17.6	17.5

48DP,DR016

Temp (F) Air Ent Cond (Edb)		Evap Air — Cfm/BF								
		4500/.086			6000/.112			7500/.136		
		Evap Air — Ewb (F)								
		72	67	62	72	67	62	72	67	62
85	TC	203	188	169	212	196	180	224	208	196
	SHC	98	122	145	120	138	167	120	158	186
	Kw	16.3	15.5	14.6	16.1	16.0	14.5	17.0	16.3	15.6
95	TC	196	178	154	204	187	168	216	199	188
	SHC	95	119	138	103	135	161	117	155	178
	Kw	17.4	16.4	15.8	17.2	16.3	15.5	18.2	17.3	16.7
105	TC	186	164	138	194	175	156	205	188	178
	SHC	91	113	130	101	130	149	115	152	168
	Kw	18.5	17.6	16.9	18.3	17.8	16.8	19.3	18.3	17.8
115	TC	175	149	126	183	158	144	194	172	166
	SHC	87	107	120	97	124	136	112	147	157
	Kw	19.5	18.8	18.2	19.4	18.5	18.8	20.4	19.3	19.1

BF — Bypass Factor
Edb — Entering Dry-Bulb
Ewb — Entering Wet-Bulb
Kw — Compressor Motor Power Input
SHC — Sensible Heat Capacity (1000 Btuh) Gross
TC — Total Capacity (1000 Btuh) Gross

NOTES:

1. Direct interpolation is permissible. Do not extrapolate.
2. The following formulas may be used:

$$t_{ldb} = t_{edb} - \frac{\text{sensible capacity (Btuh)}}{1.09 \times \text{cfm}}$$

t_{lwb} = Wet-bulb temperature corresponding to enthalpy of air leaving evaporator coil (h_{lwb}).

$$h_{lwb} = h_{ewb} - \frac{\text{total capacity (Btuh)}}{4.5 \times \text{cfm}}$$

Where:
h_{ewb} = Enthalpy of air entering evaporator coil.

3. SHC is based on 80 F edb temperature of air entering evaporator coil.
   Below 80 F edb, subtract (corr factor x cfm) from SHC.
   Above 80 F edb, add (corr factor x cfm) to SHC.

BYPASS FACTOR (BF)	ENTERING AIR DRY-BULB TEMP (F)					
	79 / 81	78 / 82	77 / 83	76 / 84	75 / 85	under 75 / over 85
	Correction Factor					
.05	1.04	2.07	3.11	4.14	5.18	Use formula shown below.
.10	.98	1.96	2.94	3.92	4.90	
.20	.87	1.74	2.62	3.49	4.36	
.30	.76	1.53	2.29	3.05	3.82	

Interpolation is permissible.
Correction Factor = 1.09 x (1 - BF) x (edb - 80).

Courtesy: Carrier Corp.

Rooftop unit data
Figure 8-11

Fan performance

UNIT MODEL 48DP,DR	CFM	0.20 Rpm	0.20 Bhp	0.40 Rpm	0.40 Bhp	0.60 Rpm	0.60 Bhp	0.80 Rpm	0.80 Bhp	1.00 Rpm	1.00 Bhp	1.20 Rpm	1.20 Bhp	1.40 Rpm	1.40 Bhp	1.60 Rpm	1.60 Bhp	1.80 Rpm	1.80 Bhp	2.00 Rpm	2.00 Bhp
012	3000	634	0.39	737	0.51	830	0.64	914	0.78	992	0.92	1066	1.08	1139	1.26	—	—	—	—	—	—
	3200	664	0.45	760	0.58	851	0.72	934	0.86	1009	1.01	1081	1.17	1150	1.34	1218	1.53	—	—	—	—
	3400	695	0.53	785	0.66	874	0.80	954	0.95	1028	1.11	1097	1.27	1164	1.44	1228	1.63	1293	1.83	—	—
	3600	726	0.61	811	0.75	897	0.90	974	1.05	1048	1.22	1115	1.38	1180	1.56	1242	1.74	1303	1.94	1364	2.15
	3800	750	0.70	839	0.85	921	1.00	997	1.16	1069	1.33	1135	1.50	1198	1.68	1260	1.87	1318	2.07	1376	2.23
	4000	792	0.80	867	0.96	945	1.12	1020	1.28	1089	1.46	1156	1.64	1217	1.82	1273	2.01	1334	2.21	1389	2.42
	4200	825	0.91	896	1.07	970	1.24	1043	1.41	1111	1.59	1176	1.78	1237	1.97	1294	2.18	1361	2.38	1405	2.57
	4400	859	1.03	927	1.20	997	1.37	1067	1.56	1135	1.74	1197	1.93	1258	2.13	1315	2.33	1369	2.53	1423	2.74
	4600	893	1.16	957	1.39	1024	1.52	1091	1.71	1157	1.90	1219	2.09	1279	2.30	1335	2.51	1380	2.71	1440	2.92
	4800	928	1.30	989	1.48	1052	1.67	1117	1.87	1181	2.07	1244	2.27	1300	2.47	1356	2.69	1410	2.91	1460	3.12
	5000	962	1.45	1020	1.64	1081	1.84	1143	2.04	1205	2.25	1265	2.46	1322	2.67	1376	2.89	1430	3.12	1480	3.34
014	3750	740	0.78	856	0.95	957	1.11	1049	1.27	1132	1.44	1202	1.62	1255	1.81	1282	2.00	—	—	—	—
	4000	773	0.92	885	1.11	983	1.27	1074	1.44	1155	1.61	1230	1.79	1293	1.99	1340	2.19	1366	2.39	—	—
	4250	807	1.07	914	1.26	1011	1.44	1099	1.61	1180	1.80	1254	1.98	1322	2.17	1379	2.38	1423	2.59	1450	2.80
	4500	842	1.23	945	1.43	1039	1.63	1124	1.80	1204	1.99	1278	2.18	1346	2.38	1409	2.59	1462	2.81	1504	3.03
	4750	878	1.40	978	1.62	1069	1.83	1152	2.01	1230	2.21	1303	2.41	1371	2.61	1435	2.82	1493	3.04	1544	3.28
	5000	915	1.60	1011	1.82	1099	2.05	1181	2.25	1256	2.44	1329	2.65	1396	2.86	1459	3.07	1519	3.29	—	—
	5250	952	1.83	1045	2.06	1129	2.30	1210	2.52	1283	2.71	1355	2.93	1422	3.15	1485	3.36	1544	3.58	—	—
	5500	990	2.06	1079	2.30	1161	2.55	1240	2.78	1313	3.00	1381	3.20	1448	3.43	1510	3.66	—	—	—	—
	5750	1028	2.31	1114	2.56	1194	2.82	1270	3.06	1342	3.30	1408	3.51	1473	3.73	1536	3.97	—	—	—	—
	6000	1067	2.59	1149	2.85	1228	3.11	1300	3.37	1372	3.62	1437	3.85	1499	4.07	—	—	—	—	—	—
	6250	1105	2.88	1184	3.15	1261	3.42	1332	3.69	1402	3.96	1466	4.20	1529	4.43	—	—	—	—	—	—
016	4500	801	1.05	890	1.26	971	1.46	1050	1.67	1125	1.88	1200	2.12	1275	2.39	1349	2.70	1421	3.03	1490	3.39
	4800	843	1.25	928	1.47	1006	1.68	1081	1.90	1153	2.13	1223	2.36	1293	2.62	1364	2.92	1433	3.24	1501	3.59
	5100	885	1.47	968	1.70	1043	1.93	1114	2.16	1183	2.40	1250	2.64	1316	2.90	1382	3.18	1448	3.49	1514	3.86
	5400	927	1.71	1008	1.95	1080	2.20	1148	2.44	1214	2.69	1279	2.94	1342	3.20	1403	3.47	1466	3.77	1529	4.16
	5700	971	1.98	1049	2.24	1118	2.50	1184	2.75	1247	3.01	1309	3.28	1370	3.54	1429	3.81	1487	4.11	1547	4.43
	6000	1016	2.28	1091	2.55	1158	2.83	1222	3.10	1282	3.36	1342	3.64	1401	3.92	1458	4.20	1514	4.50	1569	4.80
	6300	1059	2.60	1133	2.89	1198	3.17	1259	3.46	1318	3.74	1375	4.02	1432	4.31	1487	4.61	1542	4.91	—	—
	6600	1104	2.96	1174	3.26	1239	3.56	1297	3.86	1355	4.15	1410	4.45	1464	4.75	1518	5.06	—	—	—	—
	6900	1150	3.35	1218	3.67	1281	3.98	1339	4.29	1394	4.60	1447	4.91	1500	5.22	1550	5.54	—	—	—	—
	7200	1194	3.77	1260	4.10	1321	4.42	1377	4.75	1431	5.07	1482	5.40	—	—	—	—	—	—	—	—
	7500	1238	4.23	1303	4.57	1362	4.91	1418	5.25	1470	5.59	—	—	—	—	—	—	—	—	—	—

EXTERNAL STATIC PRESSURE (in. wg.)

☐ Standard motor and drive.

☐ Optional motor and drive (016 unit offered with optional drive only).

▓ Field-supplied motor and/or drive.

Rpm — Revolutions per minute
Bhp — Brake horsepower input to fan

Conversion — Bhp to watts

$$\text{Watts input} = \frac{\text{Bhp} \times 746}{\text{Motor efficiency}}$$

Motor efficiencies:
1 hp — 0.81
1-1/2 hp — 0.78
2 hp — 0.83
3 hp — 0.84

NOTES:
1. Fan performance is based on wet coils, clean filters and casing losses.
2. Motors have service factors as shown in Physical Data table on page 8.
3. Condenser fan watts input:
   48DP,DR012 1385 watts
   48DP,DR014 1400 watts
   48DP,DR016 2672 watts

Courtesy: Carrier Corp.

Rooftop unit data
Figure 8-11 (continued)

Electrical data

UNIT MODEL VOLTS-PH-HZ	VOLTAGE RANGE		COMPR*		OUTDOOR FAN MOTOR		INDOOR FAN MOTOR		COMBUSTION FAN MOTOR	POWER SUPPLY	
	Min	Max	RLA	LRA	Qty	FLA	Hp	FLA	FLA	Min Ckt Amps	MOCP (Amps)
48DP,DR012 208/230-3-60	187	253	24.0 (ea)	136 (ea)	2	4.6	1.0 1.5	3.45 5.50	.90	70 70	80 80
48DP,DR012 460-3-60	414	508	10.5 (ea)	49 (ea)	2	2.1	1.0 1.5	1.60 2.60	.57	30 30	35 35
48DP,DR012 575-3-60	518	632	8.3 (ea)	41 (ea)	2	1.8	1.0 1.5	1.35 1.70	.90	25 25	25 25
48DP,DR014 208/230-3-60	187	253	27.0 (ea)	137 (ea)	2	4.6	2.0 3.0	7.40 10.40	.90	78 78	90 90
48DP,DR014 460-3-60	414	508	11.6 (ea)	69 (ea)	2	2.3	2.0 3.0	3.30 4.70	.57	35 35	40 40
48DP,DR014 575-3-60	518	632	9.3 (ea)	55 (ea)	2	1.8	2.0 3.0	2.30 3.80	.90	30 30	30 30
48DP,DR016 208/230-3-60	187	253	61.6	266	2	7.7	3.0	10.40	.90	105	125
48DP,DR016 460-3-60	414	508	32.0	120	2	3.3	3.0	4.70	.57	55	60
48DP,DR016 575-3-60	518	632	25.6	96	2	2.6	3.0	3.80	.90	50	45

FLA — Full Load Amps **MOCP** — Maximum Overcurrent Protection *48DP,DR012 and 014 units contain 2 compressors. Values apply
LRA — Locked Rotor Amps **RLA** — Rated Load Amps   to each.

Typical piping and wiring

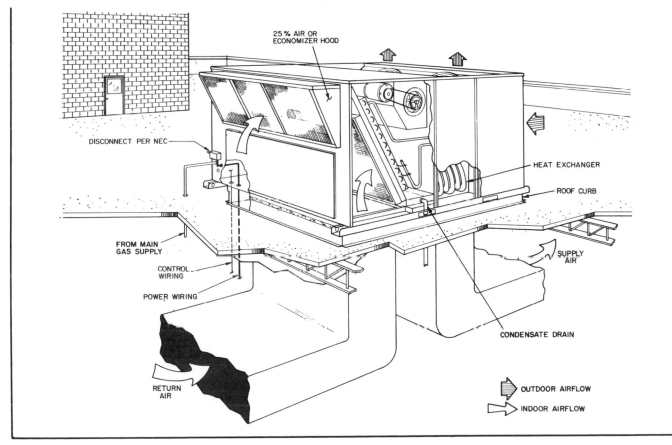

Rooftop unit data
Figure 8-11 (continued)

Instead, we can put the outlets and returns in the ceiling. The ductwork will run between the roof and the dropped ceiling. These installations are quick and easy if you use a sheet metal main duct with flexible circular metal ducts running to the individual outlets. Use insulated ductwork for best efficiency.

The remaining steps, like wiring, piping and the like, can be done as described in Chapter 6. Remember to check your design carefully to make sure it's complete.

Auto Repair Shop

This building is similar in size and construction to the convenience store. But as you'll see, the way a building is used can make major changes in design conditions. Figure 8-12 shows the floor plan and elevation for the shop.

We'll group the rooms with similar conditions into zones. The office, toilet and waiting room (Zone 2) will be heated and air conditioned, using a gas-fired rooftop unit with electric air conditioning. We won't do this part of the design since it's almost identical to the work we did on our convenience store.

The repair area will be heated but not cooled. We'll use gas-fired unit heaters mounted 12 feet above the floor.

Design Conditions

Our building is located in Omaha, Nebraska. Look back to Figure 5-10 in Chapter 5 for the outside design conditions:

Summer: 91° Fdb, 75° Fwb, 50% rh
Winter: -3° Fdb, 15 mph wind, 20% rh

For Zone 1: Summer: None
　　　　　　 Winter: 60° F

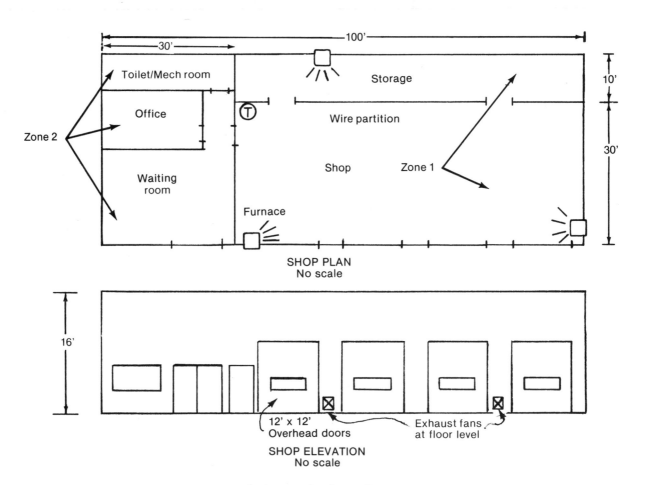

Auto repair shop plan
Figure 8-12

Heat Loss and Gain

Since we won't be cooling the shop, we won't figure a heat gain. To figure heat loss, use the same steps we applied to the store. But there's something else to take into consideration. Frequent opening of the large overhead doors in Zone 1 will cause a major infiltration of outside air. We estimate that there will be an average of three complete air changes per hour. This outside air will have to be heated.

Zone 1

Walls: $[(180 \times 16) - (4 \times 144)] \times .1 \times 63 = 14,515$
Windows and glass door: $77 \times 1.1 \times 63 = 5,336$
Overhead doors: $4 \times 144 \times .5 \times 63 = 18,144$
Floor (perimeter): $180 \times 25 = 4,500$
Roof: $2,800 \times .08 \times 63 = 14,112$
Infilt: $40 \times 70 \times 16 \times 3 \times .075 \times .24 \times 63 = 152,410$

Total Btu/hr $\quad$ 209,017

Outside Air

You can find the amount of outside air required for repair shops in the table in Chapter 9. According to the building code in Omaha, a power exhaust system must also be installed to remove an equal amount of inside air from the shop. We'll have more on this design later. For now, we only need to know the amount of outside air to be heated.

For the repair area, we'll use 1.5 cfm/sq.ft. of floor space. This is over and above any infiltration. Despite being separated from the repair area by only a wire partition, the storage area is not required to take in any outside air.

Zone 1: $30 \times 70 \times 1.5$ cfm = 3,150 cfm

This air must be heated to the outside design temperature for the shop, 60°F:

$3,150$ cfm $\times 1.1 \times 63 = 218,295$ Btu/hr

Total Zone 1 heating load:
$209,017 + 218,295 = 427,312$ Btu/hr

Heater Size and Location

We'll use manufacturer's data to help us decide how many heating units to use. We need enough units to get a good air distribution throughout the area. But keep in mind that each unit installation will increase job cost. If you find yourself losing out to a competitor who prices down his bid by installing one large unit, here's the argument that may convince the owner to accept your bid. Explain to him how better air distribution leads to more efficient heating and happier, more productive employees.

The units in the shop will take in outside air through the wall. The storage room unit is not required to take in any outside air. We'll expect this unit to cover about 25% of our total heat loss without outside air — about 52,870 Btu/hr. The remaining load of 374,442 Btu/hr is divided between the two shop heaters. Each heater will need a capacity of about 190,000 Btu/hr. It will also have to heat 1800 cfm of outside air.

Several manufacturers make furnaces suitable for our job. Look at Figure 8-13. We'll choose a Reznor model 250 and locate two in the shop. For the storage area, we'll use a model 75. Figure 8-14 is a diagram showing how we'll install each shop unit. The shop unit installation is similar except we don't need any outside air.

Each unit has a throw of 30 feet. We've included an inlet louver, along with a damper to cut off outside air when the unit is shut off. The discharge outlet has louvers for manual adjustment. The unit has a duct system that places the return at floor level and the discharge 12 feet above the floor, where the air distribution won't be blocked by cars.

Remember that we must have a power exhaust unit equal to our intake air to comply with the code. A typical application is described in the industry as a sidewall propeller fan. These can be direct or belt drive, mounted to a sized wall opening. Backdraft dampers and wall-mounting collars are available to make installation easier and to prevent outside air from leaking in. Actual model selection is based on unit size, cfm capacity and operating pressure. In our shop we'll use two units. Each unit will remove 1,800 cfm. They'll be wired to operate when the heating units run. We'll mount them at floor level between the overhead doors, with a wire guard to protect people from the fan and moving parts. But we'll need to have the general contractor install heavy steel guard posts to prevent damage from cars backing out of the stalls.

PACKAGED FORCED AIR FURNACE FOR HEATING AND MAKE-UP AIR, COMMERCIAL / INDUSTRIAL

TECHNICAL DATA

SIZE	75	100	125	150	175	200	225	250	300	350	400
BTUH Input	75,000	100,000	125,000	150,000	175,000	200,000	225,000	250,000	300,000	350,000	400,000
*Thermal Output Capacity	57,750	77,000	96,250	115,500	134,750	154,000	173,250	192,500	231,000	269,500	308,000
Unit Amps 115V (less motor)	0.4	0.4	0.4	0.4	0.4	0.4	0.4	0.4	0.4	0.4	0.4
Control Amps (24V)	0.4	0.4	0.4	0.4	0.4	0.4	0.4	0.4	0.4	0.4	0.4
Maximum CFM (XE)	925	1230	1540	1850	2150	2460	2770	3080	3700	4310	4930
Maximum CFM (HXE)****	1390	1850	2310	2770	3230	3700	4160	4620	5550	5180	5920
Minimum CFM (XE)	620	820	1030	1230	1440	1640	1850	2050	2460	2870	3280
Minimum CFM (HXE)****	925	1230	1540	1850	2150	2460	2770	3080	3700	4310	4930
Approx. Net Wt. (lbs.)	221	228	285	332	332	386	386	494	494	515	576
Approx. Ship Wt. (lbs.)	265	275	336	407	412	468	468	594	594	645	716
**Gas Connection	1/2"	1/2"	1/2"	1/2"	1/2"	1/2"	1/2"	3/4"***	3/4"	3/4"	3/4"
Filter Size (Filters optional)	(2) 20x20	(2) 20x20	(2) 20x20	(2) 20x25	(2) 20x25	(2) 16x20 (2) 16x25	(2) 16x20 (2) 16x25	(2) 20x20 (2) 20x25	(2) 20x20 (2) 20x25	(1) 20x20 (3) 20x25	(3) 16x25 (2) 20x25

*   AGA ratings for altitudes to 2000 feet. Above 2000 feet de-rate by orifice change, 4% for each 1000 feet above sea level.

**   Sizes shown are for natural gas connections, **not** supply line size.

*** Gas connections for propane change to 1/2 inch where indicated.

**** Prefix (H) indicates high CFM units.

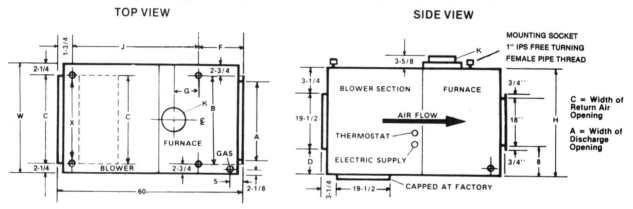

TOP VIEW   SIDE VIEW

B = Hanger ₵ Furnace Section

X = Hanger ₵ Blower Cabinet

C = Width of Return Air Opening

A = Width of Discharge Opening

DIMENSIONS (Accurate within plus or minus 1/4")

MODEL	A	B	C	D	F	G	H	J	K	W	X
75	12½	13½	17¼	9½	18	3¾	32¼	40¼	5" Rd	22	16¼
100	12½	13½	17¼	9½	18	3¾	32¼	40¼	6" Rd	22	16¼
125	15¼	16¼	17¼	9½	18	3¾	32¼	40¼	7" Rd	22	16¼
150	20¾	21¾	22¾	9½	19¼	2½	32¼	39	8" Ov	27½	21¾
175	20¾	21¾	22¾	9½	19¼	2½	32¼	39	8" Ov	27½	21¾
200	26¼	27¼	28¼	12½	18	2¼	35¼	40¼	8" Ov	33	27¼
225	26¼	27¼	28¼	12½	18	2¼	35¼	40¼	8" Rd	33	27¼
→ 250	34½	35½	36½	12½	18	2¼	35¼	40¼	8" Rd	41¼	35¼
300	34½	35½	36½	12½	18	2¼	35¼	40¼	10" Ov	41¼	35¼
350	40	41	42	12½	19¼	2¼	35¼	39	12" Ov	46¾	41
400	45½	46½	47½	12½	18	2¼	35¼	40¼	12" Ov	52¼	46½

NOTE: Model 75 and 100 furnace modules are 3" narrower than blower cabinet.

CLEARANCES FROM COMBUSTIBLES:
1. Top, flue, side opposite controls — 6"
2. Control side — width of unit plus 6"
3. Bottom — 0"

Courtesy: Reznor Corp.

Repair shop furnace data
Figure 8-13

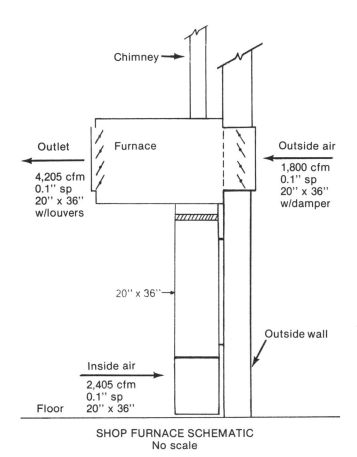

Outlet
4,205 cfm
0.1" sp
20" x 36"
w/louvers

Furnace

Chimney

Outside air
1,800 cfm
0.1" sp
20" x 36"
w/damper

20" x 36"

Outside wall

Inside air
2,405 cfm
0.1" sp
20" x 36"

Floor

SHOP FURNACE SCHEMATIC
No scale

**Repair shop furnace schematic
Figure 8-14**

Ductwork

It'll be up to the HVAC contractor to mount the unit heaters and install the ductwork. Size the ductwork using the methods we studied in Chapter 6 on warm air design. The size will depend on velocity, cfm and pressure drops.

To calculate cfm for duct sizing, we need to take a slightly different approach from what we've used before. Recall that of our total cfm, 1,800 cfm from each unit must be outside air. This will amount to about one-half of the unit's total cfm and could lead to problems. Normally the air mix entering the heating unit should be above freezing. Where outside air makes up only a small part of the total mix, this isn't a problem. But if we design the ducts for this job to allow 1,800 cfm of outside air on a winter day, we have to be sure to mix enough inside air through our return duct to keep the total air temperature above freezing. Here's how to handle it.

We know our inside return air has a design

temperature of 60°F. Our outside design temperature for winter is -3°F. We select a mix temperature of 33°F or above.

Using the following formula, we can find the percentage of outside air we'll need in the total mix.

$$Ti (1 - x) + To (x) = Tmix$$
$$60 (1 - x) + [-3(x)] = 33$$
$$60 - 60x - 3x = 33$$
$$-60x - 3x = 33 - 60$$
$$-63x = -27$$
$$x = .428 = 42.8\%$$

Since there's 1,800 cfm of outside air, the total cfm of the mix will equal:

$$\frac{1,800}{.428} = 4,206 \text{ cfm}$$

Before continuing our sizing steps, we can use these figures to check if we have enough furnace capacity to insure heat on a -3°F winter design day. We'll use this data to find the temperature rise, delta T, of the furnace.

$$delta \ T = \frac{furnace \ Btu}{1.1 \ x \ total \ cfm}$$

$$delta \ T = \frac{192,500}{1.1 \ x \ 4,205}$$

$$delta \ T = 41.6° \ F$$

This means our 192,500 Btu-capacity furnace can heat the 4,205 cfm of mixed air 41.6°F.

Since the air mix enters at 33°F and the temperature rise is 41.6°F, the discharge temperature on a winter design day will equal 74.6°F. The inside design temperature is 60°F, so on a winter design day of -3°F, we would still have 14.6°F per unit to handle building heat loss.

The inside air cfm will equal:

$$4,205 - 1,800 = 2,405 \text{ cfm inside air}$$

The return air ductwork will have to handle 2,405 cfm. The outside air ductwork must handle 1,800 cfm. We'll size grilles and ducts to these capacities.

But we also have to figure in our pressure drops. By consulting the manufacturer's data, we select the louvers for both the outside air intake and the hot air outlet with a pressure drop of 0.1" each.

TOTAL ADJUSTED EXTERNAL STATIC PRESSURE

Size	Temp Rise	CFM	.1 RPM	.1 BHP	.2 RPM	.2 BHP	.3 RPM	.3 BHP	.4 RPM	.4 BHP	.5 RPM	.5 BHP	.6 RPM	.6 BHP	.7 RPM	.7 BHP	.8 RPM	.8 BHP	.9 RPM	.9 BHP	1.0 RPM	1.0 BHP	1.1 RPM	1.1 BHP	1.2 RPM	1.2 BHP
175	90	1440	615	.29	675	.32	725	.36	770	.40	810	.44	860	.47	910	.51	945	.56	1010	.68	1040	.72	1080	.76	1185	1.08
	81	1600	670	.35	720	.38	770	.42	810	.46	860	.50	900	.54	940	.58	975	.63	1080	.88	1115	.94	1150	1.00	—	—
	68	1900	780	.50	820	.54	850	.58	900	.62	940	.67	975	.72	1010	.77	1045	.82	—	—	—	—	—	—	—	—
H175	68	1900	720	.44	760	.48	800	.52	840	.56	880	.60	900	.65	960	.70	995	.75	1030	.80	1065	.86	1100	.92	1135	.98
	54	2400	875	.78	910	.83	945	.88	980	.93	1010	.98	1045	1.05	1080	1.12	1110	1.19	1140	1.26	1170	1.33	1200	1.40	1240	1.47
	43	3000	1080	1.42	1105	1.50	1130	1.58	1155	1.66	1180	1.74	1205	1.82	1230	1.91	1255	2.00	1280	2.10	1305	2.20	1325	2.31	1345	2.42
	36	3600	1290	2.56	1310	2.64	1330	2.73	1350	2.82	1370	2.91	1390	3.00	1410	3.09	1430	3.19	—	—	—	—	—	—	—	—
200	90	1640	520	.26	590	.30	650	.34	705	.38	745	.42	785	.46	825	.50	865	.54	905	.58	960	.66	1000	.70	1070	.96
	82	1800	570	.31	630	.35	680	.39	730	.43	765	.47	800	.50	840	.54	880	.58	920	.62	1000	.84	1035	.90	—	—
	70	2100	660	.44	705	.48	750	.52	790	.56	830	.60	870	.64	900	.68	935	.72	970	.78	—	—	—	—	—	—
H200	70	2100	600	.39	645	.42	690	.46	735	.50	775	.54	815	.58	850	.62	885	.66	920	.70	955	.75	990	.81	1025	.87
	53	2800	790	.76	820	.80	850	.85	880	.90	910	.95	940	1.00	970	1.06	1000	1.12	1030	1.19	1060	1.26	1090	1.33	1120	1.40
	42	3500	955	1.40	990	1.46	1015	1.52	1045	1.58	1070	1.66	1095	1.74	1120	1.82	1145	1.90	1170	1.98	1195	2.06	1220	2.17	1245	2.27
	35	4200	1140	2.54	1165	2.63	1190	2.72	1215	2.81	1240	2.90	1265	2.99	1290	3.09	1315	3.19	—	—	—	—	—	—	—	—
225	90	1850	585	.34	645	.38	695	.42	740	.46	775	.50	810	.53	850	.57	895	.62	935	.67	970	.71	1010	.76	1040	.81
	79	2100	665	.45	710	.49	755	.53	795	.57	835	.61	875	.65	905	.69	940	.73	975	.80	1005	.86	1040	.92	1075	.97
	69	2400	740	.58	780	.63	815	.68	850	.73	885	.77	920	.82	950	.87	980	.92	1010	.97	1040	1.04	1070	1.11	1100	1.18
H225	69	2400	680	.53	720	.57	760	.61	800	.65	835	.69	870	.74	900	.79	930	.84	960	.90	990	.96	1025	1.02	1060	1.09
	59	2800	790	.76	820	.80	850	.85	880	.90	910	.95	940	1.00	970	1.06	1000	1.12	1030	1.19	1060	1.26	1090	1.33	1120	1.40
	49	3400	940	1.32	970	1.38	1000	1.44	1030	1.50	1055	1.58	1080	1.66	1105	1.74	1130	1.82	1155	1.90	1180	1.99	1205	2.09	1230	2.19
	39	4200	1140	2.54	1165	2.63	1190	2.72	1215	2.81	1240	2.90	1265	2.99	1290	3.09	1315	3.19	—	—	—	—	—	—	—	—
250	90	2050	560	.35	630	.40	700	.46	755	.46	815	.53	875	.65	925	.80	975	.85	1025	.90	1075	.95	1115	1.00	—	—
	80	2300	630	.43	690	.48	740	.58	790	.58	850	.68	910	.77	960	.87	1000	.92	1050	.97	1100	1.02	1150	1.10	—	—
	70	2650	710	.57	760	.66	810	.77	860	.77	910	.87	955	.98	995	1.02	1045	1.10	1095	1.18	1145	1.27	1185	1.35	1225	1.43
H250	70	2650	640	.45	700	.55	750	.65	800	.75	850	.80	900	.90	945	.95	990	1.00	1035	1.08	1080	1.16	1125	1.24	1170	1.32
	53	3500	815	.89	860	.96	905	1.03	945	1.10	985	1.17	1025	1.24	1060	1.31	1100	1.40	1135	1.48	1175	1.55	1215	1.64	1255	1.73
	42	4400	1010	1.51	1040	1.58	1080	1.66	1115	1.74	1145	1.82	1175	1.90	1205	2.00	1235	2.09	1270	2.18	1305	2.27	1340	2.36	1375	2.45
	35	5200	1170	2.24	1200	2.36	1230	2.48	1260	2.60	1290	2.72	1320	2.86	1350	3.00	1300	3.13	—	—	—	—	—	—	—	—
300	90	2460	665	.48	710	.54	760	.65	810	.76	865	.80	920	.85	965	.90	1010	.96	1060	1.01	1110	1.09	1155	1.17	1195	1.25
	79	2800	740	.70	780	.77	830	.82	880	.87	930	.92	980	.97	1020	1.02	1065	1.10	1110	1.18	1150	1.26	1190	1.34	1230	1.42
	69	3200	830	.87	875	.92	920	.97	965	1.02	1010	1.10	1050	1.18	1090	1.26	1130	1.34	1165	1.42	1200	1.50	1240	1.58	1280	1.66
H300	69	3200	760	.80	800	.85	850	.90	900	.90	945	.95	990	1.07	1035	1.15	1080	1.23	1115	1.32	1155	1.41	1190	1.50	1225	1.60
	60	3700	860	.97	900	1.03	940	1.11	980	1.19	1020	1.19	1055	1.35	1090	1.43	1125	1.52	1160	1.60	1200	1.68	1240	1.76	1280	1.84
	50	4400	1010	1.50	1040	1.58	1070	1.66	1100	1.74	1130	1.82	1160	1.90	1190	1.98	1220	2.06	1250	2.16	1280	2.26	1315	2.36	—	—
	41	5350	1200	2.56	1230	2.62	1260	2.75	1290	2.87	1320	2.99	1350	3.12	—	—	—	—	—	—	—	—	—	—	—	—
350	89	2900	760	.75	810	.80	860	.80	910	.90	955	.90	1000	1.00	1040	1.08	1080	1.16	1120	1.24	1165	1.32	1210	1.40	—	—
	80	3200	830	.86	880	.91	930	.96	975	1.04	1015	1.12	1050	1.20	1090	1.28	1130	1.36	1170	1.44	1210	1.52	1250	1.60	—	—
	70	3700	950	1.16	990	1.24	1030	1.32	1070	1.40	1110	1.48	1150	1.56	1185	1.64	1220	1.72	1260	1.80	1300	1.88	—	—	—	—
H350	70	3700	860	.98	905	1.06	950	1.14	990	1.22	1030	1.30	1070	1.38	1110	1.46	1150	1.54	1185	1.62	1220	1.70	1255	1.78	1280	1.86
	60	4300	980	1.42	1020	1.50	1060	1.58	1100	1.66	1135	1.74	1170	1.82	1200	1.91	1230	2.00	1265	2.09	1300	2.19	1335	2.30	1370	2.41
	54	4800	1090	1.84	1125	1.92	1160	2.00	1195	2.11	1225	2.22	1255	2.33	1285	2.44	1315	2.55	1345	2.66	1375	2.77	1405	2.88	—	—
	50	5300	1170	2.28	1200	2.40	1230	2.52	1260	2.64	1290	2.76	1320	2.88	1350	3.00	1375	3.15	—	—	—	—	—	—	—	—
400	92	3200	610	.61	670	.68	720	.75	770	.83	820	.83	860	1.00	900	1.08	935	1.16	970	1.24	1005	1.32	1040	1.40	—	—
	80	3700	700	.86	750	.94	800	1.02	840	1.10	880	1.18	920	1.26	955	1.34	990	1.42	1020	1.50	1060	1.58	—	—	—	—
	70	4200	800	1.21	845	1.29	890	1.37	925	1.45	960	1.56	995	1.67	1030	1.78	1065	1.89	1100	2.00	1135	2.14	—	—	—	—
H400	70	4200	705	.98	750	1.07	795	1.15	835	1.23	875	1.32	910	1.41	945	1.50	980	1.62	1015	1.74	1050	1.86	1085	1.98	1180	2.12
	60	4900	810	1.40	850	1.50	890	1.60	930	1.70	960	1.80	990	1.90	1025	2.00	1060	2.12	1095	2.26	1125	2.40	1155	2.54	1220	2.68
	55	5400	890	1.80	925	1.90	960	2.00	995	2.14	1025	2.28	1055	2.42	1085	2.56	1115	2.70	1145	2.84	1175	2.98	1200	3.12	—	—
	49	6000	975	2.65	1010	2.72	1040	2.79	1070	2.86	1095	2.93	1120	3.00	1150	3.15	—	—	—	—	—	—	—	—	—	—

Courtesy: Reznor Corp.

**Furnace fan data
Figure 8-15**

Even though this is an auto repair shop, we still want to avoid excessive system noise and drafts. The inside return grille will also use this drop. Keep in mind that you can select higher drops on your jobs but this will mean larger fans and motors to do the work, and a noisier system.

There's also a pressure drop in the ductwork running from the unit to the floor. Again we'll use 0.1'' per 100' of duct. We'll figure our 12' length in this system as we did in the ductwork in Chapter 6.

$$12' \; x \; 1.5 \; x \; \frac{0.1}{100} \; = \; 0.018''$$

Our total pressure drops for the system are:

.10   discharge louvers
.018   ductwork
.10   filters
.10   return grille
0.318 total

By applying the pressure drops, cfm and a velocity limit of 500 fpm to manufacturer's data as we did in Chapter 6, we're able to size our intake and outlets louvers and ductwork for both the air returns and supply. We'll apply the total pressure drop to the furnace data to make sure we've chosen the correct size fan to move the air through the system. Since our external static pressure is 0.318, we need a 2 HP motor for 4,205 cfm. See Figure 8-15.

An outside air intake 20'' x 36'' will work best with our unit, since it matches the unit's intake opening. It gives us 720 square inches of area, well over the minimum. The return duct and intake grille inside the building must handle 2,405 cfm. We'll use another rectangular duct sized 20'' x 36'' to allow ample capacity and match the duct connections on the furnace.

Although it isn't shown in our data, Reznor can supply many temperature control options. We can choose a damper system to supply the correct ratio of outside and inside air per our system design. This and other options allow you to order a unit tailored to the job specifications. This provides easier installation and start-up, and increased owner satisfaction.

Controls

Our repair shop will normally only operate about 10 hours a day. There's little inside the shop that's sensitive to cold temperatures, so we'll want a manual switch to shut each unit on and off. This allows the owner flexibility in saving fuel. He can shut off one, two or all three units overnight, depending on outside temperatures. Each furnace will also be controlled by a thermostat for automatic operation. Both manual and thermostat switches control the gas burner, the exhaust fan and the outside damper.

The unit for the storage area doesn't have an outside air intake. But it does have a thermostat control with a manual override. Unless it's very cold outside, the owner can shut off his two shop units and run the storage room unit overnight at a very low temperature. This should provide enough heat in the storage and shop areas, while using very little fuel.

Since our furnaces are gas-fired, they'll need a chimney through the roof. Check state and local codes and the manufacturer's data for the correct installation. If you're working on a new building, check your contract to see who's responsible for chimney installation. Coordinate your work with the general contractor as needed.

Codes will also require a separate system for removing exhaust gases that can accumulate when mechanics run the engine they are working on. A simple system uses hoses run from the car exhausts through a hole in the overhead door. More elaborate systems use fans and ductwork run overhead or below the floor. We'll describe these systems in Chapter 9.

Ventilation Systems and System Balancing

Dealing with air pollutants is a critical part of HVAC work. Pollution is everywhere. Even the outdoor air contains as many as 30 million dust or pollutant particles in a single cubic foot. When this "fresh air" is drawn into a building, indoor pollutants, like fumes, smoke, liquid particulates and gases, are added to the load. It's up to you to design a ventilation system that will clean this air to an acceptable level.

The condition of the air is probably as important to the occupants as the temperature. Dusty, dirty or smelly air will always provoke complaints. All systems, even small ones, are now required by code to have some method of filtering or exchanging the air. Residential warm air systems have the advantage of adapting easily to filtration and humidity control. But many commercial businesses, like restaurants and welding shops, require extensive custom cleaning and circulating systems to comply with codes. Figure 9-1 gives the ventilation requirements for both residential and commercial installations. This section will help you in planning this kind of work.

In ventilation systems, as in all HVAC work, your goal is customer satisfaction. Start out with careful design and installation. Then sit down with

the owner and set up a regular maintenance plan to make sure the system operates at peak efficiency.

Types and Sizes of Pollutants
Pollutants vary greatly in particle size. The particles are measured in microns; a micron is 1/25,400 of an inch. The period at the end of this sentence is 400 microns across.

Particles 10 microns and larger are normally visible with the naked eye. Pollutants in this category include dust, lint and pollen. Particles less than 10 microns are visible only with a microscope. Examples of these are tobacco smoke, smudging particles and fine dust.

Particles larger than 10 microns settle out fairly rapidly and can be found suspended in air only near their source or under strong wind conditions. As a result, their control usually isn't too difficult. Particles of less than two microns are of more concern, since these smaller particles are most likely to remain in the lungs. These particles also remain suspended in the air for a long time and can travel long distances unless forcibly removed.

All airborne particles can be damaging to those who come in contact with them. Besides lung damage — which may not be apparent right away

— unclean air can lead to throat and eye irritation, fatigue, drowsiness, odor complaints, chronic colds and other health problems.

Methods of Cleaning the Air

There are two basic methods of air cleaning. One or the other will cover almost every job you'll come across. The first method is the air exchange system, in which contaminated air is forced out of the building and replaced with fresh air. These systems are often required in commercial, industrial and shop locations. Many of the new superinsulated and airtight homes also require air exchange systems.

The second method uses some type of mechanical filter or cleaner to clean the air that circulates through the building. In these systems, it's assumed that natural infiltration will supply enough fresh air to avoid problems. The fiberglass panel filters found in many residential hot air systems are typical. Although these filters do clean the circulating air somewhat, their main purpose is to prevent dust and dirt from causing equipment malfunction. For this reason, air exchange systems also include filters. They do a good job of removing the larger particles they're designed to trap — about 15% of all the particles found in the air. They keep the equipment safe. The microscopic particles they miss only harm people.

Ventilation Requirements for Occupants

	Estimated persons/ 1000 ft² floor area.[g]	Required ventilation air, per human occupant	
		Minimum cfm	Recommended cfm
RESIDENTIAL			
Single Unit Dwellings			
General Living Areas, Bedrooms, Utility Rooms	5	5	7-10
Kitchens, Baths, Toilet Rooms[a]	—	20	30-50
Multiple Unit Dwellings and Mobile Homes			
General Living Areas, Bedrooms, Utility Rooms	7	5	7-10
Kitchens, Baths, Toilet Rooms[a]	—	20	30-50
Garages[b]	—	1.5[b]	2-3[b]
COMMERCIAL			
Public Rest Rooms	100	15	20-25
General Requirements—Merchandising (Apply to all forms unless specially noted)			
Sales Floors (Basement and Ground Floors)	30	7	10-15
Sales Floor (Upper Floors)	20	7	10-15
Storage Areas (Serving Sales Areas and Storerooms)	5	5	7-10
Dressing Rooms	—	7	10-15
Malls and Arcades	40	7	10-15
Shipping and Receiving Areas	10	15	15-20
Warehouses	5	7	10-15
Elevators	—	7	10-15
Meat Processing Rooms[c]	10	5	5
Pharmacists' Workrooms	10	20	25-30
Pet Shops[b]	—	1.0[b]	1.5-2[b]
Florists[d]	10	5	7
Greenhouses[d,e]	1	5	7-10
Bank Vaults	—	5	5
Dining Rooms	70	10	15-20
Kitchens[f]	20	30	35
Cafeterias, Short Order; Drive-Ins, Seating Areas	100	30	35
Bars (Predominantly Stand-Up)	150	30	40-50
Cocktail Lounges	100	30	35-40

Ventilation requirements
Figure 9-1

Ventilation Requirements for Occupants

	Estimated persons/ 1000 ft² floor area.[g]	Required ventilation air, per human occupant	
		Minimum cfm	Recommended cfm
Hotels, Motels, Resorts			
Bedrooms	5	7	10-15
Living Rooms (Suites)	20	10	15-20
Baths, Toilets (attached to bedrooms)[a]	—	20	30-50
Corridors	5	5	7-10
Lobbies	30	7	10-15
Conference Rooms (Small)	70	20	25-30
Assembly Rooms (Large)	140	15	20-25
Cottages (treat as single-unit dwellings) (See also Food Services, Industrial, Merchandising, Barber and Beauty Shops, Garages for associated Hotel/Motel Services)			
Dry Cleaners and Laundries			
Commercial[f,g]	10	20	25-30
Storage/Pickup Areas	30	7	10-15
Coin-Operated[g]	20	15	15-20
Barber, Beauty, and Health Services			
Beauty Shops (Hairdressers)	50	25	30-35
Reducing Salons (Exercise Rooms)	20	25	30-35
Sauna Baths and Steam Rooms	—	5	5
Barber Shops	25	7	10-15
Photo Studios			
Camera Rooms, Stages[h]	10	5	7-10
Darkrooms	10	10	15-20
COMMERCIAL			
Shoe Repair Shops (Combined Workrooms/Trade Areas)	10	10	15-20
Garages, Auto Repair Shops, Service Stations			
Parking Garages (enclosed)[b]	—	1.5[b]	2-3[b]
Auto Repair Workrooms (general)[b,i]	—	1.5[b] ←	2-3[b] ←
Service Station Offices	20	7	10-15
Theaters			
Ticket Booths	—	5	7-10
Lobbies, (Foyers and Lounges)	150	20	25-30
Auditoriums (No Smoking)	150	5	5-10
Auditoriums (Smoking Permitted)	150	10	10-20
Stages (with Proscenium and Curtains)[h,j]	70	10	12-15
Workrooms	20	10	12-15
Ballrooms (Public)	100	15	20-25
Bowling Alleys (Seating Area)	70	15	20-25
Gymnasiums and Arenas			
Playing Floors-Minimal or No Seating	70	20	25-30
Locker Rooms[k]	20	30[k]	40-50[k]
Spectator Areas	150	20	25-30
Ramps, Foyers, and Lobbies	150	10	15-20
Amusement Parlors and Pool Rooms	25	20	25-30
Tennis, Squash, Handball Courts	—	20	25-30
Swimming Pools	25	15	20-25
Ice-Skating, Curling, and Roller Rinks	70	10	15-20
Transportation			
Waiting Rooms	50	15	20-25
Ticket and Baggage Areas, Corridors, and Gate Areas	50	15	20-25
Control Towers	50	25	30-35
Hangars[l]	2	10	15-20
Platform	150	10	15-20
Concourses	150	10	15-20
Repair Shops	—	10	15-20

**Ventilation requirements
Figure 9-1 (continued)**

Ventilation Requirements for Occupants

	Estimated persons/ 1000 ft^2 floor area.[g]	Required ventilation air, per human occupant	
		Minimum cfm	Recommended cfm
Offices			
General Office Space	10	15	15-25
Conference Rooms	60	25	30-40
Drafting Rooms, Art Rooms	20	7	10-15
Doctors' Consultation Rooms	—	10	10-15
Waiting Rooms	30	10	15-20
Lithographing Rooms[g]	20	7	10-15
Diazo Printing Rooms[g]	20	7	10-15
Computer Rooms	20	5	7-10
Keypunching Rooms	30	7	10-15
Communication			
TV/Radio Broadcasting Booths, or Studios[h]	20	30	35-40
Motion Picture and TV Stages	20	30	35-40
Pressrooms	100	15	20-25
Composing Rooms	30	7	10-15
Engraving Shops	30	7	10-15
Telephone Switchboard Rooms (Manual)	50	7	10-15
Telephone Switchgear Rooms (Automatic)	—	7	10-15
Teletypewriter/Facsimile Rooms	—	5	7-10

INDUSTRIAL (including agricultural processing)

(Occupational safety laws in the various states usually regulate the ventilation requirements. These are almost always far in excess of the ventilation requirements for the occupants.) ASHRAE Standard 62-73 lists requirements for occupants only. In general, 25 cfm per occupant is recommended except for mining or metalworking, where 40 cfm is recommended.

INSTITUTIONAL

Schools			
Classrooms	50	10	10-15
Multiple Use Rooms	70	10	10-15
Laboratories[m]	30	10	10-15
Craft and Vocational Training Shops[m]	30	10	10-15
Music, Rehearsal Rooms	70	10	15-20
Auditoriums	150	5	5-7.5
Gymnasiums	70	20	25-30
Libraries	20	7	10-12
Common Rooms, Lounges	70	10	10-15
Offices	10	7	10-15
Lavatories	100	15	20-25
Locker Rooms[k]	20	30[k]	40-50[k]
Lunchrooms, Dining Halls	100	10	15-20
Corridors	50	15	20-25
Utility Rooms	3	5	7-10
Dormitory Bedrooms	20	7	10-15
Hospitals, Nursing and Convalescent Homes			
Foyers	50	20	25-30
Hallways	50	20	25-30
Single, Dual Bedrooms	15	10	15-20
Wards	20	10	15-20
Food Service Centers	20	35	35
Operating Rooms, Delivery Rooms[n]	—	20	—
Amphitheatres	100	10	15-20
Physical Therapy Areas	20	15	20-25
Autopsy Rooms	10	30	40-50
Incinerator Service Areas[o]	—	5	7-10
Ready Rooms, Recovery Rooms[n]	—	15	—
(For Shops, Restaurants, Utility Rooms, Kitchens, Bathrooms, and Other Service Items, see Hotels)			
Research Institutes			
Laboratories[m]	50	15	20-25
Machine Shops	50	15	20-25
Darkrooms, Spectroscopy Rooms	50	10	15-20
Animal Rooms[n]	20	40	45-50

Ventilation requirements
Figure 9-1 (continued)

Ventilation Requirements for Occupants

	Estimated persons/ 1000 ft² floor area.[g]	Required ventilation air per human occupant	
		Minimum cfm	Recommended cfm
Military and Naval Installations			
Barracks	20	7	10-15
Toilets/Washrooms	100	15	20-25
Shower Rooms	100	10	15-20
Drill Halls	70	15	20-25
Ready Rooms, MP Stations	40	7	10-15
Indoor Target Ranges[p]	70	20	25-30
Museums			
Exhibit Halls	70	7	10-15
Workrooms	10	10	15-20
Warehouses	5	5	7-10
Correctional Facilities, Police and Fire Stations (see also Gymnasiums, Libraries, Industrial Areas)			
Cell Blocks	20	7	10-15
Eating Halls	70	15	20-25
Guard Stations	40	7	10-15
Veterinary Hospitals			
Kennels, Stalls, Operating Rooms[n]	20	25	30-35
Reception Rooms	30	10	15-20
ORGANIZATIONAL			
Churches, Temples (see Theaters, Schools and Offices)			
Legislative Halls			
Legislative Chambers	70	20	25-30
Committee Rooms and Conference Rooms	70	20	25-30
Foyers, Corridors	50	20	25-30
Offices	10	10	15-20
Press Lounges	20	20	25-30
Press/Radio/TV Booths	20	20	25-30
Public Rest Rooms	20	15	20-25
Private Rest Rooms (for Food Service, Utilities, etc. see Hotels)	—	20	30-50
Survival Shelters[n]	—	5	2.5	—	—

[a] Installed capacity for intermittent use.
[b] cfm per sq ft of floor area.
[c] Spaces maintained below 50 F are not covered by these requirements unless the occupancy is continuous. Ventilation from adjoining spaces is permissable. When occupancy is intermittent, infiltration will normally exceed ventilation requirement.
[d] Maximum allowable concentration (MAC) for sulfur dioxide is 30 mmg/m³.
[e] Ventilation to optimize plant growth, temperature, and humidity will almost always be greater than shown.
[f] Exhaust to outside; source control as required.
[g] Installed equipment must incorporate positive exhaust and control (as required) of undesirable contaminants (toxic or otherwise).
[h] Thermal effects probably determine requirements.
[i] Stands where engines are run must incorporate systems for positive exhaust withdrawal.
[j] Special ventilation will be needed to eliminate stage effect contaminants.
[k] cfm/locker.
[l] Special solvent and exhaust problems handled separately.
[m] Special requirements control systems may be required.
[n] Special requirements or codes may determine requirements.
[o] Special exhaust systems required.
[p] Floor area behind firing line only.

**Ventilation requirements
Figure 9-1 (continued)**

Air cleaners are a more efficient method of cleaning the air. There are several types. The first is electronic. Electricity is used to place a positive charge on the particles that pass between two plates. The plates are alternately charged. The positive particles are repelled by the positive plates and attracted to the negative plates, where they're collected.

Other cleaners are nonelectric. They use one or more filtering principles such as straining, diffusion or interception, to trap a wide range of particle sizes.

Both types of cleaners have the same average efficiency, removing about 65% of airborne particles. An electronic cleaner normally has greater efficiency at the beginning of its service period and drops off as it loads. A nonelectric cleaner's efficiency increases with use, but if allowed to go too long without service, it will drop off quickly.

Selling Air Cleaning Systems

Whenever you design or install a heating or cooling system, you should try to sell an efficient method of air cleaning. For commercial jobs, codes make it a necessity. They'll also give a detailed description of the requirements. But for other jobs you'll have to sell the owner on the idea.

That shouldn't be too hard. Most people today value clean air. Ask them for specific complaints about the air quality of their old system. Point out that clean air cuts down on fatigue, disease and odor. Proper air cleaning can eliminate up to 90% of allergy-causing pollen from the air. Clean air can reduce cleaning and painting bills on walls and ceilings.

Remind the owners that it's less costly to install a filtering system now, while your crew is on the job. And explain to them that the panel filters used in most forced air systems are not really designed to clean the air, but rather to protect the equipment.

It's up to you to guide them in choosing the right method and equipment. Like anything else you install, it must do the job and satisfy the customer. An effective cleaner for a particular job will meet these three requirements: First, it'll have the ability to take out the particles that cause the customer's complaints. Second, its maintenance requirements will be acceptable to the owner. And third, the cost will be within the buyer's budget.

Maintenance is a key item. Most cleaners must be cleaned or the filters replaced on a regular basis if they're to remain effective. If maintenance is done by the homeowner, the best method will be a cleaner with easily changed, throwaway filters. If the maintenance is difficult, the owner probably won't do it, and the system won't work like it should. If the owner feels uncomfortable with the idea of doing his own maintenance, it's an ideal chance for you to suggest regular equipment checkup and maintenance by your crew.

Consult the manufacturer's data when choosing a suitable cleaner. The many different types of air cleaners on the market fit a wide range of systems. It's important to deal with a reputable company whose units will do what they say they will. Efficiency ratings are based on laboratory tests, but make sure the company used the "atmospheric dust spot test" method. This is a good method of determining what the unit will do under actual operating conditions. Look for manufacturers that have long-term warranties on their units.

For forced air systems, size and fit the units so they clean both the return air *and* any outside air coming into the system. Locate the cleaner to clean the air before it passes through the equipment.

If a forced air installation isn't possible, portable or self-contained units may be the answer. These can be installed in a central room or the one where the problem is the most serious.

So far in this chapter, we've talked about pollutants and air cleaning systems in general terms, for average commercial and residential installations. But some commercial jobs, like auto repair shops and restaurants, have special ventilation requirements. We'll discuss them in much more detail.

Vehicle Engine Exhaust

In the HVAC system for the auto repair shop in Chapter 8, we included a method of bringing in outside air and forcibly exhausting an equal amount of air to comply with the code. This was a general ventilation system. Its purpose was to take care of gasoline vapor, light particles, small exhaust leaks and vapors from chemicals such as spilled antifreeze. This system couldn't be used to remove engine exhaust since it didn't move enough cfm at a high enough rate to be effective. Although in theory a general system could be designed to handle vehicle exhaust, its size would make it uneconomical both to install and operate.

Vehicle exhaust contains the by-products of combustion: unburned fuel particles and very fine liquid drops of oil and gases. The gases are poisonous if inhaled. A safe and economical method of removing them from the shop is a system that connects directly to the engine exhaust pipe.

Figure 9-2 is a plan for a beneath-the-floor system for the auto repair shop. We have five work bays for cars or light trucks. Each bay has an exhaust inlet mounted flush with the floor. A flexible metal tube is pulled from inside the inlet and slipped over the end of the vehicle exhaust pipe. These tubes are generally 3" or 4" diameter and connect through a 60-degree "Y" fitting to the main duct, which is made of clay tile or a special noncombustible coated ductwork. The main duct runs under the floor and up the inside of the wall, high enough to be out of the way. The fan is mounted on a platform and then ducted through the wall or roof, as shown in Figures 9-3A and 9-3B.

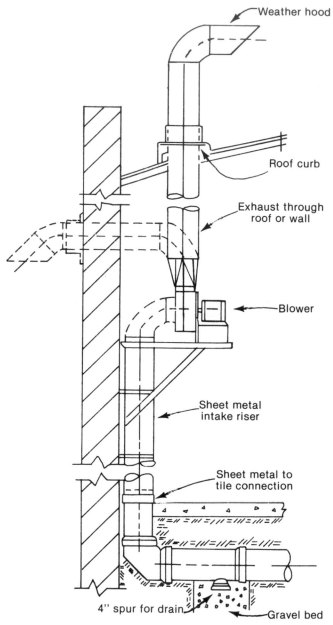

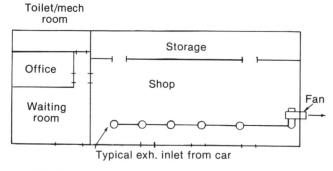

FLOOR PLAN-AUTO REPAIR SHOP EXHAUST SYSTEM
No scale

Auto shop exhaust system floor plan
Figure 9-2

Sizing the Exhaust System

Sizing the pipe exhaust system presents some new challenges. Begin by checking state and local codes to make sure you know their requirements. Since it's a forced air system, fan and duct size will depend on cfm, velocity and pressure drops. But the rough surface of the clay ductwork, the sharp changes in direction and the higher operating velocity all mean a higher pressure drop than found in a warm air system.

Before we run through these calculations, we

TYPICAL BLOWER MOUNTING & DUCT WORK
No scale

Typical exhaust system diagram
Figure 9-3A

need to review a calculation we used in Chapter 8. Recall that in designing our general ventilation system, we had 3,600 cfm of outside air entering and being exhausted, to comply with code. This outside air had to be heated. Our pipe exhaust system will also remove room air from the building

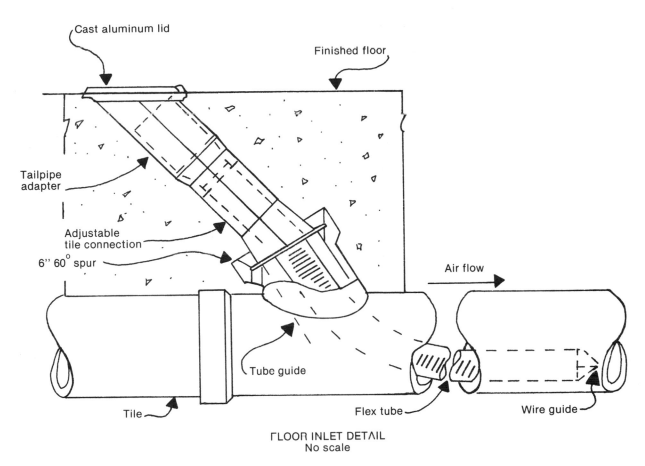

Cast aluminum lid

Finished floor

Tailpipe adapter

Adjustable tile connection

6'' 60° spur

Air flow

Tube guide

Tile

Flex tube

Wire guide

FLOOR INLET DETAIL
No scale

**Typical exhaust system diagram
Figure 9-3B**

— 300 cfm for each exhaust intake. So we have an additional 1,500 cfm of room air now being removed from the shop. The air that replaces it will come from outside, so it'll have to be heated. There are two capacities we have to check: heater size and its duct sizes.

Using our equation for outside air load:

1,500 cfm x 1.1 x 63° F	103,950 Btu/hr
Original load	+427,312 Btu/hr
New load total	531,262 Btu/hr

Of this load, we allocated 53,000 Btu's to the storage room heater. The remainder was split between two units in the shop. Originally each shop unit had to account for 90,000 Btu/hr, and we chose a Reznor model 250. But our new load is 239,131 Btu/hr per unit. So we'll have to go to the larger unit for our two shop heaters, the model 300.

We also have to check duct and grille sizes. Originally we had 1,800 cfm that would result in a mixture temperature of 33°+F. Now we're adding 750 cfm to that.

$$1,800 + 750 = 2,550 \text{ cfm outside air}$$

To find the new inside total, divide the outside cfm by the mixture factor.

$$\frac{2550}{.428} = 5,958 \text{ total cfm}$$

$$5,958 - 2,550 = 3,408 \text{ cfm inside air return}$$

We'll look back to Figure 6-7 for the pressure drop. Using the circle formula, we can find the increase in free area needed for our inside duct and outside intake grille. In our case, the original choice of sizes was large enough to handle the increase. But be careful on jobs that you figure. As a general rule, always remember that if your original calculations were very close to the maximum or minimum values for a given capacity, any large changes later on will change those equipment sizes as well.

Now that we have enough make-up air for our exhaust pipe system, we can size the system as if it were a warm air system. Remember that the static pressure losses are generally higher due to the duct surface, high velocity, and sharp changes in direction — 0.2'' to 0.3'' per 100' of duct.

But instead of doing all the calculations, it's simpler to rely on the exhaust system manufacturers' assistance. Figures 9-4A and 9-4B are typical of the data in the manufacturers' catalogs. Use the data for sizing the system. If you have additional questions, a call to the manufacturer should bring plenty of help.

For our shop, we'll use Example 2 in Figure 9-4A. There are two optional layouts. We'll use the one on the left since it's closer to our layout. We need 1500 cfm at a total static pressure of 2.0''. From the data in Figure 9-4B, we choose blower model D-150 with a 1½ hp motor and direct-drive fan. It can supply 1600 cfm at 2.0'' static pressure, ample for our system.

Restaurant Hoods

The air in a commercial kitchen has a heavy load of pollutants given off from food preparation. The heated air contains vapors of fats and greases, particles of food and carbon, smoke and sometimes combustion gases from gas-fired cooking equipment. Removing these pollutants is essential for occupant health and to avoid unpleasant cooking odors leaking into customer areas.

Remove air from the cooking area by exhaust hoods mounted over the cooking equipment. The exact amount of air and the acceptable method of removal is controlled by the building code in force for that area. The National Fire Protection Association #96 (NFPA), the National Sanitation Foundation (NSF), the National Electrical Code (NEC), the Building Officials and Code Administrators (BOCA), the Uniform Mechanical Code (UMC), and the U.S. Department of Agriculture (USDA) are some of the regulatory agencies that set the guidelines for manufacture and installation of kitchen hood exhaust systems. State, county and local officials use their guidelines to make interpretations about acceptable systems for their area. The local code official can, and occasionally does, render official interpretations contrary to the national code's intent. Also, some national codes may be in direct conflict with each other. It's imperative that you, the contractor, know your code administrator and follow his interpretations.

The air needed to replace the air exhausted by the kitchen hood is called *secondary* or *tempered* air. In most cases, the volume of tempered air is large enough to require the addition of a separate unit to heat and cool it. If the demand is small enough, however, the tempered air may be supplied through the regular heating and cooling units.

The remainder of the air, called *untempered air*, may be introduced directly into the hood. This air doesn't need to be heated or cooled. It's brought in from outside by a supply fan and fed directly into the hood.

The total amount of air required for the hood depends on hood design and local codes. Typically, the total amount of air exhausted is 100 cfm per square foot of hood opening.

Types of Hoods

The two types of hoods commonly used in restaurant installations are shown in Figure 9-5A. The wall canopy is wall-mounted and exposed on three sides. It has a single row of filters. The island canopy is hung from the ceiling, exposed on all four sides. It has a double row of filters, back to back. There's also another type, the backshelf, which is similar to the wall-mounted. Hoods come painted on both sides, with paint on the exterior and stainless steel on the interior, or stainless steel throughout.

All the hoods are installed to extend down to about 30'' to 36'' above the cooking surfaces *where the heat is generated,* depending on the code. The hoods must also extend to cover a code-determined amount of horizontal area of the cooking surface.

Begin the hood design job by selecting the style that fits the kitchen, local codes and the owner's budget. Most often, the codes will be the primary guideline in determining which type is acceptable.

Examples of Underfloor Systems

1) Single inlets with 3" diameter flexible tubes and 6" diameter spurs

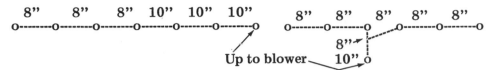

200 cfm each inlet
Total 1,200 cfm @ 2.5" static pressure
Blower: 1½ hp; check mfg's data
Ducts: 10" diameter up to blower; 10" diameter exhaust

2) Single inlets with 4" diameter flexible tube and 8" diameter spurs

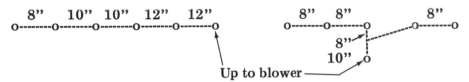

300 cfm each inlet
Total 1,500 cfm @ 2.0" static pressure
Blower: 1½ or 1 hp; check mfg's data
Ducts: 12" diameter up to blower; 12" diameter exhaust

3) Dual inlets with 3" diameter flexible tubes and 10" diameter spurs

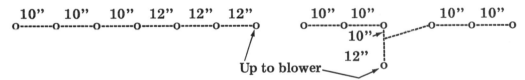

300 cfm each dual inlet
Total 1,800 cfm @ 2.5" static pressure
Blower: 1½ hp; check mfg's data
Ducts: 12" diameter up to blower; 12" diameter exhaust

4) Dual inlets with 4" diameter flexible tubes and 12" diameter spurs

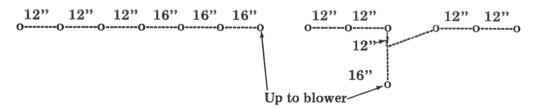

400 cfm ea. dual inlet for trucks, 300 cfm for cars
Total 2,400 cfm at 2.5" static pressure
Blower: 2 hp; check mfg's data
Ducts: 16" diameter up to blower; 16" diameter exhaust

Courtesy: National System of Garage Ventilation, Inc.

Garage system design data
Figure 9-4A

BLOWERS

AMCA RATED

SPECIAL FEATURES AVAILABLE:
ACID RESISTANT PAINT ON WHEEL AND INSIDE OF HOUSING ● NON-FERROUS WHEEL ● BELT GUARD ● DRIVE COVER ● VIBRATION PADS ● FLEXIBLE CONNECTION AT INTAKE AND EXHAUST

DIRECT DRIVE—forward curve fan wheel

Forward curve steel fan wheel for quieter operation. Baked enamel finish. 16 GA housing. 1750 R.P.M. clockwise rotation. 8 discharge positions. Open drip-proof motor.

MODEL	HP.	C. F. M. 2.0"SP	2.5"SP	3.0"SP	A	B	C	D	E	F	G	H	I	J	K	L	M	N
D-50	½	600			9	6½	10¾	12⅛	8	9⅞	6¾	9¼	7⅛	7¾	3⅜	14½	7	5⅝
D-75	¾	1100	600		9	6½	10¾	12⅛	8	9⅞	6¾	9¼	7⅛	7¾	3⅜	14½	7	5⅝
D-100	1	1280	910		10	5⅞	11¾	14¾	8	10¾	8	10¾	8⅛	9⅛	4⅞	13⅞	7	6⅞
→ D-150	1½	1600	1280	1040	10	5⅞	11¾	14¾	11½	11¼	8	10¾	8⅛	9⅛	4⅞	17⅜	10	6⅞
D-200	2	2130	1850	1660	10	8	11¾	14¾	11½	11¼	8	10¾	8⅛	9⅛	4⅞	19½	10	6⅞

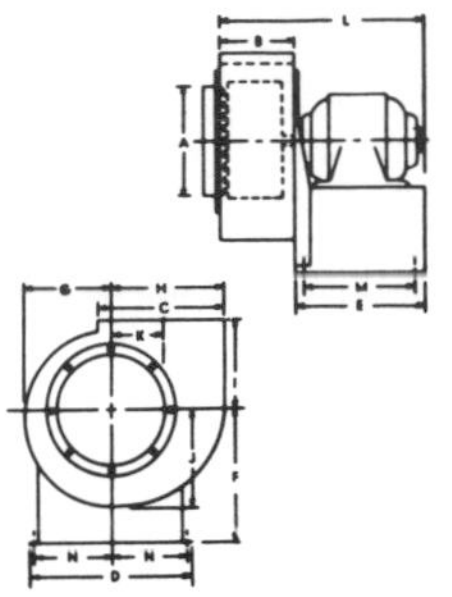

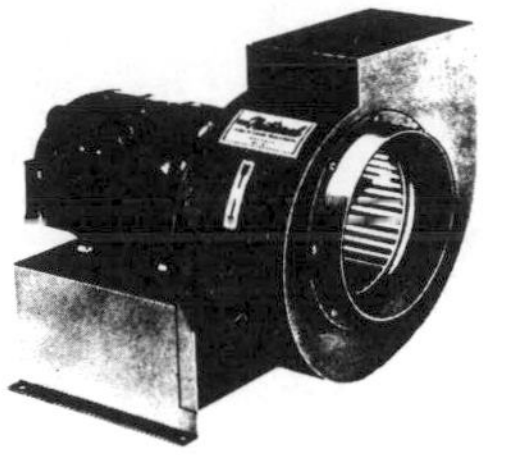

BELT DRIVE—forward curve fan wheel

Forward curve steel fan wheel for quieter operation. Baked enamel finish. 16 GA housing. Variable speed motor pulley. Clockwise rotation. 8 discharge positions. Open drip-proof motor.

MODEL	HP.	C. F. M. 2.0"SP	2.5"SP	3.0"SP	A	B	C	D	E	F	G	H	I	J	K	L	M	N
BN-1	¾	1439			13¼	9⅝	13¼	16⅜	26¼	13½	11¼	14¾	1 3/16	10½	11½	17	10⅛	19 1/16
BN-2	1	1716			13¼	9⅝	13¼	16⅜	26¼	13½	11¼	14¾	1 3/16	10½	11½	17	10⅛	19 1/16
BN-3	1½	2059	1880		13¼	9⅝	13¼	16⅜	26¼	13½	11¼	14¾	1 3/16	10½	11½	17	10⅛	19 1/16
BN-4	2	2574	2402	2231	13¼	9⅝	13¼	16⅜	26¼	13½	11¼	14¾	1 3/16	10½	11½	17	10⅛	19 1/16
BT-5	1½	2816	2432		16	11¾	16¼	19¼	30⅜	15½	13⅜	17⅝	1 3/16	12⅜	14⅛	17⅞	12⅜	21
BT-6	2	3328	3072	2560	16	11¾	16¼	19¼	30⅜	15½	13⅜	17⅝	1 3/16	12⅜	14⅛	17⅞	12⅜	21
BT-7	3	4352	3840	3584	16	11¾	16¼	19¼	30⅜	15½	13⅜	17⅝	1 3/16	12⅜	14⅛	17⅞	12⅜	21
BX-9	5	6460	6080	5697	19⅝	14¼	19⅝	23⅛	35¼	17½	15¼	21½	2	14⅝	17⅛	21⅞	15	23½

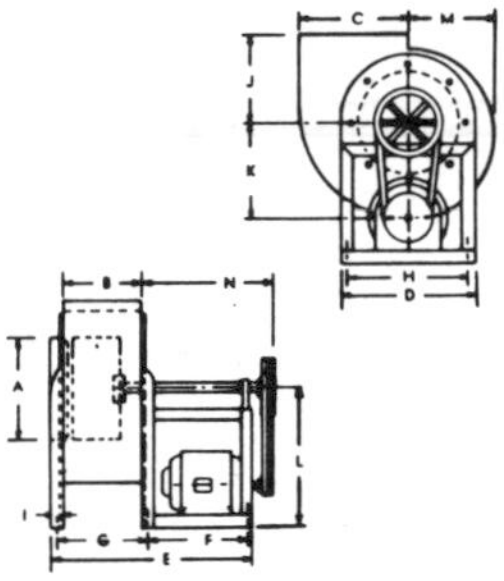

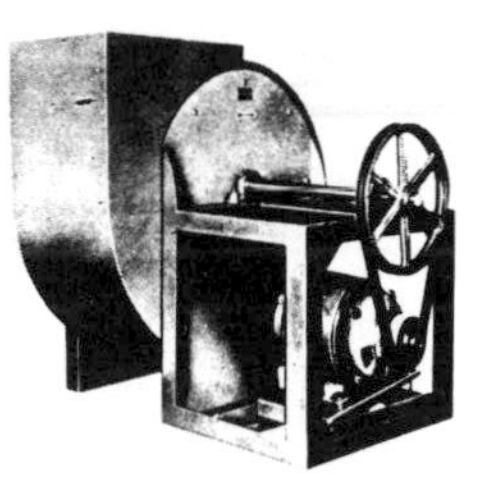

BELT DRIVE—backward curve fan wheel

Backward curve fan wheel for non-overloading design. Baked enamel finish. 16 GA housing. Variable speed motor pulley. Clockwise rotation. 8 discharge positions. Open drip-proof motor.

MODEL	H.P.	1"S.P.	1½"	2"	2½"	3"	3½"	4"	4½"	5"
105	½	1317	1129	1003	700	467				
	¾	1630	1379	1254	1129	914	827	510		
122	1	2223	1967	1710	1539	1283	905	667	452	
	2	2993	2822	2651	2394	2223	2052	1967	1733	1440
135	1½	2937	2727	2413	2203	1888	1595	1229	913	662
	2	3357	3147	2832	2623	2413	2237	1965	1575	1251
150	2	3712	3456	3072	2816	2688	2245	1713	1325	983
	3	4480	4224	3968	3712	3328	3200	2944	2631	2169
182	3	5859	5481	5103	4536	4158	3454	2672	2139	1683
	5	7182	6804	6615	6237	5859	5481	5103	4725	3941
200	5	8113	7649	7186	6722	6259	5795	5331	4421	3600
	7½	9272	9040	8577	8345	7881	7418	6954	6722	6259

MODEL	A	B	C	D	E	F	G	H	I	J	K	L	M	N
105	11¼	8½	11¾	13	14¾	7 13/16	8	9 1/16	15¾	25½	14¾			½
122	13¼	9¾	13	14½	17½	9 5/16	10¼	10 3/16	18½	27⅝	15⅛	11½	13¼	½
135	14 3/16	10 13/16	14 5/16	15¾	19	10¼	11¼	11⅝	20	29⅜	15⅞	12½	14½	½
150	16 3/16	11 13/16	15⅞	17¾	20½	11 7/16	12½	12 13/16	21½	31⅛	16⅝	13½	16	½
182	19½	14⅜	19⅜	21	24¼	13 13/16	14½	15 11/16	25½	35⅛	17¾	16⅛	19¼	⅝
200	21⅜	16	21¼	22¾	26¾	15 3/16	16	17¼	28	37⅛	18⅛	17¾	21¼	⅝

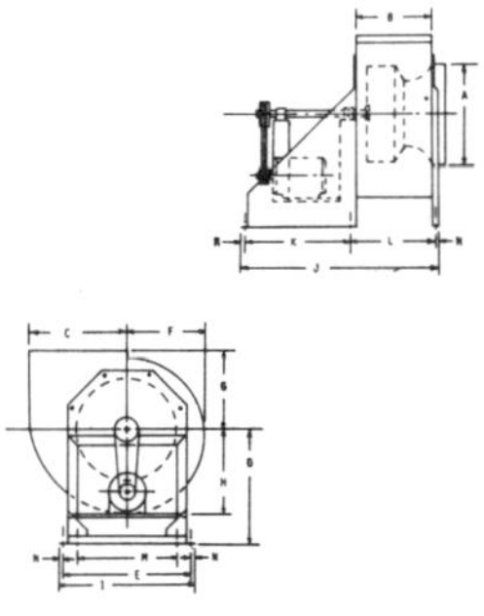

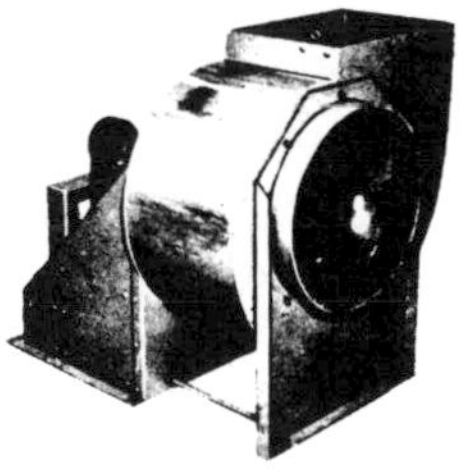

SYSTEM CAPACITY CHART

Based on Total C.F.M. of system.

Cars—3" & 4" tubes		Trucks—4" & 4½" tubes	
C.F.M.	S.P.	C.F.M.	S.P.
200- 600	2.0"	300- 400	3.0"
700-2400	2.5"	500-1600	3.5"
2500-3500	3.0"	1700-3600	4.0"
3600-5600	3.5"	3700-4400	4.5"
5700-6900	4.0"	4500-6200	5.0"

Courtesy: National System of Garage Ventilation, Inc.

Garage system design data
Figure 9-4B

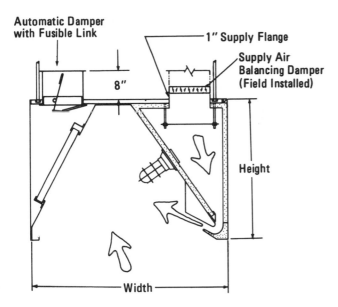

WALL HOOD

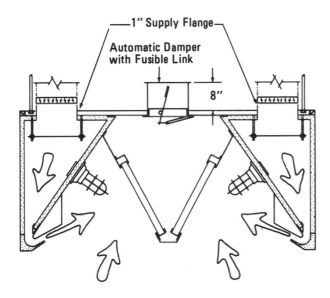

ISLAND HOOD

Courtesy: Econovent Systems, Inc.

**Wall and island exhaust hoods
Figure 9-5A**

Sizing the Hood

From the kitchen floor plan and specs, check the sizes of the cooking equipment. Scaling the appliances off the floor plan does not always guarantee actual equipment sizes. Confirm in writing with the owner or his architect what the in-

stalled cooking gear will be and its layout. Also, check with the equipment supplier to verify equipment sizes and capacities.

Electric cooking equipment— For electric cooking, the canopy hood must have a minimum of 6'' overhang. This means a 6'' overhang from the front edge of the cooking surface to the edge of the supply air discharge lip. This discharge lip is approximately 6'' from the outside front face of the hood. So it will be 12'' from the cooking gear to the outside edge of the hood. This minimum also applies at the sides. Allow for utility hook-up at the rear of the appliances, usually 6''.

Be aware of floor to ceiling height. The standard dimension from floor to underside of the canopy hood is 6'6'' to 7'. Check the code to make sure the installation height will be within the maximum distance allowed above the cooking surface.

Gas-fired cooking equipment— Hood installation for gas-fired equipment requires special equipment. With exhaust gas flue temperatures ranging from 450°F to 1,500°F, there's a high rate of exhaust air expansion above the cooking surfaces. Cooking vapors and fumes can escape the hood capture area into the kitchen, creating health and cleaning problems. Use the following design procedures.

Gas-fired griddles: Frying-type griddles have a high exhaust gas flue discharge temperature and heavy grease loading from fatty meats. Their high surface temperature also contributes to potential problems. Whatever the number of griddles, follow these hood sizing steps:

1) For wall canopies, provide an 18'' overhang of the equipment at the front and 12'' at the sides. Specify a canopy at least 4'6'' wide. This dimension is based on a 30''-deep griddle surface, 6'' from the rear wall, with a front edge 18'' from the outside edge of the wall canopy hood.

2) Use stainless steel-lined hoods.

3) We recommend a height of 2'10'', but never go less than 2'8''.

Hood maintenance: When the griddle will cause a heavy grease-loading condition, inform the owner that a daily cleaning, filter washing and wiping down of the hood will be necessary. This will insure good system efficiency and prevent grease fires in the unit.

Gas-fired fryers: Fryers are generally smaller than griddles, but their flue discharge temperatures are two to three times higher than the gas griddle. Heat expansion is a problem here, also. Follow these guidelines:

1) Have a 12'' overhang at the front and a 6'' overhang at the sides with a minimum 4'0'' wide canopy hood.

2) Use stainless steel-lined hoods.

3) Again, we recommend a height of 2'10'', but not less than 2'8''.

Open-flame charbroilers: Open-flame broilers have conditions similar to gas griddles.

1) Because the open-flame style of cooking causes heavy grease loading, you need an exhaust canopy hood for each unit.

2) Maximum dimensions of the cooking unit should be 36'' wide and 30'' deep.

3) Use a canopy hood at least 4'6'' wide. Provide for an 18'' overhang at the front and 12'' at the sides.

4) Use a stainless steel-lined hood.

Upright broilers: This type of appliance has a roll-out broiling tray. The only way to insure capture of vapors is to have the proper overhang.

1) Use a stainless steel-lined hood.

2) Use a 5' wide canopy hood with an 18'' overhang. The unit is 36'' deep, and will be set 6'' from the wall.

If two or more hoods are placed together, a butting trim strip is required. A good-quality silicone sealer may be used if approved by local codes.

Sizing Ductwork and Fans

We'll begin sizing the air handling part of the system by finding the total supply air. This is based on our hood dimensions. For all hoods, allow 100 cfm per square foot of hood. This air will be 70% to 80% untempered, depending on the code. Here's an example, based on a wall canopy 8' long, 3'6'' wide, and 2'10'' deep:

At 80% untempered:

100 cfm x 8' x 3.5' =	2,800	cfm
2,800 cfm x .8 =	2,240	cfm untempered
2,800 − 2,240 =	560	cfm tempered

Tempered air supply— Be sure to check the building heating and cooling unit and distribution system to make sure it has the capacity to handle the tempered air load. Review the auto repair shop sections in this and the previous chapter.

If the building's HVAC unit doesn't supply air to the kitchen, you'll have to provide transfer grilles in the walls or doors between the kitchen and other rooms. If the walls between the kitchen and other rooms are fire walls, you may have to install fire dampers. This is easier to do if you install a transfer duct above the ceiling with a grille on each side of the wall. Install the damper in the duct. Size the duct for about 1,000 fpm and the grilles at 500 fpm.

If the building's HVAC unit isn't large enough to supply the tempered air increase, you'll need to install a separate heating unit. If the kitchen is air conditioned, you'll also need a larger cooling unit. Place the unit so it supplies the air directly to the kitchen. Follow the hood manufacturer's recommendations for maximum hood efficiency. Duct routing and the type of diffuser are important. Size the unit and its ductwork for the tempered air cfm using the methods described in Chapters 7 and 8.

Untempered air supply— The untempered air system will have a fan, ductwork and manual balancing damper. You can easily size the ductwork to supply the air from the manufacturer's data showing the hood supply opening. These openings are sized based on supplying the required cfm at a velocity of 1,500 fpm within an acceptable pressure drop.

In the above example, the wall canopy is 8' long and 3'6'' wide. Its supply opening, from Table II in Figure 9-5B, is 10'' x 30''. It'll handle 2,240 cfm (80 cfm per square foot of canopy), as shown in the manufacturer's data.

HOOD DIMENSIONAL DATA

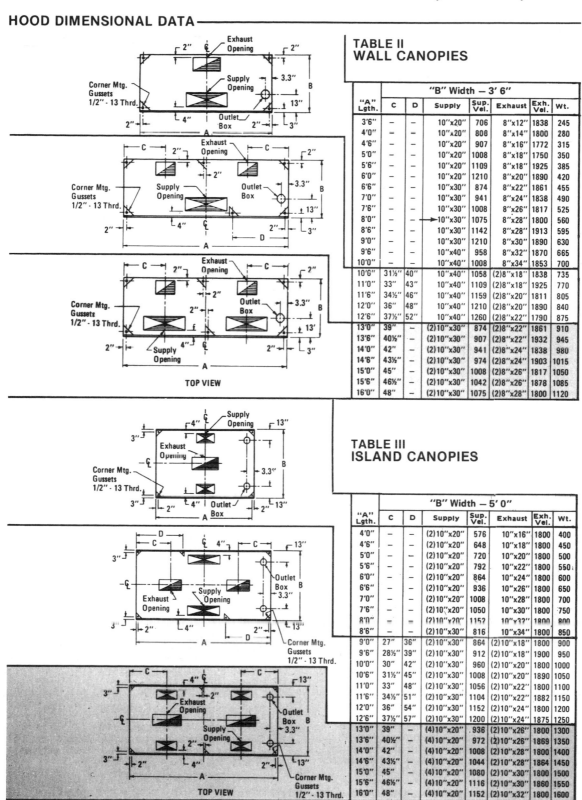

TABLE II
WALL CANOPIES

"A" Lgth.	"B" Width — 3' 6"						
	C	D	Supply	Sup. Vel.	Exhaust	Exh. Vel.	Wt.
3'6"	–	–	10"x20"	706	8"x12"	1838	245
4'0"	–	–	10"x20"	808	8"x14"	1800	280
4'6"	–	–	10"x20"	907	8"x16"	1772	315
5'0"	–	–	10"x20"	1008	8"x18"	1750	350
5'6"	–	–	10"x20"	1109	8"x18"	1925	385
6'0"	–	–	10"x20"	1210	8"x20"	1890	420
6'6"	–	–	10"x30"	874	8"x22"	1861	455
7'0"	–	–	10"x30"	941	8"x24"	1838	490
7'6"	–	–	10"x30"	1008	8"x26"	1817	525
8'0"	–	–	→10"x30"	1075	8"x28"	1800	560
8'6"	–	–	10"x30"	1142	8"x28"	1913	595
9'0"	–	–	10"x30"	1210	8"x30"	1890	630
9'6"	–	–	10"x40"	958	8"x32"	1870	665
10'0"	–	–	10"x40"	1008	8"x34"	1853	700
10'6"	31½"	40"	10"x40"	1058	(2)8"x18"	1838	735
11'0"	33"	43"	10"x40"	1109	(2)8"x18"	1925	770
11'6"	34½"	46"	10"x40"	1159	(2)8"x20"	1811	805
12'0"	36"	48"	10"x40"	1210	(2)8"x20"	1890	840
12'6"	37½"	52"	10"x40"	1260	(2)8"x22"	1790	875
13'0"	39"	–	(2)10"x30"	874	(2)8"x22"	1861	910
13'6"	40½"	–	(2)10"x30"	907	(2)8"x22"	1932	945
14'0"	42"	–	(2)10"x30"	941	(2)8"x24"	1838	980
14'6"	43½"	–	(2)10"x30"	974	(2)8"x24"	1903	1015
15'0"	45"	–	(2)10"x30"	1008	(2)8"x26"	1817	1050
15'6"	46½"	–	(2)10"x30"	1042	(2)8"x26"	1878	1085
16'0"	48"	–	(2)10"x30"	1075	(2)8"x28"	1800	1120

TABLE III
ISLAND CANOPIES

"A" Lgth.	"B" Width — 5' 0"						
	C	D	Supply	Sup. Vel.	Exhaust	Exh. Vel.	Wt.
4'0"	–	–	(2)10"x20"	576	10"x16"	1800	400
4'6"	–	–	(2)10"x20"	648	10"x18"	1800	450
5'0"	–	–	(2)10"x20"	720	10"x20"	1800	500
5'6"	–	–	(2)10"x20"	792	10"x22"	1800	550
6'0"	–	–	(2)10"x20"	864	10"x24"	1800	600
6'6"	–	–	(2)10"x20"	936	10"x26"	1800	650
7'0"	–	–	(2)10"x20"	1008	10"x28"	1800	700
7'6"	–	–	(2)10"x20"	1050	10"x30"	1800	750
8'0"	–	–	(2)10"x20"	1152	10"x32"	1800	800
8'6"	–	–	(2)10"x30"	816	10"x34"	1800	850
9'0"	27"	36"	(2)10"x30"	864	(2)10"x18"	1800	900
9'6"	28½"	39"	(2)10"x30"	912	(2)10"x18"	1900	950
10'0"	30"	42"	(2)10"x30"	960	(2)10"x20"	1800	1000
10'6"	31½"	45"	(2)10"x30"	1008	(2)10"x20"	1890	1050
11'0"	33"	48"	(2)10"x30"	1056	(2)10"x22"	1800	1100
11'6"	34½"	51"	(2)10"x30"	1104	(2)10"x22"	1882	1150
12'0"	36"	54"	(2)10"x30"	1152	(2)10"x24"	1800	1200
12'6"	37½"	57"	(2)10"x30"	1200	(2)10"x24"	1875	1250
13'0"	39"	–	(4)10"x20"	936	(2)10"x26"	1800	1300
13'6"	40½"	–	(4)10"x20"	972	(2)10"x26"	1869	1350
14'0"	42"	–	(4)10"x20"	1008	(2)10"x28"	1800	1400
14'6"	43½"	–	(4)10"x20"	1044	(2)10"x28"	1864	1450
15'0"	45"	–	(4)10"x20"	1080	(2)10"x30"	1800	1500
15'6"	46½"	–	(4)10"x20"	1116	(2)10"x30"	1860	1550
16'0"	48"	–	(4)10"x20"	1152	(2)10"x32"	1800	1600

Hood dimensional data
Figure 9-5B

HOOD DIMENSIONAL DATA

TABLE II (Continued)
WALL CANOPIES

"B" Width — 4' 0"

C	D	Supply	Sup. Vel.	Exhaust	Exh. Vel.	Wt.
–	–	10"x20"	806	8"x14"	1800	280
–	–	10"x20"	922	8"x16"	1800	320
–	–	10"x20"	1037	8"x18"	1800	360
–	–	10"x20"	1152	8"x20"	1800	400
–	–	10"x30"	845	8"x22"	1800	440
–	–	10"x30"	922	8"x24"	1800	480
–	–	10"x30"	998	8"x26"	1800	520
–	–	10"x30"	1075	8"x28"	1800	560
–	–	10"x30"	1152	8"x30"	1800	600
–	–	10"x30"	1229	8"x32"	1800	640
–	–	10"x40"	979	8"x34"	1800	680
27"	31"	10"x40"	1037	(2)8"x18"	1800	720
28½"	34"	10"x40"	1094	(2)8"x18"	1900	760
30"	37"	10"x40"	1152	(2)8"x20"	1800	800
31½"	40"	10"x40"	1210	(2)8"x20"	1890	840
33"	–	(2)10"x30"	845	(2)8"x22"	1800	880
34½"	–	(2)10"x30"	883	(2)8"x22"	1882	920
36"	–	(2)10"x30"	922	(2)8"x24"	1800	960
37½"	–	(2)10"x30"	960	(2)8"x24"	1875	1000
39"	–	(2)10"x30"	998	(2)8"x26"	1800	1040
40½"	–	(2)10"x30"	1037	(2)8"x26"	1869	1080
42"	–	(2)10"x30"	1075	(2)8"x28"	1800	1120
43½"	–	(2)10"x30"	1114	(2)8"x28"	1864	1160
45"	–	(2)10"x30"	1152	(2)8"x30"	1800	1200
46½"	–	(2)10"x30"	1190	(2)8"x30"	1860	1240
48"	–	(2)10"x30"	1229	(2)8"x32"	1800	1280

"B" Width — 4' 6"

C	D	Supply	Sup. Vel.	Exhaust	Exh. Vel.	Wt.
–	–	10"x20"	907	8"x16"	1772	315
–	–	10"x20"	1037	8"x18"	1800	360
–	–	10"x20"	1166	8"x20"	1823	405
–	–	10"x30"	864	8"x22"	1841	450
–	–	10"x30"	950	8"x24"	1856	495
–	–	10"x30"	1037	8"x26"	1869	540
–	–	10"x30"	1123	8"x30"	1755	585
–	–	10"x30"	1210	8"x32"	1772	630
–	–	10"x40"	972	8"x34"	1787	675
24"	25"	10"x40"	1037	(2)8"x18"	1800	720
25½"	28"	10"x40"	1102	(2)8"x18"	1913	765
27"	31"	10"x40"	1166	(2)8"x20"	1823	810
28½"	–	(2)10"x30"	821	(2)8"x20"	1924	855
30"	–	(2)10"x30"	864	(2)8"x22"	1841	900
31½"	–	(2)10"x30"	907	(2)8"x24"	1772	945
33"	–	(2)10"x30"	950	(2)8"x24"	1856	990
34½"	–	(2)10"x30"	994	(2)8"x26"	1791	1035
36"	–	(2)10"x30"	1037	(2)8"x28"	1869	1080
37½"	–	(2)10"x30"	1080	(2)8"x28"	1808	1125
39"	–	(2)10"x30"	1123	(2)8"x30"	1880	1170
40½"	–	(2)10"x30"	1166	(2)8"x30"	1822	1215
42"	–	(2)10"x30"	1210	(2)8"x30"	1890	1260
43½"	–	(2)10"x40"	940	(2)8"x32"	1835	1305
45"	–	(2)10"x40"	972	(2)8"x34"	1787	1350
46½"	–	(2)10"x40"	1004	(2)8"x34"	1846	1395
48"	–	(2)10"x40"	1037	(2)8"x34"	1906	1440

"B" Width — 5' 0"

C	D	Supply	Sup. Vel.	Exhaust	Exh. Vel.	Wt.
–	–	10"x20"	1008	8"x18"	1750	350
–	–	10"x20"	1152	8"x20"	1800	400
–	–	10"x30"	864	8"x22"	1841	450
–	–	10"x30"	960	8"x24"	1875	500
–	–	10"x30"	1056	8"x28"	1768	550
–	–	10"x30"	1152	8"x30"	1800	600
–	–	10"x40"	936	8"x32"	1828	650
–	–	10"x40"	1008	8"x34"	1853	700
22½"	22"	10"x40"	1080	(2)8"x18"	1875	750
24"	25"	10"x40"	1152	(2)8"x20"	1800	800
25½"	28"	10"x40"	1224	(2)8"x20"	1912	850
27"	–	(2)10"x30"	864	(2)8"x22"	1841	900
28½"	–	(2)10"x30"	912	(2)8"x24"	1781	950
30"	–	(2)10"x30"	960	(2)8"x24"	1875	1000
31½"	–	(2)10"x30"	1008	(2)8"x26"	1817	1050
33"	–	(2)10"x30"	1056	(2)8"x26"	1904	1100
34½"	–	(2)10"x30"	1104	(2)8"x28"	1848	1150
36"	–	(2)10"x30"	1152	(2)8"x30"	1800	1200
37½"	–	(2)10"x30"	1200	(2)8"x30"	1875	1250
39"	–	(2)10"x40"	936	(2)8"x32"	1828	1300
40½"	–	(2)10"x40"	972	(2)8"x32"	1898	1350
42"	–	(2)10"x40"	1008	(2)8"x34"	1853	1400
43½"	–	(2)10"x40"	1044	(2)8"x36"	1812	1450
45"	–	(2)10"x40"	1080	(2)8"x36"	1875	1500
46½"	–	(2)10"x40"	1116	(2)8"x38"	1865	1550
48"	–	(2)10"x40"	1152	(2)8"x38"	1895	1600

SUPPLY VELOCITIES BASED ON 80 CFM PER SQUARE FOOT OF CANOPY

EXHAUST VELOCITIES BASED ON 100 CFM PER SQUARE FOOT OF CANOPY

TABLE III (Continued)
ISLAND CANOPIES

"B" Width — 5' 6"

C	D	Supply	Sup. Vel.	Exhaust	Exh. Vel.	Wt.
–	–	(2)10"x20"	634	10"x16"	1980	440
–	–	(2)10"x20"	713	10"x18"	1980	495
–	–	(2)10"x20"	792	10"x20"	1980	550
–	–	(2)10"x20"	871	10"x22"	1980	605
–	–	(2)10"x20"	950	10"x24"	1980	660
–	–	(2)10"x20"	1030	10"x26"	1980	715
–	–	(2)10"x20"	1109	10"x28"	1980	770
–	–	(2)10"x20"	1188	10"x30"	1980	825
–	–	(2)10"x30"	845	10"x32"	1980	880
–	–	(2)10"x30"	898	10"x34"	1980	935
27"	36"	(2)10"x30"	950	(2)10"x18"	1980	990
28½"	39"	(2)10"x30"	1003	(2)10"x20"	1881	1045
30"	42"	(2)10"x30"	1056	(2)10"x20"	1980	1100
31½"	45"	(2)10"x30"	1109	(2)10"x22"	1890	1155
33	48	(2)10"x30"	1162	(2)10"x22"	1980	1210
34½"	–	(4)10"x20"	911	(2)10"x24"	1897	1265
36"	–	(4)10"x20"	950	(2)10"x24"	1980	1320
37½"	–	(4)10"x20"	990	(2)10"x26"	1904	1375
39"	–	(4)10"x20"	1030	(2)10"x26"	1980	1430
40½"	–	(4)10"x20"	1069	(2)10"x26"	1909	1485
42"	–	(4)10"x20"	1109	(2)10"x28"	1980	1540
43½"	–	(4)10"x20"	1148	(2)10"x30"	1914	1595
45	–	(4)10"x20"	1188	(2)10"x30"	1980	1650
46½"	–	(4)10"x30"	818	(2)10"x32"	1918	1705
48"	–	(4)10"x30"	845	(2)10"x32"	1980	1760

"B" Width — 6' 0"

C	D	Supply	Sup. Vel.	Exhaust	Exh. Vel.	Wt.
–	–	(2)10"x20"	691	12"x16"	1800	480
–	–	(2)10"x20"	778	12"x18"	1800	540
–	–	(2)10"x20"	864	12"x20"	1800	600
–	–	(2)10"x20"	950	12"x22"	1800	660
–	–	(2)10"x20"	1037	12"x24"	1800	720
–	–	(2)10"x20"	1123	12"x26"	1800	780
–	–	(2)10"x20"	1210	12"x28"	1800	840
–	–	(2)10"x30"	864	12"x30"	1800	900
–	–	(2)10"x30"	922	12"x32"	1800	960
–	–	(2)10"x30"	979	12"x34"	1800	1020
27"	36"	(2)10"x30"	1037	(2)12"x18"	1800	1080
28½"	39"	(2)10"x30"	1094	(2)12"x18"	1900	1140
30"	42"	(2)10"x30"	1152	(2)12"x20"	1800	1200
31½"	45"	(2)10"x30"	1210	(2)12"x20"	1890	1260
33"	–	(4)10"x20"	950	(2)12"x22"	1800	1320
34½"	–	(4)10"x20"	994	(2)12"x22"	1882	1380
36"	–	(4)10"x20"	1037	(2)12"x24"	1800	1440
37½"	–	(4)10"x20"	1080	(2)12"x24"	1875	1500
39"	–	(4)10"x20"	1123	(2)12"x26"	1800	1560
40½"	–	(4)10"x20"	1166	(2)12"x26"	1809	1620
42"	–	(4)10"x20"	1210	(2)12"x28"	1800	1680
43½"	–	(4)10"x30"	835	(2)12"x28"	1864	1740
45"	–	(4)10"x30"	864	(2)12"x30"	1800	1800
46½"	–	(4)10"x30"	893	(2)12"x30"	1860	1860
48"	–	(4)10"x30"	922	(2)12"x32"	1800	1920

"B" Width — 6' 6"

C	D	Supply	Sup. Vel.	Exhaust	Exh. Vel.	Wt.
–	–	(2)10"x20"	749	12"x16"	1950	520
–	–	(2)10"x20"	842	12"x18"	1950	585
–	–	(2)10"x20"	936	12"x20"	1950	650
–	–	(2)10"x20"	1030	12"x22"	1950	715
–	–	(2)10"x20"	1123	12"x24"	1950	780
–	–	(2)10"x20"	1217	12"x26"	1950	845
–	–	(2)10"x30"	874	12"x28"	1950	910
–	–	(2)10"x30"	936	12"x30"	1950	975
–	–	(2)10"x30"	998	12"x32"	1950	1040
–	–	(2)10"x30"	1061	12"x34"	1950	1105
27"	36"	(2)10"x30"	1123	(2)12"x18"	1950	1170
28½"	39"	(2)10"x30"	1186	(2)12"x20"	1852	1235
30"	–	(4)10"x20"	936	(2)12"x20"	1950	1300
31½"	–	(4)10"x20"	983	(2)12"x22"	1861	1365
33"	–	(4)10"x20"	1030	(2)12"x22"	1950	1430
34½"	–	(4)10"x20"	1076	(2)12"x24"	1869	1495
36"	–	(4)10"x20"	1123	(2)12"x24"	1950	1560
37½"	–	(4)10"x20"	1170	(2)12"x26"	1875	1625
39"	–	(4)10"x20"	1217	(2)12"x26"	1950	1690
40½"	–	(4)10"x30"	842	(2)12"x28"	1880	1755
42"	–	(4)10"x30"	874	(2)12"x28"	1950	1820
43½"	–	(4)10"x30"	905	(2)12"x30"	1885	1885
45	–	(4)10"x30"	936	(2)12"x30"	1950	1950
46½"	–	(4)10"x30"	967	(2)12"x32"	1889	2015
48"	–	(4)10"x30"	998	(2)12"x32"	1950	2080

Hood dimensional data
Figure 9-5B (continued)

HOOD DIMENSIONAL DATA

TABLE II (Continued) WALL CANOPIES

		"B" Width — 5' 6"					
C	D	Supply	Sup. Vel.	Exhaust	Exh. Vel.	Wt.	"A" Lgth.
–	–	10"x20"	1109	10"x16"	1732	385	3'6"
–	–	10"x30"	845	10"x18"	1760	440	4'0"
–	–	10"x30"	950	10"x20"	1782	495	4'6"
–	–	10"x30"	1066	10"x22"	1800	550	5'0"
–	–	10"x30"	1162	10"x24"	1815	605	5'6"
–	–	10"x40"	950	10"x26"	1828	660	6'0"
–	–	10"x40"	1030	10"x28"	1839	715	6'6"
–	–	10"x40"	1109	10"x30"	1848	770	7'0"
–	–	10"x40"	1188	10"x32"	1856	825	7'6"
24"	–	(2)10"x30"	845	(2)10"x18"	1760	880	8'0"
26½"	–	(2)10"x30"	898	(2)10"x18"	1870	935	8'6"
27"	–	(2)10"x30"	950	(2)10"x20"	1782	990	9'0"
28½"	–	(2)10"x30"	1003	(2)10"x20"	1881	1045	9'6"
30"	–	(2)10"x30"	1056	(2)10"x22"	1800	1100	10'0"
31½"	–	(2)10"x30"	1109	(2)10"x22"	1890	1155	10'6"
33"	–	(2)10"x30"	1162	(2)10"x24"	1815	1210	11'0"
34½"	–	(2)10"x40"	911	(2)10"x24"	1898	1265	11'6"
36"	–	(2)10"x40"	950	(2)10"x26"	1828	1320	12'0"
37½"	–	(2)10"x40"	990	(2)10"x28"	1768	1375	12'6"
39"	–	(2)10"x40"	1030	(2)10"x28"	1839	1430	13'0"
40½"	–	(2)10"x40"	1069	(2)10"x30"	1782	1485	13'6"
42"	–	(2)10"x40"	1109	(2)10"x30"	1848	1540	14'0"
43½"	–	(2)10"x40"	1148	(2)10"x32"	1794	1595	14'6"
45"	–	(2)10"x40"	1188	(2)10"x32"	1856	1650	15'0"
46½"	–	(2)10"x40"	1228	(2)10"x34"	1805	1705	15'6"
48"	–	(2)10"x40"	1267	(2)10"x34"	1864	1760	16'0"

TABLE IV BACKSHELF HOODS

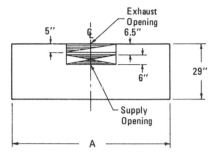

"A" Lgth.	Supply	Sup. Vel.	Exhaust	Exh. Vel.	Wt.
3'0"	6"x14"	1028	5"x14"	1543	195
3'6"	6"x16"	1050	5"x16"	1575	228
4'0"	6"x18"	1067	5"x18"	1600	260
4'6"	6"x20"	1080	6"x20"	1620	293
5'0"	6"x22"	1091	5"x22"	1636	325
5'6"	6"x24"	1100	5"x24"	1650	358
6'0"	6"x26"	1108	5"x26"	1662	390
6'6"	6"x28"	1114	5"x28"	1671	423
7'0"	6"x30"	1120	5"x30"	1680	455
7'6"	6"x32"	1125	5"x32"	1688	488
8'0"	6"x34"	1129	5"x34"	1694	520
8'6"	6"x36"	1133	5"x36"	1700	553
9'0"	6"x38"	1137	5"x38"	1705	585
9'6"	6"x40"	1140	5"x40"	1710	618
10'0"	6"x42"	1143	5"x42"	1714	650

SUPPLY VELOCITIES BASED ON 200 CFM PER LINEAR FOOT OF BACKSHELF

EXHAUST VELOCITIES BASED ON 250 CFM PER LINEAR FOOT OF BACKSHELF

PATENTED: U.S. Numbers — 3,978,777; 4,047,519; 4,153,044 Canadian Numbers — 1,051,710; 1,069,749.

TABLE III (Continued) ISLAND CANOPIES

"B" Width — 7' 0"

C	D	Supply	Sup. Vel.	Exhaust	Exh. Vel.	Wt.
–	–	(2)10"x20"	806	14"x16"	1800	560
–	–	(2)10"x20"	907	14"x18"	1800	630
–	–	(2)10"x20"	1008	14"x20"	1800	700
–	–	(2)10"x20"	1109	14"x22"	1800	770
–	–	(2)10"x20"	1210	14"x24"	1800	840
–	–	(2)10"x30"	874	14"x26"	1800	910
–	–	(2)10"x30"	941	14"x28"	1800	980
–	–	(2)10"x30"	1008	14"x30"	1800	1050
–	–	(2)10"x30"	1075	14"x32"	1800	1120
–	–	(2)10"x30"	1143	14"x34"	1800	1190
27"	36"	(2)10"x30"	1210	(2)14"x18"	1800	1260
28½"	–	(4)10"x20"	958	(2)14"x18"	1900	1330
30"	–	(4)10"x20"	1008	(2)14"x20"	1800	1400
31½"	–	(4)10"x20"	1058	(2)14"x20"	1890	1470
33"	–	(4)10"x20"	1109	(2)14"x22"	1800	1540
34½"	–	(4)10"x20"	1159	(2)14"x22"	1882	1610
36"	–	(4)10"x20"	1210	(2)14"x24"	1800	1680
37½"	–	(4)10"x30"	840	(2)14"x24"	1875	1750
39"	–	(4)10"x30"	874	(2)14"x26"	1800	1820
40½"	–	(4)10"x30"	907	(2)14"x26"	1869	1890
42"	–	(4)10"x30"	941	(2)14"x28"	1800	1960
43½"	–	(4)10"x30"	974	(2)14"x28"	1864	2030
45"	–	(4)10"x30"	1008	(2)14"x30"	1800	2100
46½"	–	(4)10"x30"	1042	(2)14"x30"	1860	2170
48"	–	(4)10"x30"	1075	(2)14"x32"	1800	2240

"B" Width — 7' 6"

C	D	Supply	Sup. Vel.	Exhaust	Exh. Vel.	Wt.
–	–	(2)10"x20"	864	14"x16"	1929	600
–	–	(2)10"x20"	972	14"x18"	1929	675
–	–	(2)10"x20"	1080	14"x20"	1929	750
–	–	(2)10"x20"	1188	14"x22"	1929	825
–	–	(2)10"x30"	864	14"x24"	1929	900
–	–	(2)10"x30"	936	14"x26"	1929	975
–	–	(2)10"x30"	1008	14"x28"	1929	1050
–	–	(2)10"x30"	1080	14"x30"	1929	1125
–	–	(2)10"x30"	1152	14"x32"	1929	1200
–	–	(2)10"x40"	810	14"x34"	1929	1275
27"	–	(4)10"x20"	972	(2)14"x18"	1929	1350
28½"	–	(4)10"x20"	1026	(2)14"x20"	1832	1425
30"	–	(4)10"x20"	1080	(2)14"x20"	1929	1500
31½"	–	(4)10"x20"	1134	(2)14"x22"	1841	1575
33"	–	(4)10"x20"	1188	(2)14"x22"	1929	1650
34½"	–	(4)10"x30"	828	(2)14"x24"	1848	1725
36"	–	(4)10"x30"	864	(2)14"x24"	1929	1800
37½"	–	(4)10"x30"	900	(2)14"x26"	1854	1875
39"	–	(4)10"x30"	936	(2)14"x26"	1929	1950
40½"	–	(4)10"x30"	972	(2)14"x28"	1860	2025
42"	–	(4)10"x30"	1008	(2)14"x28"	1929	2100
43½"	–	(4)10"x30"	1044	(2)14"x30"	1864	2175
45"	–	(4)10"x30"	1080	(2)14"x30"	1929	2250
46½"	–	(4)10"x30"	1116	(2)14"x32"	1868	2325
48"	–	(4)10"x30"	1152	(2)14"x32"	1929	2400

"B" Width — 8' 0"

C	D	Supply	Sup. Vel.	Exhaust	Exh. Vel.	Wt.	"A" Lgth.
–	–	(2)10"x20"	922	16"x16"	1800	640	4'0"
–	–	(2)10"x20"	1037	16"x18"	1800	720	4'6"
–	–	(2)10"x20"	1152	16"x20"	1800	800	5'0"
–	–	(2)10"x30"	845	16"x22"	1800	880	5'6"
–	–	(2)10"x30"	922	16"x24"	1800	960	6'0"
–	–	(2)10"x30"	998	16"x26"	1800	1040	6'6"
–	–	(2)10"x30"	1075	16"x28"	1800	1120	7'0"
–	–	(2)10"x30"	1152	16"x30"	1800	1200	7'6"
–	–	(2)10"x40"	922	16"x32"	1800	1280	8'0"
–	–	(2)10"x40"	970	16"x34"	1800	1360	8'6"
27"	–	(4)10"x20"	1037	(2)16"x18"	1800	1440	9'0"
28½"	–	(4)10"x20"	1095	(2)16"x18"	1800	1520	9'6"
30"	–	(4)10"x20"	1152	(2)16"x20"	1800	1600	10'0"
31½"	–	(4)10"x20"	1210	(2)16"x20"	1890	1680	10'6"
33"	–	(4)10"x30"	845	(2)16"x22"	1800	1760	11'0"
34½"	–	(4)10"x30"	883	(2)16"x22"	1882	1840	11'6"
36"	–	(4)10"x30"	922	(2)16"x24"	1800	1920	12'0"
37½"	–	(4)10"x30"	960	(2)16"x24"	1875	2000	12'6"
39"	–	(4)10"x30"	998	(2)16"x26"	1800	2080	13'0"
40½"	–	(4)10"x30"	1037	(2)16"x26"	1869	2160	13'6"
42"	–	(4)10"x30"	1075	(2)16"x28"	1800	2240	14'0"
43½"	–	(4)10"x30"	1114	(2)16"x28"	1864	2320	14'6"
45"	–	(4)10"x30"	1152	(2)16"x30"	1800	2400	15'0"
46½"	–	(4)10"x30"	1190	(2)16"x30"	1860	2480	15'6"
48	–	(4)10"x30"	1229	(2)16"x32"	1800	2560	16'0"

Courtesy: Econovent Systems, Inc.

**Hood dimensional data
Figure 9-5B (continued)**

Typical data for the supply air pressure drops may read something like this:

Wall and Island Canopy Hoods
1) Supply side static pressure drops:
a) internal: 3/8" to 1/2"
b) total: 3/4" (Assume 20' supply duct run at 1,500 fpm velocity and two 90-degree elbows with turning vanes and gravity backdraft dampers.)

This means you can use a total static pressure of 3/4" to size your supply fan, as long as your installation doesn't exceed the guidelines above. Figure 9-6 lists typical supply fan and performance data based on the above limitations. Select the supply fan at the calculated cfm and static pressure. Choose a model that will operate at its middle range of rpm. Never undersize the horsepower of the motor. A good rule of thumb is to size for 125% capacity. We'll choose model AS-12.

If you have a job where the installation is more involved, call the manufacturer to enlist his help in selecting the right size equipment.

When installing the untempered air fan and ductwork, follow these guidelines:

• Place the supply fan a minimum of 10' (or more — check local code) from any exhaust unit, flue or vent stack.

• Select the best possible route for ductwork — as straight and as short as possible. Be aware of beams, joists and other obstacles.

• Remember to locate the balancing damper at or near the hood.

• Insulate exterior air supply ductwork with insulation that meets these standards: minimum 4-pound density, 200 PSF at 10% deformation compressive strength, K-factor of 0.24 at 75°F, with factory-applied aluminum facing.

• Fasten the insulation with mechanical fasteners on 18" centers, with a minimum of two rows of fasteners on all sides of the ducts. Seal all joints and punctures with vapor barrier mastic.

Exhaust Air

Select the exhaust air system from the hood in the same way you selected the supply air system. Size the ductwork according to the hood exhaust opening, which is based on a 2,000 fpm velocity. The pressure drops listed will be based on an assumed straight duct length of 15', backdraft damper if code required, and mesh or panel filters. The total static pressure is usually between 3/4" and 1¼", depending on the installation. Choose the exhaust fan based on your full calculated cfm and appropriate static pressure. Use the manufacturer's data to choose the correct size fan and motor. Again, if you have a job that's different from the manufacturer's guidelines, give the company a call and they'll be glad to help.

When installing the exhaust ductwork, locate the roof outlet as close as possible to the hood. This will keep your run short. Avoid sharp bends if possible. Be aware that many codes require welded seams, access doors for cleaning, and a two-hour fire-rated enclosure for exhaust ducts. They often require exhaust velocities of at least 1,500 to 2,000 fpm.

For the entire installation — hood, supply and exhaust work — coordinate with the inspectors, manufacturer, owner, architect, engineer, general contractor, and other trades. Make sure they all have the information necessary for a proper installation. Determine the correct sizing and placement of roof openings, electrical connections, structural anchors, and equipment location and clearances. If this is bid work, make sure it's clearly stated who is responsible for work such as roof openings and curbs, exhaust duct fire-rated enclosure, supply duct insulation and the like. Keep a checklist so you don't overlook any items. Figure 9-7 is a checklist we use. Use it as it is, or adopt any ideas from it to make your own. Remember, you'll always bid an item you overlook as free.

Hoods also need lights and fire extinguishing systems. Some hood manufacturers furnish these. Or they can be installed by a local contractor. Be sure you find out who's going to furnish these items.

The filters in the hoods must meet code requirements. Your filter supplier will know about these.

Finally, get all local code variances and other trade agreements for related work in writing.

Balancing the System

Under ideal conditions, it's possible for you to design a ventilation, warm air or hydronic system

SUPPLY FAN

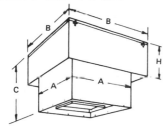
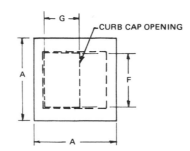

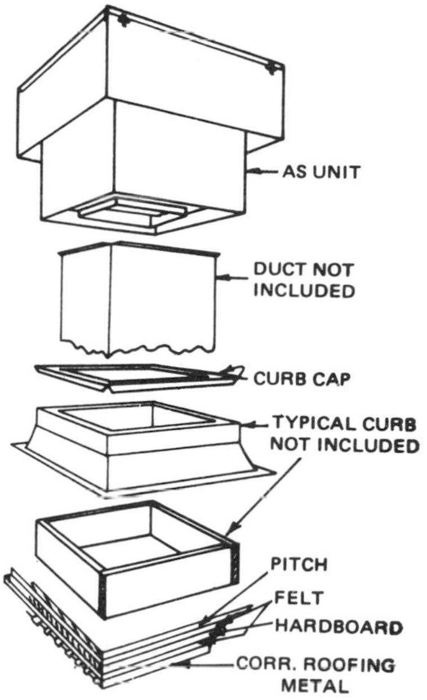

MODEL	A	B	C	BLOWER DISCHARGE D	E	F	G	H	ROOF CURB O.D.	ROOF OPN'G	Filter Size 4 Req'd	WT. LBS.
AS 10(*)	28-3/8	38-1/2	30	11-3/8	13-1/8	14	12	12-1/4	27x27	22x22	12x24	188
AS 12(*)	23-3/8	42-5/8	33-1/8	13-7/16	15-5/8	16	14	16-1/2	31x31	23x23	16x25	234
AS 15(*)	32-3/8	42-5/8	33-1/8	15-7/8	18-5/8	19	17	16-1/2	31x31	23x23	16x25	260
AS 18(*)	30-3/8	45-3/4	40-1/4	18-7/8	21-7/8	23	20	20-3/4	35x35	27x27	20x25	350

* Denotes motor size.

The supply air fan is a forward curve utility blower mounted in a down-blast discharge position. The fan is enclosed in a galvanized sheet metal housing, with (4) permanent cleanable filters, disconnect switch (1 or 3 phase), curb cap, vibration isolators, and all motors, drive sets and belts. Easy access to fan is provided by a lift-off cover and a side access plate.

PERFORMANCE

MODEL	Air Flow CFM	Out Vel. FPM	.125 in S.P. RPM	MHP	.250 in S.P. RPM	MHP	.375 in S.P. RPM	MHP	.500 in S.P. RPM	MHP	.625 in S.P. RPM	MHP	.750 in S.P. RPM	MHP
AS-10(*)	600	588	350											
	700	686	355		440									
	800	784	360	1/4	450	1/4	570	1/4						
	1000	980	375		460		575							
	1200	1176	395		470		580		660	1/3	745	1/3	810	1/3
	1400	1372	430		500		600	1/3	670		750		815	
	1600	1569	480		530	1/3	620		680		760	1/2	830	1/2
	1800	1765	520	1/3	590		645	1/2	705	1/2	785		840	
	2000	1961	575		620		690		740		810		890	
	2200	2157	620		680	1/2	730		790		860	3/4	905	3/4
	2400	2353	660	1/2	700		760	3/4	810	3/4	880		920	
AS-12(*)	1600	1111	350		400	1/4	490		550	1/3	605	1/3		
	1800	1250	360	1/4	410		495	1/3	560		610			
	2000	1389	370		430	1/3	500		570		615	1/2		
	2200	1528	380		450		510		575	1/2	620		655	1/2
	2400	1667	400	1/3	460	1/2	530	1/2	590		630		660	
	2600	1806	430		490		540		600		640	3/4	680	3/4
	2800	1945	450	1/2	510		560		610	3/4	650		700	
	3000	2084	470		540	3/4	580	3/4	625		680		715	
	3200	2223	505	3/4	560		605		650		690	1	730	1
	3400	2362	530		580		625		675	1	710		750	
AS-15(*)	3000	1493	340		400	1/2	440	1/2	490		530		575	3/4
	3200	1592	360	1/2	410		460		500	3/4	540	3/4	580	
	3400	1692	380		425		470		510		550		585	
	3600	1791	400		430		480	3/4	520		560	1	590	1
	3800	1891	410		450	3/4	490		525	1	565		590	
	4000	1990	430	3/4	460		505	1	530		570		595	
	4200	2089	440		470	1	520		540		575		605	
	4400	2189	450		500		525		560		585	1-1/2	620	1-1/2
	4600	2289	475	1	510		540	1-1/2	575	1-1/2	600		635	
	4800	2388	500		525	1-1/2	545		580		605		640	
	5000	2488	525		540		575		600		620		650	
AS-18(*)	3000	1045	250		275	1/2	325	1/2	380		415	3/4		
	3500	1219	255	1/2	280		340		385	3/4	420		460	1
	4000	1394	260		285		345		390		425	1	470	
	4500	1568	270		295	3/4	350	3/4	395	1	435	1-1/2	480	1-1/2
	5000	1742	275	3/4	305		355		400	1-1/2	450		490	
	5500	1916	290		330	1	360	1	420		460	2	500	2
	6000	2090	320	1	350		380	1-1/2	430	2	480		510	
	6500	2265	340		370	1-1/2	395		440					
	7000	2440	360	1-1/2	390		410	2						

* Recommended Minimum Motor Horse Power (MHP)

3 – 1/4 HP   5 – 1/2 HP   7 – 1 HP   9 – 2 HP
4 – 1/3 HP   6 – 3/4 HP   8 – 1-1/2 HP   0 – 3 HP

Courtesy: Econovent Systems, Inc.

Supply fan data
Figure 9-6

Building location: _____

Name, address and phone number of:

   Owner: _____

   Architect: _____

   HVAC consultant: _____

   Building inspector: _____

   Fire inspector: _____

   Health inspector: _____

   Electrical inspector: _____

Check applicable codes:

   National Fire Protection Association Chapter 96 _____

   Building Officials Code Administration _____

   Uniform Mechanical Code _____

   U.S. Dept. of Agriculture _____

   National Electrical Code _____

   Local code _____

Permit applications and shop drawing submitted to: Approval received:

   Name _____ Date _____ Date _____

   Name _____ Date _____ Date _____

   Name _____ Date _____ Date _____

   Name _____ Date _____ Date _____

Equipment manufacturer; Name and phone number of local representative: _____

_____

Type of equipment needed: _____

_____

_____

_____

1)   Length, width and height of exhaust hood(s)? _____
      Field welding of separate modules required? _____

2)   Is the floor to ceiling height adequate for the installation? _____
      6'6" minimum from floor to underside of exhaust hood, 7'0" maximum height? _____

Kitchen venting system checklist
Figure 9-7

3) Is there room above ceiling to run supply and exhaust ductwork? _____
Does the building structure interfere with hood or ductwork installation? _____
Who's responsible for roof openings? _____

4) Wooden or steel joists or rafters? _____
Check local codes for fire protection clearances and protection between ductwork and joists. _____
_____

5) Determine the basic exhaust system layout, in the kitchen and on the roof. _____
Check its location with existing or planned other HVAC equipment. _____
Is supply air intake sized for 20% make-up air? _____
Is it located a minimum of 10' from any exhaust fan or vent stack? (or local code-required distance?)
_____

6) Obtain spec sheets of cooking equipment. _____
Determine equipment layout with architect and owner. _____
Determine fuel type - gas, electric, propane? _____

7) Review equipment size and location. _____
Make sure selected hood(s) have proper overhang and height. _____
Supply and exhaust openings are no more than 6" left or right of hood centerline? _____

8) Review return grille(s) location in kitchen. _____
A minimum of 15' from any exhaust hood? _____
Maximum velocity of 500 fpm? _____

9) Is a fire protection system required? _____
Who provides it? _____

10) Are balancing dampers installed on the supply side of the hoods? _____
Is adjustment rod exposed and accessible? _____

11) Does the local code allow backdraft dampers for supply and exhaust ducts? _____

12) Is the necessary ductwork insulated? _____
Has the electrical contractor been notified of electrical requirements? _____

13) Are duct access doors readily accessible? _____

14) Will 20% tempered air be supplied by existing or planned HVAC equipment? _____
Are the outside air intakes adjusted for this increase? _____
Are transfer grilles necessary? _____
Will they need fire dampers? _____

15) Select supply and exhaust fans. _____
Size 25% larger than required. _____

16) Installation field tested and adjusted? _____
Owner informed of maintenance and cleaning schedule? _____

Kitchen venting system checklist
Figure 9-7 (continued)

that's self-balancing. But in the real world, it's more likely that a system you design and install will need some fine-tuning. That's why it's important that dampers or balancing valves are installed in the proper locations in the system.

The purpose of balancing is to make sure the system delivers the volume of air or water necessary to provide satisfactory heat distribution. You'll size the duct or piping components to deliver air or water without excessive pressure at the fan or pump. That is, the pressure loss from the fan or pump to the supply outlet will be sufficient to leave just enough pressure to overcome the pressure loss through the supply outlets along the entire system. But it's not possible to predict these losses exactly at the design stage. When the system is installed, actual conditions will change pressure drops along the way. For that reason, you'll rely on dampers or balancing valves for the final system adjustment.

Before reading the next section, you may find it helpful to review Chapter 2 on testing instruments.

Air System Balancing

Air system fans operate at a constant speed. For a given air flow rate, a certain pressure will be developed at the fan discharge. If the pressure changes, the quantity of air delivered will also change. Each damper adjustment affects the pressure available at the fan and the air flow through the system. The air flow from an individual outlet cannot be adjusted without affecting the rest of the system. Therefore, you must balance the system using a systematic procedure. We recommend a method called *proportional balancing*.

Figure 9-8A shows a fan and duct system schematic. Its pressure and air flow characteristics are graphed on the chart in Figure 9-8B. The duct system has four supply outlets in a branch line with a damper at each outlet. There's also a damper in the branch duct between the fan and supply outlets.

Each part of the system has this characteristic: *The pressure required to overcome the resistance of that section of the system varies as the square of the volume of air flowing through it.*

For example, Curve A shows the pressure characteristic of the last section of duct from Outlet 2 to Outlet 1. (We'll assume Outlet 1 is at the end of the system). As the flow to supply Outlet 2 doubles, from a value of 8 units to 16 units, the

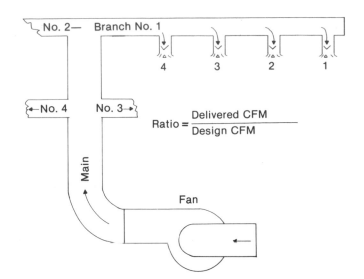

$$\text{Ratio} = \frac{\text{Delivered CFM}}{\text{Design CFM}}$$

Fan and duct system schematic
Figure 9-8A

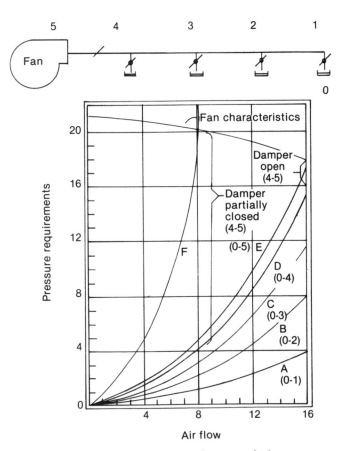

Fan and duct system characteristics
Figure 9-8B

pressure requirement becomes four times as great. It increases from one unit to four units of pressure. As more sections are added to the duct system, the same process applies. Curve C represents the system including the last three supply outlets. The flow increased from 8 to 16 units and the pressure requirement went up from 3 units to 12 units.

Curve D shows the pressure of the system from Outlet 4 on to the end of the system. This again has the same square relationship between the pressure needed and the air flow. The entire system, from the fan to the end, is represented by Curve E. This includes the open damper in the branch duct between points 4 and 5.

If that damper is partially closed, a new system curve, Curve F, will be defined. The intersection of this curve with the fan curve shows the new air flow through the system. Note that the relationship between successive parts of the system downstream from the Curve F damper was not changed by varying the damper position. The pressure requirement of system section 0-2 is twice the pressure requirement for section 0-1. Section 0-3 is twice that of 0-2, and so on, regardless of the position of the damper in section 4-5. This relationship is also independent of the total volume of air flow.

We took the time to go through the whole system to illustrate a key point in proportional balancing: *Once the successive sections of a system are balanced or adjusted to one another, the relationship of the two parts of the system will not be altered if those sections are not adjusted further.* For this reason, the outlet farthest from the fan is adjusted first in proportionate balancing. And it is adjusted to establish a ratio between the measured air flow at that point and the design air flow for that point. The next farthest outlet from the fan is adjusted until a ratio between actual and design flow rate is established that is the same as that established at the first outlet.

The same relationship between successive sections of the system also applies to branches of the system. That is, once two successive branches are adjusted, the pressure requirements of the two branches will not be altered by adjustments to the system upstream from those branches.

A critical point in balancing is that outlets in the same branch must be adjusted to the same ratio of *actual* to *design* air flow rate. Any ratio can be chosen as long as the same one is used for each outlet. However, the industry standard is a ratio of

1.0. With a properly designed system, this will result in the smallest adjustment of fan speed.

Steps for Proportional Balancing
Follow these steps each time you're using proportional balancing to adjust a system:

Preparation before going to the job—

1) Make sure the system is finished and started up.

2) Obtain the design drawings showing each fan and outlet. Note the total flow for each system and outlet.

3) Locate on the drawings each main duct damper and each branch duct damper. A *main duct* is defined as a duct that serves two or more branch ducts. A *branch duct* is defined as a duct that serves two or more outlets.

4) Obtain an anemometer. Get the proper K-factors from the manufacturer for the returns and exhausts. (K-factors are explained in Chapter 2.) The K-factor depends on the design of the diffuser. Keep in mind that similar designs by different manufacturers will have different K-factors. Don't guess. Get the right one.

It's also important to get a schematic showing the proper location of the anemometer's probe for testing, along with the serial numbers of the probe and anemometer. Check the instrument data to be sure you have the right scale on the instrument for the job, and that its probe matches the instrument. Remember, if you don't record K-factor velocities in the correct manner and location, with the correct instrument and probe, your field data will be worthless.

5) Prepare calculation and report sheets. Include columns for room location, outlet number, model and size, design flow rate, flow factor or K-factor, and the K-factor velocity for each outlet.

Preparation on the job—

1) Set all dampers in the system in the open position.

2) Lock the outside air or recirculating air damper in the open position.

3) Start the fan and check the direction of its rotation.

4) Adjust the fan speed if necessary to obtain the maximum air flow without overloading the motor. The air flow should be equal to or greater than the design air flow. Check this flow with your instrument at the main return duct close to the fan. If it's not equal to or greater than the design flow, you may have to increase your motor size. But first check the duct system for defects or changes made on the job and not shown on plan revisions. If the air flow is insufficient because of a defect or change, there's no point in trying to balance the system. Determine the cause of the problem and notify the owner of your findings.

Testing and Adjustment

Use a balancing worksheet like the one in Figure 9-9. It's very important that the outlets be numbered, tested and adjusted in the correct sequence. The farthest branch and the farthest outlet on the branch will be Outlet 1 on Branch 1. The remaining outlets and branches are numbered consecutively, working back toward the fan. Test and adjust the outlets and branches in the same sequence.

Follow these steps:

1) Spot check outlets along the branch to determine if enough air is being delivered to give measurable readings.

2) Measure velocities at Outlet 1 on the farthest branch.

3) Find the average velocity and record it on the worksheet. Make sure you place the number on the correct line and column on the sheet.

4) Find the ratio (R) of measured to design velocity. The formula is:

$$R = \frac{\text{Measured velocity}}{\text{Design K-factor velocity}}$$

Record the ratio on the worksheet.

5) Find the velocity and ratio for Outlet 2 and write it down. Compare ratios for Outlets 1 and 2. If they aren't equal, adjust Outlet 2 so its R will approach Outlet 1's R. *Do not* adjust Outlet 1.

6) After adjustment, record the new velocity and ratio for Outlet 2.

7) Go back to Outlet 1. Measure the velocity and figure the ratio. If the ratios between Outlet 1 and 2 vary by more than 0.1, they aren't proportionately balanced. Further adjustment to Outlet 2 will be needed. After each adjustment, make sure you record the new values for both Outlets 1 and 2. Continue the adjusting procedure until the R values fall within 0.1. *Remember to only adjust Outlet 2.*

8) After Outlets 1 and 2 are balanced with each other, proceed to Outlet 3. Measure and record velocity. Figure the ratio and record it. Compare the R with the R for Outlet 2. If necessary, adjust Outlet 3 to get the same R value as shown for Outlet 2. *Do not* adjust Outlets 1 or 2.

9) Any adjustments of Outlet 3 will change the air flow through Outlets 1 and 2. However, as long as Outlets 1 and 2 are not adjusted (after they are balanced) they will remain proportionately balanced with each other. The final R value for Outlets 2 and 3 will also apply to Outlet 1, and all three will be proportionately balanced with each other.

10) Adjust Outlet 4 so its R value matches Outlet 3. Continue to each outlet in the branch until all outlets are balanced.

11) It is essential that once an outlet is proportionately balanced, you leave it alone. Don't make any further adjustments to that outlet or those previously adjusted.

12) Balance each branch in a similar manner. Measure the air flow at a selected outlet in each of the two branches farthest from the fan. Adjust branch dampers until the R values from the selected outlets are equal. The two branch ducts are now proportionately balanced.

13) Adjust each branch to balance it with the preceding branch in the same manner as you balanced the outlets.

14) Check the total air flow at the fan. Adjust the speed to obtain the correct total flow through the system. Check to be sure the fan motor is not overloaded. The fan air flow should be within plus

Job name _____ Page 1 of _____

Address _____

Balancing test by _____ Date _____

System description _____

		Required	Actual
Fan: Make _____			
Size _____	Fan rpm	_____	_____
Type _____	System cfm	_____	_____

		Rated	Actual
Motor: hp _____	Line volts	_____	_____
rpm _____	Motor amps	_____	_____

Branch no. _____

Room no.	Outlet no.	Outlet size and type	K factor	Design cfm	Design Vel_D*		Adjustment I Vel_M*	I R*	II Vel_M	II R	III Vel_M	III R
	1					B						
	2					A						
						B						
	3					A						
						B						
	4					A						
						B						

*Vel_D = Velocity at design condition
*Vel_M = Velocity measured

$R = M/D$ = Measured-to-design velocity ratio

Branch no.   cfm

Branch ___   _____

Branch ___   _____

Branch ___   _____

Branch ___   _____

Total cfm @ fan _____

Remarks: _____

_____

_____

Signed _____ Date _____

Air system balancing sheet
Figure 9-9

or minus 5% of the design flow. If you have the fan speed adjusted to give an air flow a little larger than required, use the supply damper at the fan to reduce the flow.

15) Check the air flow at the outlets at the end of each branch to be sure they meet the system requirements. Unlock the outside air or recirculating damper. Write up a final report and send a copy to the owner.

Hot Water System Balancing

Water systems are balanced like air systems, with one significant difference: The heat output of the hot water radiation unit doesn't vary directly with the water flow rate. For example, in an air system, at 50% of the rated flow, the heat output will be reduced by one-half. In a water system at 50% rated flow, heat output will be reduced by only 10%.

Reverse Return Systems

For balancing a two-pipe, reverse return system, you can determine the flow of water in each radiation loop by measuring the temperature difference between the water entering and the water leaving the loop. For this example, we'll use the baseboard serving one apartment as the loop. Using this data, the manufacturer's baseboard rating, and a conversion factor based on the design room air temperature, we can compute the flow. The formula is:

$$\text{Water flow} = \frac{\text{Baseboard length x Btu/ft mfg. rating}}{500 \times (\text{water temp in} - \text{water temp out})}$$

The conversion factor, 500, will remain constant in the formula. The rest of the data will change, depending on the actual case. For example, if we have a 10' baseboard at a catalog rating of 600 Btu/ft and a measured water temperature of 200°F in and 190°F out, actual flow will be:

$$\frac{10 \times 600}{500 \times (200 - 190)} = 1.2 \text{ gpm}$$

By comparing this with the design flow we can see if our actual flow is too large or too small. Assume the system was designed with a 20°F drop, with a flow rate of 0.6 gpm. Our actual flow is too large. The flow can be reduced by partially closing off the balancing valve in the circuit. This is the plug or globe valve located in the loop supply line. Besides being used to throttle the flow, it can be used to shut off the flow to service radiation units or the control valve in the loop.

Steps for Balancing

As we did with air system balancing, we'll divide the balancing steps into two sections, in your office and at the job site.

Preparation before going to the job—

1) Make sure the system is finished and started up.

2) Check the water design inlet and outlet temperatures used to calculate the design flow for each baseboard. In our apartment system, for example, they were 200°F entering and 180°F leaving, for a 20°F drop. We'll disregard the small temperature drops in the loop piping. Note these inlet and outlet design temperatures on the system drawing.
    Also note the design temperature of the boiler and the design flow of each circuit if they're not already on the drawing.

3) Obtain a digital electronic thermometer. It should have two temperature sensors with leads long enough to reach both ends of the longest baseboard. Make sure the thermometer has a switch so you can switch from one probe to the other.

4) Make up and fill out a worksheet like the one in Figure 9-10. Include the design pump data, design flow rates, baseboard data and room or zone designations.

5) Obtain and test a suitable multimeter for measuring pump volts and amps.

Procedures on the job—

1) Open all balancing valves fully. Open all dampers on the baseboards.

2) Start at the baseboard that's farthest from the boiler. Set the room thermostat to call for heat, at 20°F above the room temperature.

3) Check the room temperature. It must be 67°F (or room design temperature if other than 67°F) or less.

Job name _____ Page 1 of _____

Address _____

Balancing test by _____ Date _____

System description _____

Pump:	Make	_____		Required	Actual
	Size	_____	Fan rpm	_____	_____
	Type	_____	System cfm	_____	_____

				Rated	Actual
Motor:	hp	_____	Line volts	_____	_____
	rpm	_____	Motor amps	_____	_____

Branch no. _____

Room no.	Ft.	Btu/ft.	gpm	T_i	T_o	$T_i - T_o$	gpm			
							A	B	C	D

Ft. = Radiation length; Btu/ft. = Catalog rating Btu/ft.; gpm = Design flow gpm; T_i = Temp in, degree; T_o = Temp out, degree; $T_i - T_o$ = Temp difference, degree; gpm, A, B, C, D = actual flow.

Remarks: _____
_____
_____
_____

Signed _____ Date _____

Water system balancing sheet
Figure 9-10

4) Check to make sure the boiler is at design temperature.

5) Check pump volts and amps to make sure the pump isn't overloading. If there are gauges on the pump, check the pressure drop across the pump. Compare this to the manufacturer's pump curve to find the system flow.

6) Return to the baseboard farthest from the boiler. Make sure the damper is fully open and the air flow into and out of the baseboard is not blocked. Clamp or tape the temperature sensors on top of the pipe at the inlet and exit of the baseboard. Wait several minutes for the temperatures to stabilize. Read the results and record them on the worksheet. Figure the actual flow rate. Compare it with the design flow rate.

7) If the actual temperature difference is smaller than the design difference, close the balancing valve a little at a time until the temperature drop increases to within plus or minus 10% of the design value. If the temperature difference is too large, check the other loops. Make sure the room temperature in them is not above 67°F and that the thermostat is calling for heat. Reduce the flow in them to the design flow. Then recheck your first location and adjust.

If the flow at the first location farthest from the boiler still doesn't meet design values, check the pump to make sure it's the right size. Then check the system to make sure it was installed and is operating correctly. Make sure you haven't overlooked any thermostats, valves or dampers. Check pipe sizes and baseboard lengths. Make necessary changes or corrections and again balance the system. When finished, make a final check of the pump volts and amps to make sure it isn't overloaded. If so, it will need to be replaced with a larger pump or its speed adjusted to avoid system failure.

Write up a final report and send it to the owner.

Series Loop Systems
To balance a series loop single-pipe system, make sure all valves and dampers are open. Set the thermostat to call for heat. Measure the temperature drop from the outlet of the boiler to the return. Compare the measured temperature with the system's design temperature. If the system has too much heat, you can lower the boiler temperature or reduce the water flow. For too little heat, raise the boiler temperature or increase the pump size to get more flow at the baseboard units.

Balancing air or water systems isn't difficult if you follow the steps we've listed and work carefully. Stay alert for indications of system parts that are malfunctioning or changes in design that weren't part of the plan. For further information on balancing, read the chapter in the 1984 ASHRAE Systems Handbook on Testing and Balancing.

Starting Your Business

Every year thousands of hopeful entrepreneurs start businesses. Most of them fail within a year. Although there are many reasons for the failures, a government study named the two most common: lack of capital and poor bookkeeping. But this doesn't tell you much. Did the business start out with too little capital? Or did the owner squander his money on a lot of new equipment that he didn't really need? In either case, it was poor planning on the owner's part.

Perhaps another owner planned well but felt he had to be a "nice guy." Instead of going out and aggressively collecting money owed him for his jobs, he let his accounts receivable pile up. The people who owed him money left it in their bank accounts, collecting interest, because he was afraid to demand payment. Some may call this poor bookkeeping, but it's really poor management. The owner wasn't aware of the way money is used today.

When we began to write this section of the book, we talked with an HVAC contractor who had just retired. When he found out what we were after, he had this to say:

"Fellas, I've been in heating and air conditioning as a contractor for 25 years and I'm glad to get out. When I first started it was tough, but times were good. Every year the business got bigger. There was enough work and if you had a few bad jobs in a year, you could handle them. Sure, it hurt, but it didn't kill you. But nowadays, the fat's gone out of the business."

He went on to explain that with changes in the economy, profit margins on jobs are smaller. That, along with the high cost of doing business — the interest on a short-term loan to buy materials, for example — had combined to make it a lot tougher on the contractor. One or two bad jobs in a year could put a business under, especially a business just starting up. "Anybody who wants to do HVAC contracting today had better have an awfully sharp pencil," he said.

What does all of this mean? You've put in many years learning the HVAC trade and you know you can do the jobs better than a lot of companies. Should you give up your dream of running your own outfit? No, of course not. Sure, businesses fail every year. A lot of them. But some succeed. And the economy affects everybody, not just companies that are starting up. All we're saying is that getting started as a contractor is a big risk, financially, physically and mentally. It's not enough to know how to design and install HVAC jobs well. Being an owner is a whole new ball game, with a lot of new rules.

In the next two chapters we'll talk about this part of the business. You want to get all the information you need to make your business one of the successful ones. We'll cover a lot of material. But don't stop with this book. Go out and talk to people in all kinds of businesses, not just construction work. Look in your library for books on starting your own company. You'll discover that the key to being successful depends largely on your preparation.

Perhaps years ago someone could run a business by the "seat of his pants," and make it work by hunches and luck. But those were the days when money was cheap and work was plentiful. Right now, that "fat" is a lot thinner. To get your share, you'll have to work hard and be sharp.

Most economists feel that there are still good times ahead in the economy. But the steadily expanding construction boom of the last 30 years won't be repeated. Expect slow times to be a regular part of the economic cycle. If you can do well when work is slow, you'll do really well when things pick up.

Good times or bad, there's always construction work. New businesses start up to do it. You want to make sure you'll be around to reap the rewards from all the hard work of getting started.

Your Business Objective

Not everyone who thinks about going into business should do it. Some people just won't like being the boss. They won't like the paperwork, the attention to detail, the idea of being behind a desk for most of the workday. These people are better off working for someone else. There are others who might not enjoy these things, either, but they're willing to do them in return for the satisfaction gained from running their own operation.

How do you know where you fit? Take some time off by yourself. Ask yourself this question: *Why do I want to go into business?* Then give yourself an honest answer.

At first you might have a little trouble answering. Your reasons may sound a little fuzzy. That's all right in the beginning. But you have to keep thinking about it and get those reasons into focus. Finally, out of all those reasons, you have to come up with one single answer. That answer will tell you a lot about whether you'll succeed or fail as an HVAC contractor.

If, for example, you want to go into business so you can spend your winters in the Virgin Islands while your crews make money for you in the subzero weather in Minnesota, you can stop reading this chapter right now. We're not saying you won't ever be able to do it. But if you do, it will be your reward *only* after you've spent a lot of time and hard work making it possible.

Maybe your reason is that you're fed up with the way the boss treats you, or that you get laid off every winter, or that your wife and brother-in-law keep telling you to get into business for yourself. These may all be reasons to start thinking about going into business. But they shouldn't be the main reason.

Suppose, on the other hand, that you see a lot of work that isn't getting picked up, or hear customers complaining about how long it takes to get a contractor. You *know* you can do that work quickly and well. Now you're starting to zero in on a good reason. And let's not fool ourselves. Nobody would go into business, with all the attendant headaches and hard work, if he didn't expect to make more money than he could make working for someone else. As an owner, you take all the risks. It's only fair that you should get a greater share of the return. In fact, after all is said and done, that's probably the biggest motivation.

But saying your business objective is "to make money" is a little too simple. Your business objective isn't just something you'll keep to yourself. It's got to be a clear-cut statement that you can use as a tool in your business dealings.

Keep these points in mind when you formulate your business objective:

• You should be able to explain your objective, both to yourself and anyone else, in one or two sentences.

• You want a positive reason, not a negative one.

• It should reflect your understanding of existing opportunities.

After you've formulated your business objective, write it down and commit it to memory. If it's a good one, you won't want to hide it. You'll use it often, in two ways:

First, as a business owner you'll be faced with many decisions, especially when you're starting out. Many complicated decisions become simple when you weigh them against your business objective.

Second, there will be other important people who'll want to know why you decided to go into business for yourself. Anyone you approach for a loan, bankers in particular, will put considerable weight on your reason for starting your own business.

Although it's hard to believe, a banker we spoke with swore that a man came to his office and asked for a business loan because "my boss is a jerk and I want to put him out of business." Needless to say, the banker wasn't too impressed. Certainly not enough to lend him the money. He knew all he needed to know about the man asking for the money. We'll talk more on getting financing later. But keep in mind that knowing, understanding, and being able to clearly state your business objective will play an important part.

The Market Study

The amount of HVAC work for any given area will vary from year to year. If, for several years, the work in your area has increased faster than the existing contractors can get to it, you've got a good reason to think about going into business. If work stays relatively stable or declines, you can still go into business. But you'll need some way to insure that you get a share of the available work. You'll have to work faster or cheaper, or be smarter than the other contractors.

The market study is a key tool in appraising your chances of success. You may have a strong business objective, be very good at the trade and have plenty of ambition. But you need a market study to test just how practical your idea is. No matter how much time, work and money you're willing to invest, your business won't go anywhere if the work simply isn't there.

The market study will evaluate your chances of finding enough buyers for your services to make it worthwhile to start a business. Market studies for large companies experimenting with a new product like beer or detergent can be very complicated and expensive. Your study can't be that complicated, nor should it be. It can be inexpensive, simple, yet accurate. You need accurate data, which you should record in a neat, logical table or worksheet. It's the foundation for your overall business plan, and it'll be reviewed by people who are important to financing your venture.

You can begin your market study by visiting the office that issues building permits for your area. It's a good idea to call first so you know the office hours and the rules for using these files. But the permits are public record and must be made available on request. Although some offices will have the permit data summarized and make copies for you, it's a good idea to take along a notebook and pencil to jot down information you may have to dig out on your own.

Go back through the permits for at least five years and find out if the number of new and remodeled buildings in the area is increasing or decreasing. How does this compare with building activity nationwide? Can you spot significant trends, such as a decline in single-family housing but an increase in condominiums? Is a certain area a "hot spot" for building activity? Although you may not use all the information you gather in the market study, it's sure to come in handy as you proceed with your business plans.

The next step is to find out who's doing the work. Many cities require that all companies doing business within their limits take out a license. You can find out if the number of HVAC contractors working in the area is increasing or decreasing. How many of the contractors are from out of town? Sometimes these lists will also have the dollar value of the work done. This can be helpful in analyzing market conditions. For example, are big contractors going after small jobs? This could mean there aren't enough big commercial projects to keep them busy.

If you're in an area that doesn't require licensing, you can get some of this information by counting the number of contractors in the Yellow Pages of the phone book. Many phone companies and libraries will have back issues for previous years. Compare the number of contractors currently advertising in your area with the number in previous years.

Although these steps will tell you a lot about the construction business in your area, you'll want to check other sources, too. The time and energy you spend will be well worth it. You need to learn all you can before deciding to take the risk of starting your own business. Here are some other good sources of information:

• Business development agencies: These agencies, often set up by local governments or universities, are designed to help individuals get started in business. They can provide useful information and statistics on many types of business activities in the area. Many of them will also aid in arranging financing with private lenders, and offer general advice on business procedures. Keep in mind, however, that these agencies often get mixed reviews by those who have used them in the past. Some tend to emphasize special projects or geographical areas. They may slant their advice and help to those who'll fit into these plans. On the other hand, some of the agencies are very effective for almost anyone who walks into their office. Overall, they're certainly worth a close look. But to get a true picture of their effectiveness, talk with several companies that have worked with the agency.

• Local planning and permit committees: Nearly all commercial and housing projects of any size now require a lengthy permit process in most states. Planning and permit departments or committees can tell you what major jobs are in progress and where major subdivisions are being planned. Applications for sewer, water, electric, and gas utility service also indicate future building activity. Contact the local utilities or government agencies.

• Major employers may plan new or expanded facilities in your area. The new plant will attract more employees. Will a new subdivision be built to house them?

• Don't overlook residential remodeling as part of your potential market. Most cities require permits for remodeling work. If you find permits grouped in older neighborhoods, it may mean that owners are reinvesting in their properties and that the property values are increasing. Many city governments are encouraging this kind of development to preserve inner city areas. Locating the primary contractors and selling them your services can mean plenty of work over an extended time.

• Which way is the population growth headed for your market area? The local Chamber of Commerce may be able to supply details. Many real estate companies prepare surveys to sell commercial properties to investors in the area. They'll often give both population trends and economic growth estimates. Hopefully, you'll find a company in your area that's willing to part with this information free or for a small fee.

• You may want to review statistical data, especially if you're thinking about setting up operations in an unfamiliar area. The U.S. Census Bureau has a wealth of information you can use in a market survey. Besides national data, the Bureau has *Census Tracts* and *Standard Metropolitan Statistical Area* (SMSA) data available. These studies divide large cities and surrounding areas into tracts of about 5,000 residents, to show different statistics for families' income and housing.

The Bureau also compiles the number of permits issued in about 5,000 communities across the country, and publishes this information monthly in *Housing Units Authorized by Building Permit and Public Contract*. A good public library will have current and back issues on hand. A yearly subscription is $50 from:

The Superintendent of Documents
U.S. Government Printing Office
Washington, DC 20402

Every ten years the Bureau does a census on housing. Information includes the year each residence was built, the number of rooms, owner cost, value, retail costs, and the condition and type of mechanical systems. The data is available for about 300 cities and for each of the census tracts. Your state census data center will be able to supply computer printouts for the areas you're interested in. You can get your state census data center's phone number by writing or calling:

Public Information Office
Bureau of Census
Department of Commerce
Washington, DC 20233
(301) 763-4040

To find out about any other available information that may relate to your study, call the Office of the Census of Housing at (301) 763-2873, and the Construction Statistic Office at (301) 763-5631.

• Don't overlook the obvious. If you've spent any time at all in the construction business, you probably know at least one person who's worked for a general contractor in the area. An hour's conversation over a couple of beers can yield a wealth of information about possible work and local construction conditions.

• The reports published by the F. W. Dodge Company are an extremely valuable source of information. These *Dodge Reports* specialize in keeping track of many different aspects of current building projects throughout the country. One report, for example, summarizes project data for individual jobs in a given area. It will list the type of job, the owner, the architects and designers, the company that submitted the low bid, and its amount. Another report lists projects according to building types to show costs and trends for a given type of work. Still another contains information on upcoming projects.

These reports will give you a good idea of some of the jobs going on in your area and who's doing the work. If you're lucky, your local library will have back issues of the reports. Then you can get a picture of what's been happening over the last few years. If the public library doesn't have the reports, check at your nearest college or university architectural or civil engineering department. They may have them on file, or perhaps a teacher will have saved his back issues.

Besides being useful for market studies, the Dodge Reports are essential when you begin bidding on commercial jobs. For a listing of available publications, contact:

F.W. Dodge
McGraw-Hill Information Systems
P.O. Box 11354, Church Street Station
New York, NY 10249
1–800-257-5295

After you've gathered all the information you can find, you'll have to sift through it and analyze it. At worst, you'll find the area you've targeted simply isn't a good one to start up another HVAC business. You'll then have to investigate some other part of the country as a possible location.

If you find that conditions are favorable, you'll want to put down your findings on paper. This will help you check your data, but more important, it will be useful when you go to get financing. You may well argue that any banker worth his salt will know at least as much about his area's economy as you do. That's probably true. What's really important is that he sees that you know it, too. You've gone to the time and trouble to present the data to him to strengthen your loan application.

So don't try to fudge any facts or figures. Be positive but truthful. Even a fact or statistic that looks negative can be a positive factor when viewed from a different angle.

To make your market study complete, you should add a written summary and attach it to the study. Include any data and/or quotes from authoritative sources that will help define your market area. For example, you may want to add something like this:

The sharp decline in building permits and the related decline in HVAC contractors for Beaver City in 1981 may have been caused by an economic slowdown based on higher interest rates, reflecting the national trend. More important, the attached U.S. Census Bureau Tract shows that by 1981, the population growth in the area had leveled off, especially in the number of potential homeowners. But the figures for later years show that the population has once more begun to climb, indicating that people are moving into the area. In the years since 1981, however, the number of HVAC contractors declined by one more.

While the number of residential housing permits has not increased in proportion to the population growth at the rate of previous years, the number of condominiums and apartments has increased sharply.

In addition, Mr. Ralph Goodtalker of the Beaver City Chamber of Commerce points out that the city has made a concentrated effort over the last several years to make the area attractive to new businesses. Three of them — Acme Genetics, Grover Metal Stampings, and Electrochip Co. — have made substantial investments in new plants and equipment in the Beaver City industrial park. They've created 1,500 new jobs. Two other companies are now negotiating with the city for sites in the park.

Both the new businesses and the increase in population, when placed next to the decline in HVAC contractors in the area, make me believe I can succeed in my business venture.

Estimating Gross Sales

You've probably realized by now that you're putting all these steps down on paper for two reasons. First, you want to be as sure as possible that your business will succeed. Secondly, you're going to need all this paperwork to get financing. Even if you're lucky enough to have a wealthy relative, you'll need to have a firm grasp of your market and your business goals if you hope to succeed.

So, assuming you've found a favorable market for your services, the next step is making sure it will generate enough income. You'll do this by estimating your gross sales. This estimate will serve as a guide to your daily operations as well as assist you at loan time.

A quick way to estimate gross sales is to begin with your expected gross profit. A usual gross profit for the construction industry is 10%. Let's say you hope to make a $15,000 gross profit in your first year of operation. This is over and above any salary you pay yourself. Divide $15,000 by 10%:

$$\frac{\$15,000}{.10} = \$150,000 \text{ annual gross sales needed}$$

$$\frac{\$150,000}{12} = \$12,500 \text{ monthly gross sales needed}$$

This gives you a quick ballpark figure on how much business you'll need to do. But you may be better off basing your estimated sales on your operating costs. Figure 10-1 shows a typical operating cost summary. The advantage to this method is that all your operating expenses can be estimated fairly accurately, based on your own work experience and quotes from those whose services you'll purchase.

Operating expenses are also called overhead. The relationship between overhead, material cost, labor cost, and profit is explained in the next chapter. When most contracting businesses turn a 10% profit, overhead makes up 20% of the total financial outlay. So it will take five times the amount of your overhead to operate your business and make a profit.

Annual operating cost:	$60,000
	x 5
Annual sales needed:	$300,000

Now here's a step you may not want to put down on paper. Using the above example, you can see

Item	Amount
Building rent	$7,000
Heat and electric	1,400
Telephone	840
Interest on loan	12,000
Insurance	1,200
Depreciation on equipment	2,000
Owner's salary	20,800
Accountant	4,260
Legal fees	500
Truck and travel expenses	7,500
Office supplies	500
Contingency fund	2,000
Total	$60,000

Operating cost summary
Figure 10-1

you need to do enough work to generate $25,000 monthly. Compare this with your market study. Are there enough jobs available for you to make this amount every month?

If, for example, an average of only 19 building permits were issued for your area each month, and there are three other contractors, you'd probably be able to average five new construction jobs per month at the most. That means each job would have to have a value of $5,000. Basing this on your past experience and all you've found out so far, is this a realistic possibility when you're competing with other contractors? If you're not sure, you'd better have some other kind of HVAC work to fall back on to keep your sales figure where it belongs.

There are some other options. You can see that if your operating costs are low, you'll need less money coming in to survive. But you can't reduce your operating costs below a certain level. You can't, simply because your sales decline, pay less rent, telephone or loan interest. You can, of course, lay off your office help and buy an automatic phone answering machine.

The other solution, and the one we recommend, is to go into business with enough of a financial cushion to tide you over your first two years of operation. Traditionally, these are the years when your volume will need time to grow. You can get this cushion by overestimating your operating costs on paper and then holding them down when you operate. But be careful. If your estimate is too high, the banker will spot it and question it. He may not approve of your method. Some contractors have set aside some of their own cash in a separate fund to give them the security of money to fall back on. Or you may decide just to increase the size of your contingency fund and tell the banker outright what it's for.

Starting Costs
Besides operating costs, you'll have onetime start-up costs. Figure 10-2 is a list of typical starting costs in 1985 dollars for a beginning, one-man HVAC business.

Item	Amount
Building security deposit	$2,000
Tools and equipment	8,500
Office equipment and fixtures	2,000
Utility deposit	1,500
Inventory	50,000
State/Federal Employer Deposits	4,000
Truck	8,500
Legal fees	200
Accountant fees	100
Insurance	300
Loan closing costs	1,200
Miscellaneous	2,000
Contingency fund	3,000
Total	$83,300

Start-up costs
Figure 10-2

The list doesn't include accounts receivable, simply called "receivables" in bookkeeping jargon. They're actually part of your daily operations, rather than start-up. But keep in mind that there may be several months between the time you get your first jobs and the time you get paid for the work — before you get any money in your receivables. In the meantime, the rent will have to be paid, materials bills may come due, and so on.

Here's a rule of thumb: Receivables will average about 15% of the annual gross receipts. So if you expect to take in $300,000 each year, you'll be carrying about $45,000 on any given day.

Once you've established your business, you'll have the cash flow to handle this. But when you start, you may need a lump sum to cover the first three months' operating costs. Talk this over with your lender and get his advice. In our sample list, we include the money as operating costs and set aside a lump sum. However you handle it, make sure you don't forget it. And don't, under any circumstances, use it for anything except its intended purpose.

Projection and Forecast Sheets
Unlike loans for consumer items or real estate, most lenders don't have a standard form for commercial loans. Since you haven't been in business before, they'll rely a good deal on your net worth and the projected profit and loss of your business enterprise.

Net worth is explained in the next section. It simply means how much money and other assets you can put into the business. To make up your projected profit and loss, you take the information from your market study and operating cost sheet and extend it over a three-year period. Figure 10-3 is a typical sheet used by lending institutions for a business projection.

Remember to make your market study and operating cost sheets realistic. Then be sure to transfer the numbers correctly. You may find it helpful to get an accountant or tax person involved in the projections.

Personal Balance Sheet
Your market study, starting costs and operating costs together tell your lenders where you're going, and how much money you'll need to get there. The lender will also want to know where you're starting from, so you'll be asked to fill out a personal financial statement, often called a balance sheet. It may look something like the one Figure 10-4.

The sheet will have two basic sections:

Assets: things you own

Liabilities: things you owe

As the name "balance sheet" suggests, the assets and liabilities will be equal when all the figures are added. Obviously, no one owes exactly what he owns. The difference between assets and liabilities is your *net worth*. Net worth is listed as a separate entry in the liability side of the balance sheet.

─────── Projection and Forecast of Three Years Income and Expenses ───────

Net sales	$_____	$_____	$_____
Beginning inventory	_____	_____	_____
Purchase (or cost of manufacture)	_____	_____	_____
Ending inventory	_____	_____	_____
Cost of goods sold	_____	_____	_____
Gross profit	_____	_____	_____
Variable expenses			
Owners' salaries	_____	_____	_____
Employees' salaries	_____	_____	_____
Accounting and legal fees	_____	_____	_____
Advertising	_____	_____	_____
Travel and entertainment	_____	_____	_____
Delivery	_____	_____	_____
Bad debts	_____	_____	_____
Supplies	_____	_____	_____
Other expenses	_____	_____	_____
Total variable expenses	_____	_____	_____
Fixed expenses			
Rent	_____	_____	_____
Utilities	_____	_____	_____
Taxes	_____	_____	_____
Repairs	_____	_____	_____
Depreciation	_____	_____	_____
Insurance	_____	_____	_____
Interest	_____	_____	_____
Other expenses	_____	_____	_____
Total fixed expenses	_____	_____	_____
Research & development	_____	_____	_____
Total expenses	_____	_____	_____
Net profit before taxes & other income	_____	_____	_____
Other income or (expense) net	_____	_____	_____
Net profit before taxes	_____	_____	_____
Income taxes	_____	_____	_____
Net profit after taxes	_____	_____	_____
Owner's draw (if not included above)	_____	_____	_____
Net profit remaining for loan payment	_____	_____	_____
Fixed asset expenditure above annual depreciation	_____	_____	_____

(Explain assumptions made on a separate sheet.)

_____
(Signature and Date)

Profit-and-loss forecast sheet
Figure 10-3

Financial Statement & Credit Information

Name _____ Age _____

Address _____ City _____ How long _____

Employer _____ Position _____

Address _____ City _____ Length of Employment _____

Income

Salary or compensation _____

Bonuses _____

Dividends _____

Other income _____

(Alimony, child support or separate maintenance need not be revealed if you do not wish to have it considered in determining your creditworthiness.)

Assets		Liabilities	
Cash on hand	_____	Notes & Accounts Payable	
Cash in following banks		_____	_____
_____	_____	_____	_____
_____	_____	_____	_____
Stocks and Bonds	_____	Mortgage on Real Estate	_____
Real Estate	_____		
_____	_____		
Life Ins. (face value)_____		All other liabilities	
(cash value)	_____	_____	_____
Auto	_____	_____	_____
Furniture & Personal	_____	_____	_____
Other	_____	_____	_____
Total Assets	_____	**Total Liabilities**	_____
		Net Worth	_____

Contracts or Loans Payable or Credit References:

Name of Company	Address	City	Original Amt.	Present Amt.	Monthly Pymts.
1.					
2.					
3.					

Financial statement
Figure 10-4

Besides the assets and liabilities, the sheet will ask for other information, to give the lender a complete picture of your finances. He may want information on loans outstanding, loans paid off in the last several years, real estate owned or owed on, other than your home, and personal property that may have substantial value.

Be as complete as possible when filling out the form. If you have a bookkeeper or accountant who handles your taxes, you may want to get his help. If you have questions about an item, call the lender and talk it out. Assets should include anything of value which you own — cash savings, stocks and bonds, real estate, accounts receivable and so on. Liabilities are any formal debt or any other debts over $50. This includes credit cards and loans made on a casual basis. If you have any pertinent financial facts not covered on the sheet that may help your application, add a brief, typewritten summary sheet.

Avoid any omissions that will show up when the lender runs a check on your sheet. Even honest mistakes make you look bad. Too many of them will raise questions in the lender's mind about your ability to operate your own business.

How Much Do I Ask For?

There are several answers to that question.

As the business owner, your first inclination will be to ask for all you can get. In your mind, getting the largest loan possible saves you from investing your last dime. It leaves some of your assets intact in case of disaster. It will also protect you from the unexpected expenses that will crop up in your new business, no matter how carefully you've planned.

Your lender, on the other hand, will also have an amount in mind and it may be less than yours. He'll also have his reasons. First, he'll want you to invest as much of your own assets as possible. This reduces the bank's risk. It also insures that you'll do everything in your power to make the business succeed. When failure means that your wife and three small children will be penniless and on the street, the banker can feel pretty confident that you'll do the work needed.

Other limiting factors will be the interest rates, the amount of money the bank has on hand for loans, your assets to secure the loan, and the amount of money that can be paid back from your anticipated sales volume.

It's actually this last item that's most critical. Banks don't like to see businesses fail. They'd rather have you do well, so you can continue to repay your loan with its attached interest. If a business fails, the payments stop. The bank loses both the money and the interest. It's left with a group of assets to dispose of. It's all very messy for them.

Keeping all the above in mind, a good way to find a ballpark figure is to use your starting costs and operating costs. Take your starting costs and add six months' operating costs and subtract your personal investment figure. Using Figures 10-1 and 10-2, for example:

Starting costs:	$83,300
Operating costs for 6 months:	+ 360,000
	443,300
Less personal equity:	- 100,000
Total loan needed:	$343,300

As we said earlier, this is a ballpark figure. Your lender may, after reviewing your assets, offer you less. You may have to get more money from another source. On the other hand, if you have a good source of collateral — a paid-up home, for example — and if the banker has confidence in your ability and market study, he may offer you more money than you ask for.

Money Definitions

Up until now we've been talking about getting money to start your business. Now let's explain some of the different ways that money is used.

Permanent Capital

Permanent capital is money which doesn't have to be paid back. Instead it earns you, the owner, the right to share in the profits. For example, you invest $15,000 of your own money in the business. In return you receive any profit made at the end of the year.

Permanent capital is also called equity capital, or simply, equity. Your money becomes part of the business and stays with it. Equity is not a loan to the business. In essence, you buy part of the business in return for something, most often the profits. Equity can be increased by putting in more of your money or someone else's money in the business. Equity can also be generated in an existing business by selling off inventory or unwanted equipment. Bankers seldom lend equity capital. Instead, they see it as the owner's share of the risk.

Working Capital

Working capital is short-term money used by companies to bridge the gap between expenses and income. Few businesses, especially in the beginning, make money before it's needed. Until money comes in, you'll need cash on a regular basis to pay bills and meet the payroll.

For example, suppose you bid and get an HVAC contract for a condominium project with 60 units. You'll need to hire more men and order materials. But the contract stipulates that your first payment will arrive when the work is 40% completed. Until then, you'll need working capital to carry the load.

Unlike equity, working capital is relatively easy to borrow, especially if you have a contract from a reliable general contractor or developer. But you'll usually pay higher interest than you would for long-term notes. Payback may be one lump sum after a certain number of days, with interest.

Where construction work is seasonal, contractors may apply for working capital loans in the spring, after contracts have been let. If you deal with the same lender over several years and make your payments on time, getting a short-term loan may take only a phone call and a few minutes at the bank to sign the papers. On the other hand, if you're just getting started in the business, your lender may want to secure your first several short-term loans with some collateral and a payment schedule.

Make sure this schedule has some flexibility in it in case you run into a temporary cash flow problem. Often the lender will hand you a schedule that requires payment of the amounts due at the agreed-upon dates with no other options. This is fine as long as your business operates smoothly. But you're likely to hit some financial snags that leave you temporarily short of cash at the end of the month. When this happens, the schedule should allow reduced payments and/or an extension of the due date. Otherwise the lender may be able to tie up the loan collateral. If it's part of your business operations, it could turn your temporary snag into a major problem that cripples your ability to operate the business.

You may be able to reduce short-term capital amounts by relying on "trade credit." Some suppliers might give you extended payment terms for materials. But again, this may be difficult when you're first getting started, especially at today's interest rates. If you're fortunate enough to get a supplier who'll give you this advantage, don't abuse the trust. You can't afford to be marked as an undesirable credit risk.

A third alternative is to order materials according to the job schedule. That way you won't have the materials on hand — and the bill due — until they're actually needed. There's more on this in the next chapter.

Long-Term Capital

Long-term capital is used for capital expenditures. It's money marked to buy fixed assets such as major equipment or buildings. You'll usually pay it back at a fixed rate over a long period of time. The item purchased is often the security for the loan. There are many sources for long-term capital. Investment corporations, commercial banks, private investors and life insurance companies are typical. Many of those interested in lending money to beginning businesses will advertise in trade magazines and newspapers' financial sections. Be sure you get legal advice before signing with an unknown lender.

Collateral

There are many types of collateral. Collateral is a word that describes something that insures repayment, in full or in part, in case you default on your loan.

A *signature loan* is unsecured. There's no collateral for the loan other than the legally enforceable promise to repay you make when you sign the loan agreement. It is a contract, but it's very difficult to enforce in the event of default. Few commercial loans are made on the basis of the signature alone. If, in the opinion of the lender, you're a very good risk and you've repaid other notes promptly, you may be able to get working capital or other short-term loans on your signature alone.

A *savings account* is a good source of collateral. Savings-backed loans have been made popular by credit unions in recent years. Basically, you assign your savings account and turn your passbook over to the lender. In return, you're given a loan which won't exceed the amount in your account. Generally, your money stays in your account and earns interest. When the loan is repaid, the account is signed back to you. The advantage to this type of loan is that the interest you're earning on your savings negates part of the interest you're paying for the loan.

A "whole life" *insurance policy* can be used as collateral for a loan. The amount available will depend on the policy size and its cash surrender value, along with the insurance company's loan rate and other rules. Sometimes the loan can be made directly from the insurance company. But policies can also be used as collateral for other lenders. When the loan is repaid, the policy is signed back to you in its original condition. An advantage to this type of loan is that you're still covered under the policy's provisions for life insurance. The amount of the loan is deducted from the benefit paid should you die while the policy is being used as collateral.

One of the most common forms of collateral is *real estate*. It's most often used to secure long-term loans, but it can also be used for short-term lending. A loan secured by real estate involves a mortgage on the property.

Before accepting real estate as collateral, the lender will want to make sure the property is worth more than the amount he's lending you. You'll also have to provide insurance coverage adequate to cover the debt in the event of property loss.

A *chattel mortgage* is like a real estate mortgage, but the security is something other than land and buildings. In the construction business, the security is usually materials and equipment. Chattel mortgages are used for shorter time periods than real estate mortgages. Like real estate collateral, the chattel has to be worth more than the loan and insured by the borrower. The lender can make sure you don't sell the chattel by filing a *lien* against it. It's illegal for any property owner to sell an item with a lien still in force.

Money Sources

There are three general sources you can tap for money for your business:

1) Yourself, your family, business associates and friends

2) Banks, investors, and other lending institutions

3) Government agencies

Depending on your financial strength, you may use one, two or all three sources.

Yourself and Others

Very few of us have enough money of our own to start a contracting business. It's hard for any wage earner to accumulate enough excess cash after paying family expenses. It is, however, possible to save enough so that other investors will have confidence in your sincerity and ability to handle your own finances. It may take time and hard work, but many people are willing to make the sacrifices required.

If you're able to set aside a business nest egg, it will only make raising the remaining amount that much easier. If you have to rely on friends or relatives for part or all of your financing, *do it in the most businesslike way possible.*

Borrowing from friends or relatives can create problems. A refusal can mean hurt feelings or a ruined friendship. An acceptance, if done on a casual basis, can cause problems for you. If you've borrowed $10,000 from Uncle Ralph on a handshake and verbal terms, you may soon discover that he expects the return on the loan to be far different from what you thought was agreed on. He may even begin to meddle in your business affairs, on the assumption that he's now your "partner." Or, expecting a larger return on his money, he may loudly tell everyone at the annual family reunion what a lousy businessman you are.

Still, many people in business today wouldn't be there without financial help from family or friends. There are several ways you can insure that these loans will be effective and leave the relationship in good shape.

First, look for equity capital — investors rather than lenders. Invite a friend or relative to share the risk in return for a part of the profits. You contribute the technical expertise, the time and the effort. Your investor contributes the cash. Offer a fair and sporting proposition and be prepared to bargain a little. Someone who believes in you and has the spare cash might be willing to bankroll your company. The investor can always take a tax loss if it fails.

Make sure that these arrangements have some kind of control written into the contract. This will protect both you and the investor. The terms should be spelled out so you both understand them. One visit to an attorney familiar with this type of contract should be all that's necessary. Generally, the investor won't have any input in daily operations. But he may want some rights in the event of imminent failure. There are very few "silent partners" any more, willing to invest their money and then sit back and accept the outcome.

If you can't find an investor, settle for a lender. Have your attorney draw up a loan contract just like a bank's. Have your lender give it to his attorney for review. Spell out exactly what the collateral is, the interest and repayment dates, and the provisions in event of default. Above all, make sure your lender knows the risks involved. You'll be able to face friends and family with a clear conscience if you made your deal on a sound business basis.

Banks and Lending Institutions

Borrowing from a financial institution removes the potential emotional problems from your loan. On the other hand, the money may be harder to get.

We talked about some of the paperwork you'll need when you approach a bank earlier in this chapter. As a rule, bankers rely a great deal on paperwork. Your loan application should include the following, neatly typed:

- A short personal biography

- Your business objective

- Your estimated starting costs

- Your estimated operating costs

- Your market study

- Your personal balance sheet

- The total loan amount and general repayment terms you hope to get.

Before preparing all this information, *call the bank.* Make sure that the bank you want to do business with is handling your type of loan. And find out if the bank has its own forms for some of this information. If so, put the information on the bank's form.

The information that you supply in your own format should be just long enough to cover the facts and stress the positive points. Avoid long, detailed explanations. Be brief but complete. If the banker wants more information on any point, he'll ask for it.

Try to arrange your application so each financial table or major item of information is presented on a single sheet. Title each section. Combine your typed paperwork and the bank's forms in an arrangement like this:

Page 1: Include your name, address and the reason for the loan, a short personal biography, including where you've lived, your education and your business experience, and your business objective.

Pages 2 and 3: Your market study

Page 4: Estimated starting costs

Page 5: Estimated operating costs

Pages 6 and 7: Your personal balance sheet

Page 8: Loan amount and possible repayment terms

Remember that neat, attractive paperwork appeals to bankers. If you don't type well, find someone who does, even if you have to spend a few dollars. After all, you're asking for many times that amount.

When your application is ready, call the bank and ask for an appointment to discuss your loan application. Sometimes they'll ask you to drop off the application and give them several days to check it out. Then if the bank is interested in the loan, they'll call you to make an appointment for a personal interview.

For your interview, dress neatly and be on time. You're in competition for the bank's funds with other borrowers. As a rule, banks turn down more loans than they grant. It's up to you to use everything you can to make a favorable impression.

The banker, when considering your loan, will check the accuracy of your market study and estimated costs. He'll evaluate your character and business skills. He may ask questions on several areas you haven't covered in your application. Keep in mind that he may already know the answers. Finally, he'll decide how much he's willing to lend and the terms. He may tell you he'll be glad to give you the loan if you can come up with more collateral or equity.

Don't be afraid to bargain if the lender's offer is low. Tell him, in a positive way, why your figure is a better one. If your personal balance sheet shows that you're a good risk, you're in a good bargaining position. There's no reason why you have to take the first offer. If you really can't come to terms, ask for time to think it over and then shop around at other lenders.

If your application is rejected outright, ask why. Also ask the lender to suggest a second course of action. No one likes to be rejected at anything, and borrowing is an especially touchy business. But try not to get upset. Be calm and reasonable but persistent. You'll pick up information to strengthen your application when you approach a second lender.

Keep in mind that the financial industry is undergoing major changes. Deregulation has opened many new sources of financing to business borrowers, such as savings and loan associations and credit unions. Whereas one lender may make you a poor offer and refuse to bargain, another may welcome you with open arms and a better interest rate. If you're in a strong financial position, it's good business to shop around for your loan, much as you shop for a new car, home or any other major investment.

Be aware that lending institutions are controlled by state or federal commissions. These commissions make sure that each bank, savings and loan, or credit union abides by the rules governing it. At least, that's the theory. In practice, there are so many rules and regulations that the commissions rarely check on any one lender unless there's a specific complaint. In other words, don't assume that your lender is going to give you the best possible interest rate or abide by all the rules that work in your favor. Instead, he'll be looking at what's good for him. Sometimes in his enthusiasm he may stretch a point or two. It's up to you to know your rights and what to do if you feel they're being violated.

In the previous section, we listed some basic money definitions to help you get a familiar with the lending vocabulary. We strongly urge you to go to your library and study books on lending and business financing. There are many on the market, written in plain English, that explain your rights and the obligations of the lender. There are also local consumer groups you can contact for help and advice.

Finally, a word of caution: Whatever amount of money you get, use it wisely. Don't spend more than your estimates on operating and startup unless it's absolutely necessary. Even then, give it some heavy thought. Some new business owners with a fat loan in their pockets find it hard to resist splurging on new equipment, trucks, or a 50-foot sign with their name spelled out in lights. When you need equipment for your business, watch for sales from operators who gave in to the temptation. You'll find them listed in the newspaper ads under Auctions or Bankruptcy Sales. You'll get some good buys, *and* a strong reminder that there's a very fine line between making money and going broke.

Government Agencies

The Small Business Administration rarely makes loans. Instead it guarantees your loan through a private lender. Most SBA offices prefer that you work through the private lender when applying for an SBA loan. Contact your bank first and see if they work on SBA loans. Or ask if the SBA can suggest a private lender in your area.

Loan money is generally available for both long-term and working capital loans, at the market interest rate. SBA offices are located in major cities throughout the country. Find them in the phone book under U.S. Government.

The SBA has several rules for qualifying. A contracting business, for example, is considered a wholesale business. To qualify for help, the business must expect sales to be under $5 million and it must have been turned down for a loan by at least two banks or commercial lenders. These rules may vary due to policy shifts. Also, they may not apply for special situations, such as a minority-owned business or a business located in an economically depressed area. Certain SBA offices may also have funds targeted for special businesses that are needed in the area. To determine the available options, call the nearest SBA office and ask for qualifying information.

Many SBA offices hold regular meetings to give potential buyers the complete rundown on current SBA policies and money availability. If at all possible, try to attend. You can get some of the information through the mail, but it won't be as complete or up-to-date.

If you qualify, you may wish to supplement the SBA forms with some of the ones we described earlier in the chapter. This may help your application. On the other hand, the SBA will have plenty of forms of their own and you may feel this is sufficient. Figure 10-5 is the SBA loan application. Figure 10-6 is the personal financial statement.

The SBA moves relatively slowly. You may have to wait a year or more for all the paperwork to clear. Then you may have to wait until funds become available. If your application expires before the funds are available, all you can do is start over. Getting an SBA loan takes a good deal of patience and persistence. But if you haven't had any luck with other lenders, it may be your only option.

In fact, the SBA looks upon itself as a "lender of the last resort." It won't lend money if you have access to other credit. And you must have tried to self-capitalize by selling property or other assets or by using personal credit if it doesn't impose a hardship.

The SBA also puts restrictions on the money once you receive it. But these rules are flexible and can be modified if it will improve your operations or prevent a business failure.

For a point-by-point review of the SBA and other government lending programs, read *Builder's Guide to Government Loans,* published by Craftsman Book Company, Box 6500, Carlsbad, California 92008. It may be ordered from the order form in the back of this manual.

Besides the SBA, some state and municipal governments have programs that encourage businesses in their areas. Typically, these offer business advice and access to local lenders, sometimes at a favorable interest rate. But you still have to show you're capable of owning and operating the business at a profit.

Protecting Yourself and Your Business
This section isn't a substitute for professional legal counsel. But it will give you some guidelines and suggestions for setting up your business to guard against some of the common pitfalls beginning businesses fall prey to.

Business Organization
There are three common types of business organization: proprietorship, partnership and corporation. For one person, it may be best to operate as a proprietorship until the company is well established. Incorporating may cost from several hundred to a thousand dollars, plus whatever taxes your state imposes. It involves filing formal incorporation papers with a state agency. In many states, you become a proprietorship simply by acting as one. In other words, if you go through the motions typically associated with being in business, you become a proprietorship.

You can organize as a partnership when you want to begin a business but you're lacking in a certain area. For example, you may be quite good at job work and calculations but have little interest in bookwork or formal design drawing. After a few beers at the annual Christmas party, you may find yourself talking with someone who's strong in design and likes paperwork. Perhaps he wants to go into business but is leery of installation work. Or sometimes your financier wants to become your partner.

However it comes about, a partnership should be considered carefully. The obvious problem is that over time, people's opinions of one another may change, especially in a stressful business relationship. On the other hand, there are many partnerships that have worked very well for years. The best way to handle a partnership is to have a competent attorney draw up an agreement that spells out what will happen if one partner dies, or simply wants to quit the partnership. You'll also want an agreement about dissolving the partnership if it becomes unworkable. The key to all this is to get an attorney who's had experience in partnership agreements. Your local bar association can recommend one.

One disadvantage of partnerships and proprietorships is that the owners are personally responsible for all debts of the company, even in excess of the amount they're invested. In a corporation, only your investment is at risk.

Figure 10-7 outlines the ways you can organize and the advantages and disadvantages of each. Very often a contractor starts out in one mode and then, as his business grows, steps into another. Your accountant can tell you when this is advisable, since it often depends on cash flow, liability, reinvestment efforts and the like.

Insurance
Any contractor will need several types of insurance. If you have a bank loan, your lender will want the collateral insured. You'll also need property and equipment insurance to protect against losses from fire or theft. Liability insurance, to protect you from financial disaster in case of job

Form Approved
OBM No. 3245-0016

U.S. Small Business Administration

APPLICATION FOR BUSINESS LOAN

I. Applicant

30

Trade Name of Borrower	Street Address
32	34

City	County	State	Zip	Tel. No. (Inc. A/C)
36		37	39	

Employers ID Number	Date of Application	Date Application Received by SBA	Number of Employees (including subsidiaries and affiliates)
33		5	

Type of Business	Date Business Established	☐ Existing Business	At Time of Application _____
		☐ New Business	If Loan is Approved _____
Bank of Business Account		☐ Purchase Existing Business	

II. Management (Proprietor, partners, officers, directors and stockholders owning 20% or more of outstanding stock)

Name	Address	% Owned	Annual Comp.	*Military Service From	To	*Race	*Sex
			$				
			$				
			$				
			$				

* This data is collected for statistical purposes only. It has no bearing on the credit decision to approve or decline this application.

III. Use of Proceeds: (Enter Gross Dollar Amounts Rounded to Nearest Hundreds)	Loan Requested	SBA USE ONLY Approved	2 SBA Office Code	1 ① SBA Loan Number
⑤ Land Acquisition	$			
⑥ New Plant or Building Construction			**IV. Summary of Collateral:**	
⑦ Building Expansion or Repair			If your collateral consists of (A) Land and Building, (D) Accounts Receivable and/or (E) Inventory, fill in the appropriate blanks. If you are pledging (B) Machinery and Equipment, (C) Furniture and Fixtures, and/or (F) Other, please provide an itemized list (labeled Exhibit A) that contains serial and identification numbers for all articles that had an original value greater than $500. Include a legal description of Real Estate offered as collateral.	
⑧ Acquisition and/or Repair of Machinery and Equipment				
⑨ Inventory Purchase				

	Loan Requested	SBA USE ONLY Approved		Present Market Value	Present Mortgage Balance	Cost Less Depreciation
⑩ Working Capital (Including Accounts Payable)						
⑪ Acquisition of all or part of Existing Business			A. Land and Building	$	$	$
⑫ₐ Payoff SBA Loan			B. Machinery & Equipment			
⑫♭ Payoff Bank Loan (Non SBA Associated)			C. Furniture & Fixtures			
⑫c Other Debt Payment (Non SBA Associated)			D. Accounts Receivable			
⑬ All Other			E. Inventory			
⑭ Total Loan Requested	$		F. Other			
Term of Loan			Total Collateral	$	$	$

V. Previous Government Financing: If you or any principals or affiliates have ever requested Government Financing (including SBA), complete the following *

Name of Agency	Amount	Date of Request	Approved or Declined	Balance	Status
	$			$	
	$			$	
	$			$	

③ Previous SBA Financing (Check One) ☐ (1) No ☐ (2) Repaid/Other ☐ (3) Present Borrower | ④ Loan Number of 1st SBA Loan

VI. Indebtedness: Furnish the following information on all installment debts, contracts, notes, and mortgages payable. Indicate by an asterisk (*) items to be paid by loan proceeds and reason for paying same (present balance should agree with latest balance sheet submitted).

To Whom Payable	Original Amount	Original Date	Present Balance	Rate of Interest	Maturity Date	Monthly Payment	Security	Current or Delinquent
	$		$			$		
	$		$			$		
	$		$			$		
	$		$			$		

SBA Form 4 (9-81) REF SOP 50 10 1   PREVIOUS EDITIONS ARE OBSOLETE

SBA loan application
Figure 10-5

Section 3. Other Stocks and Bonds: Give listed and unlisted Stocks and Bonds *(Use separate sheet if necessary)*

No. of Shares	Names of Securities	Cost	Market Value Statement Date Quotation	Amount

Section 4. Real Estate Owned. *(List each parcel separately. Use supplemental sheets if necessary. Each sheet must be identified as a supplement to this statement and signed). (Also advises whether property is covered by title insurance, abstract of title, or both).*

Title is in name of	Type of property
Address of property (City and State)	Original Cost to (me) (us) $ _____ Date Purchased _____ Present Market Value $ _____ Tax Assessment Value $ _____
Name and Address of Holder of Mortgage (City and State)	Date of Mortgage _____ Original Amount $ _____ Balance $ _____ Maturity _____ Terms of Payment _____

Status of Mortgage, i.e., current or delinquent. If delinquent describe delinquencies

Section 5. Other Personal Property. *(Describe and if any is mortgaged, state name and address of mortgage holder and amount of mortgage, terms of payment and if delinquent, describe delinquency.)*

Section 6. Other Assets. *(Describe)*

Section 7. Unpaid Taxes. *(Describe in detail, as to type, to whom payable, when due, amount, and what, if any, property a tax lien, if any, attaches)*

Section 8. Other Liabilities. *(Describe in detail)*

(I) or (We) certify the above and the statements contained in the schedules herein is a true and accurate statement of (my) or (our) financial condition as of the date stated herein. This statement is given for the purpose of: *(Check one of the following)*

☐ Inducing S.B.A. to grant a loan as requested in application, of the individual or firm whose name appears herein, in connection with which this statement is submitted.

☐ Furnishing a statement of (my) or (our) financial condition, pursuant to the terms of the guaranty executed by (me) or (us) at the time S.B.A. granted a loan to the individual or firm, whose name appears herein.

Signature	Signature	Date

SBA FORM 413 (12-78) REF: SOP 50 50   Page 2   GPO : 1982 O—392-391

SBA loan application
Figure 10-5 (continued)

Form Approved
OMB No. 3245-0017

PERSONAL FINANCIAL STATEMENT	Return to:	For SBA Use Only
As of _____ , 19 ___.	Small Business Administration	SBA Loan No.

Complete this form if 1) a sole proprietorship by the proprietor; 2) a partnership by each partner; 3) a corporation by each officer and each stockholder with 20% or more ownership; 4) any other person or entity providing a guaranty on the loan.

Name and Address, Including ZIP Code (of person and spouse submitting Statement)	This statement is submitted in connection with S.B.A. loan requested or granted to the individual or firm, whose name appears below:
	Name and Address of Applicant or Borrower, Including ZIP Code
SOCIAL SECURITY NO. _____	
Business (of person submitting Statement)	

Please answer all questions using "No" or "None" where necessary

ASSETS		LIABILITIES	
Cash on Hand & In Banks . . . . $ _____		Accounts Payable . . . . $ _____	
Savings Account in Banks . . . . _____		Notes Payable to Banks . . . . _____	
U. S. Government Bonds . . . . _____		(Describe below - Section 2)	
Accounts & Notes Receivable . . . . _____		Notes Payable to Others . . . . _____	
Life Insurance-Cash Surrender Value Only . . _____		(Describe below - Section 2)	
Other Stocks and Bonds . . . . _____		Installment Account (Auto) . . . . _____	
(Describe reverse side - Section 3)		Monthly Payments $ _____	
Real Estate . . . . _____		Installment Accounts (Other) . . . . _____	
(Describe - reverse side - Section 4)		Monthly Payments $ _____	
Automobile - Present Value . . . . _____		Loans on Life Insurance . . . . _____	
Other Personal Property . . . . _____		Mortgages on Real Estate . . . . _____	
(Describe - reverse side - Section 5)		(Describe - reverse side - Section 4)	
Other Assets . . . . _____		Unpaid Taxes . . . . _____	
(Describe - reverse side - Section 6)		(Describe - reverse side - Section 7)	
		Other Liabilities . . . . _____	
		(Describe - reverse side - Section 8)	
		Total Liabilities . . . . _____	
		Net Worth . . . . _____	
Total . . . . $ _____		Total . . . . $ _____	

Section 1. Source of Income	**CONTINGENT LIABILITIES**
(Describe below all items listed in this Section)	
Salary . . . . $ _____	As Endorser or Co-Maker . . . . $ _____
Net Investment Income . . . . _____	Legal Claims and Judgments . . . . _____
Real Estate Income . . . . _____	Provision for Federal Income Tax . . . . _____
Other Income (Describe) . . . . _____	Other Special Debt . . . . _____

Description of items listed in Section I _____
_____
_____
_____
_____

* Not necessary to disclose alimony or child support payments in "Other Income" unless it is desired to have such payments counted toward total income.

Life Insurance Held (Give face amount of policies - name of company and beneficiaries) _____
_____
_____
_____
_____
_____
_____
_____
_____

SUPPLEMENTARY SCHEDULES

Section 2. Notes Payable to Banks and Others

Name and Address of Holder of Note	Amount of Loan		Terms of Repayments	Maturity of Loan	How Endorsed, Guaranteed, or Secured
	Original Bal.	Present Bal.			
	$	$	$		

SBA FORM 413 (12-78) REF: SOP 50 50   Edition of 8-67 May Be Used Until Stock Is Exhausted

SBA financial statement
Figure 10-6

Corporation

Advantages		Disadvantages	
1)	Limited debt liability	1)	Double taxation
2)	Easier to raise capital	2)	Subject to possible outside control and/or takeover
3)	Transferable ownership		
4)	Possible tax advantages	3)	Large amount of record keeping necessary
5)	Possible legal advantages	4)	Close government regulation
6)	Continuous existence	5)	Charter requirements
7)	Management specialization and expansion	6)	Costs to organize

Partnership

Advantages		Disadvantages	
1)	Low start-up costs	1)	Divided authority
2)	Easy to set up	2)	Unlimited debt liability
3)	Shared management	3)	Business ends if partner dies
4)	Limited regulation	4)	Difficult to find suitable partner
5)	Two sources of capital and expertise	5)	Difficult to find outside investors
6)	Possible tax advantages		

Single proprietor

Advantages		Disadvantages	
1)	Low working capital needed	1)	Difficult to raise capital
2)	All profits to owner	2)	Unlimited debt liability
3)	Owner in direct control	3)	Business ends, value drops if owner dies
4)	Low starting costs		
5)	Tax advantage to small owner		
6)	Small amount of outside regulation		

Business organization options
Figure 10-7

site or work-related accidents to employees and bystanders, is an absolute necessity. All states require workers' compensation coverage on employees, often handled by a private insurer.

You'll be too busy with your own business operations to become an expert in insurance. Besides the bare minimum mentioned above, there's also life insurance, key-man insurance, business interruption insurance, health insurance, contract bonds, automobile insurance, completed operations liability insurance and many, many more.

You won't need all types of coverage early in your business. As in many other areas, you'll need more as the business expands. But from the beginning, you'll need at least minimum coverage in key areas, *at a price you can afford*. Check with others in business to find a reliable, competent insurance agent. Rely on his advice. He'll make sure you have enough coverage without being "insurance poor."

Make sure you're comfortable with your agent and don't be afraid to shop for insurance coverages and costs. They vary a good deal. If your agent values your business, he'll try to get you the best possible coverage for the cost.

Government Regulation

Generally, the smaller the business, the fewer the regulations. But even a one-man operation can't work without a building permit in many areas. In other parts of the country, anyone doing business in a city or municipal area must have a license. Figure 10-8 summarizes the most significant government regulations and where to go for more information.

All businesses, no matter what size, are under some kind of government regulation. Despite political candidates' promises to reduce paperwork and regulation, many businesspeople say this is their biggest obstacle to efficient operation. However, we recommend that you abide by all applicable regulations, or at least those that you can't afford to fight on a legal basis. Most of the regulations are enforced by a bureaucracy with a full-time legal staff. Fighting them is often a no-win proposition.

However, you do have some rights. Government inspectors and investigators must identify themselves when they come to call. In many cases, they must notify you before they come to inspect your premises or books, and have appropriate authorization. You'll be notified of violations in writing and advised of an appeals process. Check with your attorney if you have any problems.

Legal Advice

You'll find yourself on shaky ground many times while you begin and operate your business. To clear up these uncertainties, get the advice of a competent attorney.

Only the largest contractors need full-time legal counsel. Instead, you'll probably retain your attorney on an as-needed basis. We recommend you find an attorney early in your business career. Work with your attorney on some basic, simple matters to start with, to make sure he or she suits your needs. A good working relationship is important. Don't wait until you've been slapped with a $500,000 lawsuit and then grab the phone book to begin looking for a lawyer.

You can ask your local bar association to recommend an attorney who's interested in handling new businesses. Also check with friends and other businesspeople for recommendations.

Keep in mind that the most expensive legal counsel isn't always the best. What you're after is someone with a background in construction law. Look for a younger attorney who's sharp and aggressive. He in turn may be looking for businesses that will grow, so his practice can grow with them.

When you find a suitable attorney, make an appointment to get advice on setting up your business. Here are some of the topics you may want to cover:

Personal liability— When you are personally liable for company problems — debts, injuries and the like — and how you can be protected.

Company liability— What your company is liable for and how it can be protected.

Contract liability— Although each contract you sign will be unique, you can get some general advice on where contractors get into trouble, how to protect yourself, and the remedies available. You'll find more information on contracts in the next chapter.

One final note: Keep in mind that you can often save yourself time, dollars and aggravation by settling small disputes and legal problems yourself. Use your lawyer only when absolutely necessary. If a client threatens legal action or complains about the job you did, try to settle it between yourselves. The same is true if you have a complaint. Most people recognize that it's in everyone's best interests to work it out without lawyers. You can also consider these options before going to your attorney: bring an action in small claims court, get paralegal advice, or talk to the American Arbitration Association.

On the other hand, if you're contacted by someone's lawyer, you have only two choices: contact the aggravated party directly, or turn the letter over to your lawyer. Don't contact the lawyer directly, since this may be interpreted as a legal response to the complaint. And never go to court against an attorney without your attorney, no matter how strong you feel your case may be.

Regulation	Employers affected	Basic requirements	For more information
Federal Income Tax	Employers and corporations with active cash flow; profit and loss. Any employer who pays wages (includes family members).	File Declarations of Estimated Income. Withhold taxes from employee's paychecks according to tables and applicable rate. Deposit amounts in bank at periodic intervals.	Local Federal IRS office. Also, write to: IRS
Federal Unemployment Tax	Employers with one or more employees during 20 or more weeks annually.	Contribute required percent of employees' wages and deposit. File Form 940 on or before Jan. 31 annually.	Local Federal IRS office.
Federal Wage & Hour Laws (Social Security)	Employers who pay over $50 per quarter in wages.	Obtain Social Security numbers of employees Withhold required percent and deposit.	
Title VII, Civil Rights Act - Equal Employment	Employers who pay over $50 per quarter in wages.	Treat employees and job applicants equally and fairly with no discrimination as to race, color, religion, sex, age or origin.	U.S. Dept. of Labor,* Office of Federal Contract Compliance. Local Affirmative Action Offices.
Occupational Safety & Health Act (OSHA)	All employers except family-owned/operated farms, and self-employed persons.	Work site must be kept free from hazards.	U.S. Dept of Labor, OSHA. Local OSHA office.
Employee Retirement Income Security Act	Most employers with employee benefit plans.	Requirements vary depending on plan type.	U.S. Dept. of Labor, ERISA Administration. Local office of U.S. Dept. of Health & Social Services.
State Income Tax	Varies by state.	Varies by state but similar to federal requirements. Withhold tax monies and deposit.	Nearest state revenue office.
State Unemployment Insurance	Varies by state.	Varies by state but similar to federal requirements. Contribute employer percent and withhold employee percent and deposit.	Nearest state job service/ unemployment office.
Workers' Compensation Insurance	Varies by state	Cover employees for job-related injuries by state insurance or approved private plan.	Nearest state Workers' Compensation office, or private insurance agent.
State/City/Town Licensing	Varies by area.	Filing of fees and information to insure technical skill.	Local building inspector. State/local consumer protection agency.

*U.S. Dept. of Labor, 200 Constitution Ave., NW, Washington, DC 20210

Government regulations
Figure 10-8

Other Advice

So far we've mostly talked about advice dealing with the negative side of business operations. There are sources of positive advice that cost little or nothing. Use them to help you become a better businessperson.

Small Business Administration— Besides making loan guarantees, the SBA offers free management assistance to anyone in business or planning to go into business. Ask for an appointment at your nearest SBA office. To avoid confusion, call them first and explain clearly what you have in mind. Make sure you qualify for planning assistance before setting aside time for the session.

You may also benefit from the SBA's management publications. You can get a list of what's available by requesting Form SBA 115A from the U.S. Small Business Administration, P.O. Box 15434, Fort Worth, TX 76119.

Other low cost and no cost sources— Most large universities have business departments or extension offices that hold seminars in business planning, computer technology, management techniques and other subjects. Look in your phone book under the name of the college or university. In rural areas, look under the county name for the Agricultural Extension Office. Although these offices may not have anything immediately on hand for your business, they'll be able to get suitable information for you.

Your public library is also an excellent source of information. Ask the librarian to show you the section on small business operations. Review the books to find the ones that apply to your operation. You may also want to check the section on HVAC to see if they have anything directly related to HVAC business practices.

Sometimes banks, technical colleges and local governments sponsor seminars on business financial management. Most of these are offered at little or no cost. Not only will you pick up information, you may find other tradespeople who'll be willing to discuss their problems and solutions that have worked for them.

Record Keeping

Good record keeping involves paperwork, something many tradespeople wish they could do without. But a primary reason why businesses fail is lack of bookkeeping. An up-to-date, accurate set of books will let you know exactly where you stand on any given day. It can warn you of potential problems. It can also alert you to make changes in business operations that will save you time and money. You'll see in the next chapter how accurate records on bidding and jobs can mean the difference between success and failure.

A large HVAC outfit can afford a full-time bookkeeper. That makes it easy to maintain good records, because someone has that specific responsibility. A small, one-person operation only requires bookkeeping on a part-time basis. There are several books that can help you in this area listed in the last pages of this manual.

Almost any good bookkeeping system will do, as long as it's set up correctly and updated accurately. Look at it this way. If you need a truck to haul supplies and a three-person crew to the job, you wouldn't buy a compact car. On the other hand, you wouldn't buy a fleet of flatbed semis. You'd buy a truck to fit your needs. But in a couple of years, when you have more men and more jobs, that truck may be too small.

That's how it'll be with your bookkeeping. When your company is small, you or your wife may be able to handle the chore. You may take your books to an accountant once a year for a summary and tax work. As the business expands, you'll hire a part-time bookkeeper or train someone to do it along with other office duties. You may find it makes sense to buy a small computer to maintain receivables, payables, payroll and your inventory, or to use a data processing center part-time. You may grow large enough to hire a full-time bookkeeper and use your accountant on a monthly basis.

Here's the major rule to remember as you take all these steps: Start out properly and don't delay when it's time to upgrade the system. Bookkeeping is the kind of work that has to be done every day. If you do it every day, it's a tolerable task. But leave it undone and it will soon pile into a desktop mountain that will scare you off, no matter how stout your intentions. Sure, you can hire a bookkeeper to come in and straighten out your mess. But it won't be cheap and they may discover some substantial dollar losses because of your neglect.

Someone else who won't appreciate your haphazard method is the IRS. You're required to keep a verifiable record of all the business expenses you claim at tax time. If you can't verify those expenses, they're just a pipe dream as far as the IRS is concerned.

A little later we'll discuss some basic bookkeeping terms. You won't be a bookkeeper when you've read them, but you will be able to sit down and talk intelligently with an accountant about what you want to do. And that's the first step.

Find a good accountant. He or she doesn't have to be a CPA, but you may feel more comfortable dealing with one. The CPA just means that your accountant has passed a state test and received a license. Whether a CPA or not, your accountant should immediately recognize what your business objectives are and the type of system you'll need. Basically, if you're starting out small, the system should be simple to keep up, with you doing most of the work. Most often this will be *single-entry* bookkeeping. The single-entry system uses the business checking account as the central control. Later, as your sales volume increases, you may need the *double entry* system, which shifts more of the load to the accountant.

Keep in mind these three rules when you're doing the bookkeeping:

- Be accurate
- Be complete
- Be on time

Set aside a regular time during your work day or week for bookkeeping. Make it a habit. Even after you've turned over your bookkeeping chores to someone else, you'll still have to review the books regularly. If you know how the system works, you'll be able to find the indicators we mentioned earlier and take appropriate action to protect your investment.

Remember that the financial history of your company is being written every day in the paperwork you generate. If you organize and understand that paperwork, you can use it as a tool to succeed. In today's business climate, you need every advantage you can generate. As the boss, you have the final responsibility for the success of your business, no matter how many bookkeepers and accountants end up on your payroll.

Bookkeeping Terms

These definitions will give you the basic vocabulary you need to discuss your bookkeeping setup with your accountant:

Assets— Anything your business *owns*. Cash, materials, equipment, land, buildings, and money owed you by customers (accounts receivable) are assets typically found in a construction business. The value of your assets is recorded in your bookkeeping system to offset what you owe. The market or real value of your assets is a good indication of your basic financial health and your ability to raise money when you need it.

Liabilities— Anything your business *owes*. These items might be wages, taxes, money owed to lenders (notes payable), and money owed to suppliers (accounts payable). Liabilities are recorded in your system and balanced against the assets.

Equity— Also called capital, equity is the difference between what your business owns and what it owes. In other words, if your business is healthy, you'll have value left over after you subtract the liabilities from the assets. This is the equity, and it belongs to the owner or owners.

Bookkeeping equation— This formula simplifies what we've just talked about: Assets equal Liabilities plus Equity. This easy-to-remember formula is the backbone of all accounting systems. Summary statements such as balance sheets and profit-and-loss statements use the bookkeeping equation with the assets on the left side and the liabilities and equity on the right side to allow a quick review of a company's financial history over time.

Daily summary— Your company records start with all the different size sheets of paper used for each transaction — receipts, check stubs, deposit slips, billing invoices, bank statements and the like. It's important to have a written record of all these transactions as they take place. If you do a large volume of business, you'll want to summarize this information every day. In a smaller business, it may make more sense to put all the individual records in a folder and summarize them once a week. Be sure nothing is lost, and record the information on your summary sheet accurately. Don't let the summary go longer than one week.

The journal— Information from the summary sheet is brought together in the journal. Each journal entry shows the following:

1) the date of the transaction

2) the amount of the transaction

3) a brief description of the transaction

4) the name of the account affected by the transaction.

The accounts— Accounts break down the journal information. Each account is a record of the increase or decrease of any type of asset, liability, capital, income or expense. Petty cash, repairs and maintenance, inventory, equipment, payroll and taxes are just a few of the accounts you'll need in your contracting business. Your accountant can help you set up a complete list.

Debits and credits— In double-entry bookkeeping, a credit is an account entry representing a decrease in value for that account. A debit represents an increase in value. For example, when you buy a load of ductwork from a supplier, the money paid is entered as a credit in your cash account — that is, it's subtracted from the cash balance. The same amount is entered as a debit in your materials account, since that account's value has increased by the same amount.

Debits are entered in the left column of the account, and credits in the right column. When credit entries total more than debits, the account has a credit balance. When the opposite is true, the account has a debit balance. At any time, the total debit balances of all accounts should equal the total credit balances.

The ledger— The ledger book is where you keep track of the accounts. You'll also use its columns to total all the debit and credit account entries. If the bookkeeping is free from errors, the final totals should balance. This adding and checking of accounts is called the *trial balance*. It's done on a monthly, weekly or even daily basis, to check for errors before they get buried too deeply.

Financial statements— Entries in your journal and ledger alone would give you the condition of your business at any time. But your accounts change daily and often there's a lot of information to sort out. So all this information is summarized on your financial statements.

The *balance sheet* and the *profit-and-loss* sheet are the two primary financial statements. The balance sheet is usually done once a month. It summarizes assets, liabilities, and equity, to give you your financial condition at a glance. It's called the balance sheet because it follows the bookkeeping equation: assets will equal liabilities plus equity.

The profit-and-loss statement records a business's income and expenses over time. P&L's are usually done on a quarterly or yearly basis. They summarize major income and expense accounts and let you know how much profit or loss to expect for that given time period.

In the beginning you'll probably handle the bulk of your bookwork yourself. For example, you'll keep your own daily summary, journal posting, accounts and payroll. Your accountant will do the trial balance, the P&L, and prepare your quarterly tax statements and annual income tax. As your business grows, you may want to hire someone to handle the bookwork in your office. Your accountant will recommend changes in methods and accounts to help keep your system working smoothly.

A good set of books requires the proper posting materials. Go to a good office supply store and ask for advice about books tailored to your system. Also ask your accountant for his recommendations on what books to get and where to get them.

Besides your record set, you'll need a file cabinet or lockbox to keep them in. You'll also need a desktop calculator.

Many small offices are now using personal computers to keep their books. A computer program that's tailored to your operations can save time and confusion in your bookkeeping, It can also simplify your monthly billing, inventory control, bid records, letter writing and a host of other operations. The drawback to a computerized system is that unless you've had some previous experience with computers, you can spend a long time trying to figure out what system will work best. To save time, find someone else who's operating a business like yours and has installed a computer. Most owners will be quite honest in telling you both the pluses and minuses in their system. When you do decide to buy, select *first* the software that will work for your business, and *then* buy the hardware to run it. Buy only the equipment you need, but make sure it has options for expansion later on. We also recommend you stay with a manufacturer that has a proven track record, such as Apple, IBM, Zenith or Radio Shack.

If you do use a computer for your bookwork, make sure more than one person knows how to use the equipment and system. You're the best standby person. If your office help quits or gets the flu for two weeks, you can operate it yourself or teach someone else how to do it. *Computers: The Builder's New Tool* is a manual providing invaluable guidance in this area. Again, check the order form in the back.

Inventory

Businesses that succeed today are leaner and tougher than those of previous years. Accounts receivable have to be collected promptly. High interest rates require careful juggling. Here are some suggestions to cut your inventory burden:

• Always ask about quantity discounts. Everything is available at a discount if you buy enough. On the other hand, don't buy inventory you won't use or will use very slowly, simply because it's at a discount.

• Compare shipping costs. One supplier may have a better price, but if you have to pay more in shipping charges, it'll cancel out any savings. Often if you point this out to the supplier, he'll come down enough to stay competitive.

• Have materials delivered directly to the job by the supplier, at the time when they're needed. This will save you time and money. Your crews won't have to handle the materials and you won't have to pay for them any sooner than necessary. You have to protect everything delivered to the job from weather, theft and damage as soon as it arrives. Some insurance policies require that it be kept in a locked enclosure. Know what your policy requires.

• Check suppliers' billing terms. Most now offer 15 or 30 days the same as cash. After that they begin to charge interest. If all else is equal, go with the supplier who offers 30 days. You might as well keep your money in the bank drawing interest for two extra weeks.

• Set up an inventory control system for your in-house supplies. It doesn't have to be fancy. Put the supplies in a secure area. Have a sign-out sheet on a clipboard next to the door. Assign one trustworthy person the responsibility of checking materials in and out. Make up a sign-out sheet with columns for the date, type of material and serial number,

the amount, job location, and person removing the materials. Compare this sheet on a regular basis with your master inventory file and note any shortages. Also double-check by comparing it with the appropriate job schedules. For example, if you see a lot of registers or baseboards checked out for a job that's still in the roughing-in stage, you'll want to ask why.

Advertising

Most HVAC contractors limit their advertising to the Yellow Pages in the local phone book. There's nothing wrong with this, since the bulk of your work will come through bids, trade contacts, or word-of-mouth from other customers. But you may find that it pays to be more aggressive. A well-placed ad targeted to the right audience can pay off in a big way.

The key to good advertising isn't necessarily how much money you spend — it's where you spend it. Radio, print and television all appeal to different markets. Ask your local media salespeople about their target audiences, and the best time to reach them. The rest is common sense. You probably wouldn't waste your money sponsoring local daytime television programming. But a well-written ad at an autumn sports event may bring in customers all winter long.

Advertising can be expensive. Proceed cautiously, especially in the beginning. You may have more important ways to spend your money. It's often hard to see any direct benefit when you advertise. The most any ad can do is to prompt the prospect to contact you. You still have to make the sale. An ad won't substitute for salesmanship and the ability to explain your trade in a way the customer understands and appreciates.

Keep in mind that you may be able to barter your services in return for advertising from local radio, television and magazine companies. Treat these arrangements on a businesslike basis. Draw up a contract to avoid confusion, specifying the dollar amounts of the services rendered.

Many contractors like to put their ads in programs put out by local organizations or civic groups sponsoring an event. The ad will reach only a limited audience one time. But the cost is usually low, so you can afford to advertise for several events in the course of the year. Your ad will reach only local customers and they'll begin to associate your company with events that benefit the community.

The Company Logo

A company logo isn't exactly advertising, but it can do you a lot of good. A unique, stylized sign or symbol can become strongly identified with your business in the public mind. There are many small companies that specialize in drawing up a distinctive logo at a reasonable cost. You'll get a camera-ready copy so you can reproduce the logo on stationery, business cards, and advertising.

A good logo should be easy and cheap to reproduce. It should identify your business name and the type of business you're in. It shouldn't be cluttered or difficult to read. Avoid logos that are "cute" or use a theme or design that may go out of date. Spend some time with the designer and make sure he understands what you want your logo to say about your business. Keep trying until you find one that satisfies you, then stick with it.

Professional Organizations

Anyone in business should consider the advantages of joining local business clubs and civic groups. Nearly every community has these organizations. National clubs like the Kiwanis or Lions probably have local chapters in your area. There are also local Chambers of Commerce, neighborhood associations and business clubs. Membership in these organizations enables you to exchange ideas with others in business. Membership can also make you aware of potential business opportunities. It will help you keep a handle on the local business climate. There's an advantage to being known in the community where you operate, since people like to do business with someone they know and respect.

Trade Publications and Associations

In the next chapter on Estimating and Bidding, we'll talk about trade magazines as a source of bidding information. You can also use them to keep up with innovations in products, safety and installation practices. Other articles will explain design practices and business operations. Since many of these articles are written by contractors and engineers with years in the field, their advice is usually based on firsthand experience.

One magazine that may be worth subscribing to is *Heating/Piping/Air Conditioning*. The address is P.O. Box 95759, Cleveland, OH 44101. The cost is $35 per year.

Besides reading about HVAC, you may want to join HVAC trade associations. The one most contractors feel has the biggest impact on the HVAC industry is *ASHRAE*, The American Society of Heating, Refrigerating and Air-Conditioning Engineers, Inc. It's a national organization with local chapters in most major cities in the U.S. For information on joining, write to the Membership Department at:

ASHRAE
1791 Tullie Circle NE
Atlanta, Georgia 30329

ASHRAE membership is open to anyone interested in HVAC design work. There are different membership designations, depending on education and technical experience. The initial membership fee in the U.S. is $50. This also covers the first year's dues. Annual dues are $75 thereafter. Members receive the *ASHRAE Journal,* a monthly magazine featuring articles on research, design and development of HVAC systems. You also receive the four-volume annual series of the *ASHRAE Handbook,* an excellent reference work for HVAC theory and design. Many of the tables we used in the chapters on warm air design are courtesy of ASHRAE. Their books go into the theory used in drawing up these tables. ASHRAE also offers other publications, such as research bulletins, at a reduced price to its members.

Another national organization for HVAC contractors is the Air Conditioning Contractors of America (ACCA). Although the name refers only to air conditioning, ACCA offers a wide range of services covering all areas of HVAC contracting. Membership costs $240 annually. It includes a 15-volume Environmental Systems Library, a monthly newsletter, bulletins that cover updates in both business and technical skills, public relations help, contacts with other contractors, group insurance, computer software and assistance, and many other benefits. Here's their address:

ACCA
1228 17th Street NW
Washington, DC 20036

Besides the national organizations, some areas have state or county associations for contractors. They're often made up of all types of building contractors and function as bargaining units in wage negotiations with labor unions. Some cities, however, may have an association strictly for HVAC contractors.

You'll probably want to investigate each of these organizations. Only you can determine which is of value to your business. Generally, the larger national organizations keep abreast of trends and changes in the industry. They can offer advice and training that may not be available through local groups. Local groups, on the other hand, usually have regular meetings where you can talk shop with other contractors. These meetings can be invaluable in making business contacts and learning how others have dealt with problems common to the trade.

As of this writing, money spent on membership dues, trade magazines and seminars and other work-related training is tax deductible. So is the cost of this book.

In the final chapter we'll cover the subject that makes or breaks many contractors — estimating and bidding.

Estimating, Bidding, and Contracts

This chapter could be called "Staying in Business," because that's exactly what it's all about. In the previous chapter we gave you some advice on getting started in business, and explained some business fundamentals. Earlier in the book, we stressed the technical parts of HVAC contracting and how to sell your services. But no matter how good a craftsman you are, or how solid your business foundation, you need a steady flow of work to generate the profit you deserve for the risks involved.

You might be lucky enough to get some work without effort. But you'll often be in competition with another contractor for even the smallest jobs. How well you do will depend on your company's productivity — how efficient you and your crews can be. It'll also depend on how well you can estimate and bid.

Bidding is largely a matter of paperwork. Estimating is the key to arriving at a work-producing bid. It isn't easy to become a good estimator. It takes job experience, good record keeping, informed "hunches," thorough and systematic methods, patience, and a good bit of luck.

In the following sections, we'll outline the best way we know to handle bidding and estimating. We'll give you the know-how and the methods. But you'll have to supply the patience, experience and hard work. It's a combination that will insure your continuing success as an HVAC contractor.

Estimating

All contracting estimates have three parts: cost, overhead and profit. A simple estimate may combine one, two or all three items. Large estimates may break the three categories into many smaller divisions, and include every possible job variable you can imagine. But even large estimates become simple if you remember that every item on the estimate will fit in one of the three main categories.

Costs

There are two kinds of costs for contractors: *direct job cost* and *indirect job cost*.

Direct job cost— We'll break direct job cost into two smaller categories. The first is *bare cost*, the dollar amount of materials needed for the job, without any markup for overhead and profit. The second category is *base rate*, the direct labor cost of installing the materials. This is the hourly rate, including fringe benefits but not including insurance and taxes, or overhead and profit markups.

Sometimes contractors talk about a direct job cost ratio of 40/60. This means that for every $40 spent on materials, another $60 is spent on installation. We're talking here about the bare cost and the base rate. In the past, this ratio was reversed, 60/40. Then it became 50/50. Most contractors feel 40/60 reflects current conditions. Of course it'll vary depending on your area, the local economy, and even the job.

The job cost ratio is the basis for a quick method of estimating called the *lump sum*. A contractor can find his bare cost for a job, divide it by the bare cost percent and get his total direct job cost. Then he'll add a suitable percent for profit and overhead to complete his estimate. Many contractors still use this method successfully on small jobs and on jobs where they've had a lot of experience. We'll give an example of a lump sum estimate later in the chapter, under "Preparing the Bid and Estimate."

Larger jobs, however, require tighter estimates. You'll have to use your job records and past experience to estimate the number of hours needed for job installation. Then multiply the number of hours by the base rate in your area to get an accurate estimate of labor cost.

Indirect job cost— Your estimate will also include a percentage for indirect job cost, also known as job overhead. The percent will vary greatly from job to job. For example, on a small job you may pull the materials from your building, load them into your van and head for the job to do the installation. But a larger job will include job-site insurance, special equipment which you may have to rent, job telephone, storage trailer and the like. Job overhead is the expense that you wouldn't have if you didn't have this job.

Overhead

Business or operating overhead is called simply *overhead*. It's your cost of running the business. Rent, advertising, utilities, mortgage payments and fixed payroll costs are typical overhead costs.

Most contractors figure all their overhead costs at about 22% of the total direct job cost for materials and labor. Multiply the direct job cost by 22%, and add the total to the estimate. If you figure overhead on labor alone, add a 40% markup.

Profit

The final item in the estimate is the *profit*. The most common figure used for the profit markup is 10%. After all, profit is the reason you're in business. Give yourself a reasonable return on your work and investment.

Remember that when we talk of adding in job overhead, business overhead and profits, we mean to include them in your *estimate*. You probably won't itemize them on your final bid.

Keep in mind that the percentages we used are only guidelines. Job and competitive conditions will always be the final guide when you draw up your estimate and bid. If your material, labor and overhead costs have already been cut to the bone on a specific job, the only place left to cut is your profit. But unless you're the kind of person who enjoys working for free, this isn't the place you want to cut.

If you have to reduce your price to stay competitive, look for an edge on materials, labor or job costs. You can usually find something. For example, you might be able to reduce your labor costs by using your most productive employees — your "key" people — on the job. Or sometimes your supplier can go back to the factory and get a special, one-time cost reduction on materials for this job. You in turn may promise to do substantial business with the supplier in the future. Still another possibility is substituting low-cost materials on the job in areas where quality isn't affected. But be sure you get the owner's approval of this last suggestion before you go ahead with it.

We'll describe some of these steps in more detail later in this chapter. The main idea here is for you to cut costs enough to get the work but leave your profit percentage alone, or perhaps even increase it. Often that will depend on how much inventiveness and creativity you can develop when working up the job estimate.

The Estimate

A carefully-done estimate will reflect actual project costs. It's a combination of material costs, labor cost, overhead and profit. In a sense, every estimate is a guess. Make yours a well-informed guess.

Estimates aren't legally binding on your company, but they are the basis for your bid. The bid becomes part of the job contract. We can't overemphasize the importance of a good, accurate estimate.

Organization is the key to working up a good estimate. Pulling information off a plan set and jotting it down in a haphazard manner will cost you dearly someday. Take our word for it, or find out for yourself the hard way. While no two estimators or contractors use exactly the same method, here's the general procedure we recommend:

Begin by getting a set of plans and specifications for the job. Review Chapter 4 on this procedure. If you've drawn them up yourself, so much the better. Whether it's your own work or someone else's, study the plans and specs carefully. Be alert for discrepancies between the plan and the specs. Identify room numbers and drawing numbers which coordinate sections and details to the floor plan. Measure and list everything on the drawings or in the specs.

Review the entire spec book, not just the HVAC section. Some other trade sections may have HVAC work included in them. If your estimate, and later your bid, misses these, you still have to do the work. Study the plans until you can see the entire job in your head. If it helps, draw up rough sketches to assist your review.

Doing the take-off— After you're comfortable with the plans and specs, start taking off materials and putting them on the *take-off sheet*. The take-off sheet is just one of several forms you'll need to keep things neat and orderly during estimating and bidding. The ACCA (see Chapter 10) offers a job estimate form for forced air systems that includes columns for take-off. Figure 11-1 shows a general estimating form you can buy at your local office supply store. You can also use it as a take-off sheet. Many contractors prefer lined, columnar tablets of the kind used in bookkeeping. Make a separate sheet for each type of material used for the job. The numbered lines on the sheets help keep items in order. Use the columns to record the dimensions or number of units of the supplies and materials. Remember to use only the front side of each sheet and title and number it correctly.

The purpose of the take-off sheet is not to price out the job. Instead, as its name suggests, it's used to take off all the materials from the plans and specs and gather them in a neat and orderly fashion. This material list will be the foundation for all other calculations. Like any other foundation work, if you mess it up, it'll continue to cause problems all through the job.

Here's a hint to help you avoid omissions. Make sure the plan set is yours to do with as you please. Then begin at the upper left-hand corner of the first sheet. As you calculate the piece of equipment or run of ductwork or whatever, *cross it off with a colored pencil*. We suggest you use purple or green. Save red for emergencies. Work clockwise until you've covered everything. When your take-off is finished, have someone else review it to make sure you didn't skip anything. If there's no one else who can do it, wait a day and do it yourself. Check to make sure you haven't missed any items. Then check to make sure you've listed the item correctly on your take-off sheet. When you're satisfied, take a different color pencil and make a second check mark next to the item.

Another hint: Be consistent in listing dimensions. For example, length times width times height is the preferred sequence. Take off quantities from the floor plan first, elevations second, detail drawings third, and finally the spec information.

Although many estimators like to start at a certain place on the plan and go clockwise, as we suggested, some prefer to do the take-off item by item. They'll record all the ductwork, then the accessories such as duct liner, canvas connectors, diffusers and registers, dampers and filters, and so on. This method has the advantage of grouping your items neatly on the take-off sheet. Use whichever method works best for you. You may want to do some experimenting and develop your own method. Remember that the forms used in estimating and bidding are nothing more than tools of the trade. They're designed to do a job *for* you, not *on* you. If you find yourself getting confused and disorganized, start over until you're satisfied with your system.

Whatever method you use, stick with it until you have a good reason to change. Remember, you want to end up with neat, logical groupings on the take-off sheet. But more important, you want the sheet to be complete. The more times you have to switch items from sheet to sheet, the more chance for error.

Doing the pricing— After you've finished the take-off, begin pricing. For small jobs, you can do the pricing right on the take-off sheet. If the estimate is a large one, do the pricing on a separate sheet. This is called the *pricing sheet,* or sometimes the extension sheet. Be careful in transferring items from the take-off to the pricing sheet. Don't skip any.

GENERAL ESTIMATE

BUILDING _____

LOCATION _____

ARCHITECTS _____

SUBJECT _____

ESTIMATE NO. _____

SHEET NO. _____

ESTIMATOR _____

CHECKER _____

DATE _____

DESCRIPTION OF WORK	NO. OF PIECES	DIMENSIONS			EXTENSIONS		EXTENSIONS		TOTAL ESTIMATED QUANTITY	UNIT PRICE M'T'L	TOTAL ESTIMATED MATERIAL COST	UNIT PRICE L'B'R	TOTAL ESTIMATED LABOR COST

General estimate form
Figure 11-1

It's usually best to group like materials of the same price together on the pricing sheet. Find the cost for each item and write it in the material cost column. Multiply this cost by the quantity and write the total in the extension column. Figure 11-2 shows how to use a three-column work sheet from any office supply store as a pricing sheet.

Besides the material cost, you'll want to include the labor cost on your pricing sheet. Labor costs can be handled in several ways. One method is to begin with the base labor rate. Add your markup to determine the billing rate, then estimate the time it will take for the installation. For example, assume a base hourly rate of $20. You feel that this particular job will handle a 40% labor markup and still be competitive. So you multiply $20 by 1.40 to find a billing rate of $28. Previous experience tells you that one man can install 12 x 24 ductwork at a rate of five feet per hour on this particular job. This is 42 pounds of ductwork, according to the duct weight multiplication table in Figure 11-3. Divide 42 pounds by the hourly billing rate of $28 to find a per pound cost of $1.50. Write this on the pricing sheet and multiply it by the weight of that particular ductwork to get an extended labor cost.

Another method uses time as the labor unit. Some contractors prefer this since it permits flexibility when they're unsure of labor costs. This is often the case if an estimate is done on a job that may be some months in the future and where labor negotiations are still pending. Using the example above, we know that 42 pounds of ductwork will take one hour to install. That breaks down into 0.7 pounds per minute. Let's assume we have 600 pounds to install:

$$\frac{600}{.7} = 857 \text{ minutes}$$

$$\frac{857}{60} = 14.3 \text{ hours}$$

Multiply 14.3 hours by the going labor rate to get a final quote on the 12 x 24 ductwork.

Consider the variables— Before we go on to discuss other parts of the estimate, there are several important items we should look at here. All of them point to the same thing. Your estimate is not simply a matter of pulling off materials and pricing them. Instead, you'll need to adjust it to reflect the job variables that only you — the boss — are aware

of. For example, we took a labor mark-up of 40%. But if you felt that the bidding on this particular job was going to be very close, you might want to cut that to 37%. But would you really be cutting your labor markup? Realistically, you'd be cutting your profit. But perhaps you have 600 pounds of new 12 x 24 duct stashed away that you bought last year when prices were down. Maybe your supplier has mentioned that your competition has been calling for quotes on that size ductwork. You're reasonably sure they'll go with the market price, which is higher than the price you paid for it. So you could afford to cut your labor markup and use the market price of the ductwork. Your profit margin would stay intact because of the difference in what you paid for the ductwork and what you're selling it for. And your bid would be very competitive because of the lower labor markup.

Another item is the *productivity* of your crew. Perhaps when we estimated that one man would install only five feet of ductwork in an hour, you think he's far too slow. Perhaps you or any man on your crew could do twice that much. But stop and think a moment. That rate reflects the figure for that particular job. There's 600 pounds of 12 x 24, which is about 75 linear feet. That ductwork will have to be moved from your shop to the job. Your man will have to lay it out, get his tools set up and get organized. Then he can begin hanging the ductwork from the joists, running it through the walls, jointing and bracing. Sure, perhaps on the second day your man will be able to install 15 feet per hour. But on the first day he may install only 10 feet all day. That set-up time on the first day will happen, whether you account for it in your bid or not. If you don't, you'll pay for it out of your profit.

Here's something else to consider: What will happen if you assign two men on that job? Productivity should increase by at least one and a half times, or perhaps almost double. Would this justify the extra labor cost? And suppose you had 6,000 pounds of ductwork to install, instead of 600. Again productivity would increase, since the set-up time is the same for 600 pounds as for 6,000 pounds. Also, the men would become more efficient as they grew accustomed to the work. The point we're trying to make with all this is that estimating is a difficult and tricky process. You have to know your suppliers, your crews, and your local competition. You also need to know how well you've done in the past.

ACME CONDOS

PRICING SHEET

JOB #54

6/1/85

SHEET: 1 OF 2

	MATERIALS	UNIT	COST PER UNIT	TOTAL COST
2	600 lbs 12 X 16 DUCTWORK	600 —	75	450 00
3	250 lbs 12 X 12 DUCTWORK	250 —	1 20	300 00
4	22 EA 8 X 10 REGISTERS	22 —	6 90	151 80
5	14 EA 10 X 6 GRILLES	14 —	4 85	67 90
6	4 EA 10 X 12 DUCT ACCESS DOORS	4 —	13 25	53 00
8	ETC.			

	LABOR	UNIT	COST PER UNIT	TOTAL COST
24	850 lbs DUCTWORK	850 —	1 50	1275 00
25	22 EA REGISTERS, 8 X 10	22 —	7 50	165 00
26	14 EA GRILLES, 10 X 6	14 —	6 85	95 90
27	4 EA DUCT ACCESS DOORS, 10 X 12	4 —	15 00	60 00
29	ETC.			

Pricing sheet
Figure 11-2

Based on 24 Gauge Plus 20%

TOTAL DUCT WEIGHT (STEEL) MULTIPLICATION TABLE (Length In Feet)

SEMI-PERIMETER (INCHES)	5	10	15	20	25	30	35	40	45	50	55	60	65	70	75	80	85	90	95	100
24	28	56	84	112	140	168	196	224	252	280	308	336	364	392	420	448	476	504	532	560
25	29	58	87	116	145	174	203	232	261	290	319	349	377	406	435	464	493	522	551	580
26	30	60	90	120	150	180	210	240	270	300	330	360	390	420	450	480	510	540	570	600
27	32	64	96	128	160	192	214	256	288	320	352	384	416	448	480	512	544	576	608	640
28	33	66	99	132	165	198	231	264	297	330	363	396	429	462	495	528	561	594	627	660
29	34	68	102	136	170	204	238	272	306	340	374	408	442	476	510	544	578	612	646	680
30	35	70	105	140	175	210	245	280	315	350	385	420	455	490	525	560	595	630	665	700
31	36	72	108	144	180	216	252	288	324	360	396	432	463	504	540	576	612	648	684	720
32	37	74	111	148	185	222	259	296	333	370	407	444	481	518	555	592	629	666	705	740
33	39	78	114	156	195	234	273	304	342	380	418	456	494	532	570	608	646	684	722	760
34	40	80	120	160	200	245	280	320	360	400	440	480	520	560	600	640	680	720	760	800
35	41	82	123	164	205	246	287	320	369	410	451	492	533	574	615	656	697	738	779	820
36	42	84	126	168	210	252	294	336	378	420	462	504	546	588	630	672	714	756	798	840
37	43	86	129	172	215	258	301	344	387	430	473	516	559	602	645	688	731	774	817	860
38	44	88	132	176	220	264	308	352	396	440	484	528	572	616	660	704	748	792	836	880
39	45	90	135	180	225	270	315	360	405	450	495	540	585	630	675	720	765	810	855	900
40	46	92	138	184	230	276	322	368	414	460	506	552	598	644	690	736	782	828	874	920
41	47	94	141	188	235	282	329	376	423	470	517	564	611	658	705	752	799	846	893	940
42	48	96	144	192	240	288	336	384	432	480	528	576	624	672	720	768	816	864	912	960
43	49	98	147	196	245	294	343	392	441	490	539	585	637	686	745	784	833	882	931	980
44	50	100	150	200	250	300	350	400	450	500	550	600	650	700	750	800	850	900	950	1000
45	52	104	156	208	260	312	364	416	468	520	572	624	676	728	780	832	884	936	998	1040
46	53	106	159	212	265	318	371	424	447	530	583	636	689	742	795	848	901	954	1007	1060
47	54	108	162	216	270	324	378	432	486	540	594	648	702	756	810	864	918	972	1026	1080
48	55	110	165	220	275	330	385	440	495	550	605	660	715	770	825	880	935	990	1045	1100
49	56	112	168	224	280	336	392	448	504	560	616	672	728	784	840	896	952	1008	1064	1120
50	57	114	171	228	285	342	399	456	513	570	627	684	741	798	855	912	969	1026	1089	1140
51	59	118	177	236	295	354	413	472	531	590	649	708	676	826	885	944	1003	1062	1121	1180
52	60	120	180	240	300	360	420	480	540	600	660	720	780	840	900	960	1020	1080	1145	1200
53	61	122	183	244	305	366	427	488	549	610	671	732	793	854	915	976	1037	1098	1159	1220
54	62	124	186	248	310	372	434	496	558	620	682	744	806	868	930	992	1054	1116	1178	1240
55	63	126	189	252	315	378	441	504	567	630	693	756	819	882	945	1000	1071	1134	1197	1260
56	64	128	192	256	320	384	448	512	576	640	704	768	832	896	960	1024	1088	1152	1216	1280
57	66	132	198	264	330	396	462	528	594	660	725	792	858	924	990	1056	1122	1188	1254	1320
58	67	134	201	268	335	402	469	536	603	670	737	824	871	938	1005	1072	1139	1206	1273	1340
59	68	136	204	272	340	408	476	544	612	680	748	816	884	952	1020	1088	1156	1224	1292	1360
60	69	138	207	276	345	414	483	552	621	690	759	828	897	966	1035	1104	1173	1242	1311	1380
61	70	140	210	280	350	420	490	560	630	700	770	840	910	980	1050	1120	1190	1260	1330	1400
62	71	142	213	284	355	426	497	568	639	710	781	852	923	994	1065	1136	1207	1278	1329	1420
63	73	146	219	292	365	438	511	584	657	730	803	876	949	1022	1095	1168	1241	1314	1387	1460
64	74	148	222	296	370	444	518	592	666	740	814	888	962	1036	1100	1184	1258	1332	1406	1480
	5	10	15	20	25	30	35	40	45	50	55	60	65	70	75	80	85	90	95	100

Duct weight multiplication table
Figure 11-3

TOTAL DUCT WEIGHT (STEEL) MULTIPLICATION TABLE (Length in Feet)

Based on 22 Gauge Plus 20%

SEMI-PERIMETER (INCHES)	5	10	15	20	25	30	35	40	45	50	55	60	65	70	75	80	85	90	95	100
65	90	180	270	360	450	540	630	720	810	900	990	1080	1170	1260	1350	1440	1540	1630	1720	1800
66	92	184	276	368	460	552	644	736	828	920	1012	1104	1196	1288	1380	1472	1564	1656	1748	1840
67	93	186	279	372	465	558	651	744	837	930	1023	1116	1209	1302	1395	1488	1581	1674	1767	1860
68	95	190	285	380	475	570	665	760	855	950	1045	1140	1235	1330	1425	1520	1615	1710	1805	1900
69	97	194	291	388	485	582	679	776	873	970	1067	1164	1261	1358	1455	1552	1649	1746	1843	1940
70	98	196	294	392	490	588	686	784	882	980	1078	1176	1274	1372	1470	1568	1666	1764	1862	1960
71	100	200	300	400	500	600	700	800	900	1000	1100	1200	1300	1400	1500	1600	1700	1800	1900	2000
72	101	202	303	404	505	606	707	808	909	1010	1111	1212	1313	1414	1515	1616	1717	1818	1919	2020
73	102	204	306	408	510	612	714	816	918	1020	1122	1224	1326	1428	1530	1632	1734	1836	1938	2040
74	104	208	312	416	520	624	728	832	936	1040	1144	1248	1352	1456	1560	1664	1768	1872	1976	2080
75	106	212	318	424	530	636	742	848	954	1060	1166	1272	1378	1484	1590	1696	1802	1908	2014	2120
76	107	214	321	428	535	642	749	856	963	1070	1177	1284	1391	1498	1615	1722	1829	1936	2043	2150
77	108	216	324	432	540	648	756	864	972	1080	1188	1296	1404	1512	1620	1728	1836	1944	2052	2160
78	109	218	327	436	545	654	763	872	981	1090	1199	1308	1417	1526	1635	1744	1853	1962	2071	2180
79	110	220	330	440	550	660	770	880	990	1100	1210	1320	1430	1540	1650	1760	1870	1980	2090	2200
80	112	224	336	448	560	672	784	896	1008	1120	1232	1344	1456	1568	1680	1792	1904	2016	2128	2240
81	114	228	342	456	570	684	798	912	1026	1140	1254	1368	1482	1596	1710	1824	1938	2062	2166	2280
82	115	230	345	460	575	690	805	920	1035	1150	1265	1380	1495	1610	1725	1840	1955	2070	2185	2300
83	116	232	348	464	580	696	812	928	1044	1160	1276	1392	1508	1624	1740	1856	1972	2088	2204	2320
84	117	234	351	468	585	702	819	936	1053	1170	1287	1404	1521	1638	1755	1872	1989	2106	2223	2340
85	118	236	354	472	590	708	826	944	1062	1180	1298	1416	1534	1652	1770	1888	2006	2124	2242	2360
86	120	240	360	480	600	720	840	960	1080	1200	1320	1440	1560	1680	1800	1920	2040	2160	2280	2400
87	122	244	366	488	610	732	854	976	1098	1220	1342	1464	1586	1708	1830	1952	2074	2196	2318	2440
88	123	246	369	492	615	738	861	984	1107	1230	1353	1476	1599	1722	1845	1968	2091	2214	2337	2460
89	124	248	372	496	620	744	868	992	1116	1240	1364	1488	1612	1736	1860	1984	2108	2232	2356	2480
90	126	252	378	504	630	756	882	1008	1134	1260	1386	1512	1638	1764	1890	2016	2142	2268	2394	2520
91	128	256	384	512	640	768	896	1024	1152	1280	1408	1536	1664	1792	1920	2048	2176	2304	2432	2560
92	130	260	390	520	650	780	910	1040	1170	1300	1430	1560	1690	1820	1950	2080	2210	2340	2470	2600
93	131	262	393	524	655	786	917	1048	1179	1310	1441	1572	1703	1834	1965	2096	2227	2358	2489	2620
94	132	264	396	528	660	792	924	1056	1188	1320	1452	1584	1716	1848	1980	2112	2244	2376	2508	2640
95	133	266	399	532	665	798	931	1064	1197	1330	1463	1596	1729	1862	1995	2128	2261	2394	2527	2660
96	134	268	402	536	670	804	938	1072	1206	1340	1474	1608	1742	1876	2010	2144	2278	2412	2546	2680
97	136	272	408	544	680	816	952	1088	1224	1360	1496	1632	1768	1904	2040	2176	2312	2448	2584	2720
98	137	274	411	548	685	822	959	1096	1233	1370	1507	1644	1781	1918	2055	2192	2329	2466	2603	2740
99	139	278	417	556	695	834	973	1112	1251	1390	1529	1668	1807	1946	2085	2224	2363	2502	2641	2780
100	141	282	423	564	705	846	987	1128	1269	1410	1551	1692	1833	1974	2115	2256	2397	2538	2679	2820
101	142	284	426	568	710	852	994	1136	1278	1420	1562	1704	1846	1980	2130	2272	2414	2556	2698	2840
102	143	286	429	572	715	858	1001	1144	1287	1430	1573	1716	1859	2002	2145	2288	2431	2574	2717	2860
103	145	290	435	580	725	870	1015	1160	1305	1450	1595	1740	1885	2030	2175	2320	2465	2610	2755	2900
104	146	292	438	584	730	876	1022	1168	1314	1460	1606	1752	1898	2044	2190	2336	2482	2628	2774	2920

Duct weight multiplication table
Figure 11-3 (continued)

Job Records

One absolute essential for good estimating is a file of *job comparison records*. These are your best source of information on your crew productivity. They'll tell you why one job made money and another didn't.

Each job record should contain a list of the estimated manhours for a job, broken down to show the different job parts. Often these divisions will closely follow your take-off divisions. For example, ductwork, furnaces, roof vents, air conditioning equipment, and chimneys could all be separate categories. As the job proceeds, add the actual manhours used for the installations, along with the names of the crew. Make notes to explain the differences between estimated and actual hours, and how you might make improvements for the next job.

Another part of the job record will show material, labor and overhead costs at regular intervals. The form shown in Figure 11-4 is handy for this part of the record. At the end of the job, summarize these costs and compare them to the estimated costs to show the profit and loss for the job. Summarizing the material and labor cost on a regular basis also tells you if the job is exceeding the estimate as it progresses. If it is, you can step in and take corrective measures before it becomes a total loss.

Your job records should also include notes to remind you why any one job made or lost money: bad weather, a low bid, lack of supervision, quick installation, materials substitution or any other variable that made a difference.

Estimate Summary Sheets

Depending on the job, you may need more than the take-off and pricing sheets. For large jobs, you'll usually have an estimate summary sheet. This sheet breaks down job-related costs when they're so large they can't be absorbed by normal labor and material markup. Listings on the summary will include pricing sheet data, job overhead, material handling, labor cost breakdown, overhead, profit, job insurance, job financing, performance bond cost, and sales or excise taxes. Figure 11-5 shows an estimate summary sheet from the National Electrical Contractors' Association.

Contract Documents

Contract documents is the legal name for all the paperwork that ties together the different parts of a contracting job. Generally, these documents fall into three broad areas: bidding requirements, specifications, and plans and drawings.

Bidding Requirements

The bidding requirements section is usually the first part of the specifications. As shown in Figure 11-6, the instructions cover such areas as bid requirements, performance bonds, material substitution, site examinations, contracts and billing forms, special permits and the like. It's important that you read and understand this section well. Acceptance of your bid will rely on your following the information found here. For example, the section on definitions will tell you exactly what is meant by many of the words or terms used in the contract documents. Ignorance will not exempt you from your legal obligations if your bid is accepted and a contract signed. Failure to comply with any bid requirement, such as the site inspection, could prove to be very costly later on.

Sometimes information in the contract section will refer to American Institute of Architects (AIA) documents. For example, you might read, under the *Performance Bond and Labor and Material Bond* section of the contract:

Subparagraph 7.5.1 of the AIA General Conditions shall be supplemented as follows . . .

There will then be a paragraph of specific instructions. Here's what's happening: The basic conditions relating to bonding used in the contract are taken directly from volumes published by the AIA. The further instructions listed in the contract supplement these general conditions. In some contracts the AIA paragraphs are copied and included. In other contracts, they're just referred to. It's up to you to know what the AIA conditions are. Two publications by the AIA cover bidding and contracts. They are AIA Document A70, *Instructions to Bidders* and AIA Document A701, *General Conditions of the Contract for Construction.* You can get copies from:

American Institute of Architects
1735 New York Avenue NW
Washington, DC 20006

Plans and Specifications

In Chapter 5 we reviewed plans and specifications as a design tool. In this chapter we need to under-

COST SUMMARY AND ESTIMATE SHEET										

NAME _____ JOB NO._____

ADDRESS_____ DATE BEGUN_____

DATE ENDED_____

MATERIALS			LABOR			OVERHEAD OR SUB-CONTRACTS			SUMMARY	
DATE	DESCRIP.	AMT.	DATE	DESCRIP.	AMT.	DATE	DESCRIP.	AMT		
									TOTAL MATERIALS	
									TOTAL LABOR	
									TOTAL DIRECT OVERHEAD	
									TOTAL INDIRECT OVERHEAD	
									TOTAL SUB-CONTRACTS	
									TOTAL COST	
									ESTIMATE	
									COST	
									PROFIT	
									(LOSS)	
TOTALS										

TOPS FORM 3019 LITHO IN U.S.A.

Cost summary and estimate sheet
Figure 11-4

stand how they're used in bidding and estimating.

You've already seen how the plan sheets are the basis for your materials take-off sheet and the estimate that follows. It's obvious that a good estimate relies on an accurate plan. But what happens if there's an error in the plan? That depends on who drew up the plan.

If you're responsible for the plan, of course you'll want to correct the mistake as soon as possible. If it's something you catch when you're doing your take-off sheet and estimate, the problem can be corrected without too much trouble. But if you get a call from your foreman on the job and he says he's got 100 feet of duct installed but it doesn't go anywhere, that's another story. Correcting it will eat into the job profit. But no matter what the

stage of the job or how much it costs you, you're honor-bound to correct any mistakes caused by errors in your plans. There's no question it's a painful process when the extra cost eats up the job profit. But if you try to cover up a mistake and keep going, you'll face legal problems and an unhappy customer later on. It's better to bite the bullet and chalk it up to experience. It'll be one mistake you won't make again for a very long time.

But what happens if it's someone else who's drawn up the plans? Again, the answer isn't simple. Most engineers, architects and owners try to cover themselves against error by placing the full responsibility for plan and specification interpretation on the contractor. Following is a quote from an actual bid requirement document used on a job:

4. Inadequacies and Omissions

A. No oral explanation in regard to meaning of drawings and specifications will be made and no oral instructions will be given before awarding of contract. Bidders shall bring discrepancies, omissions, conflictions or doubt as to true meaning of any part of contract documents to Owner or Owner's representative's attention at least 10 days before due date for bids. Prompt clarification will immediately be supplied to bidders by Addendum and each Addendum shall be acknowledged on the Bid Form. All Addenda so issued shall become part of Contract Documents. Addenda issued after receipt of bids will be mailed or delivered only to selected bidders.

B. Failure to request clarification or interpretation of Contract Documents will not relieve contractor of responsibility. Signing of contract will be considered as implication that Contractor has thorough comprehension of full intent and scope of Contract Documents.

C. Neither the Owner nor his representative will be responsible for oral instructions. Only a written interpretation or correction by Addendum shall be binding. No bidder shall rely upon any interpretation or correction given by any other method.

These sections are clear. The contractor bears the final responsibility for making sure his bid reflects the total costs of the work required in contract documents prepared by someone else — specifically the plans and specs. If you have a doubt or find a mistake, get the clarification or correction *in writing*. Second, once your bid is accepted and the contract signed, you're responsible for performing all the work in the manner designated by the plans and specs *at your bid cost*. If you've overlooked something in the plans or specs, tough luck. You eat the extra cost. So read and study all contract documents, especially plans and specifications. Understand them all, not just those that refer to HVAC work.

Addenda and Change Orders

As mentioned in our sample bid instruction above, addenda are changes in the plans and specs that are made after the contract documents are sent out, but *before* the bids are submitted. It's understood your bid covers work described by the plan and specs. It should also specify what addenda are included. Some bid forms will have room for this information. If not, you may need to include a statement like, "Price includes all Addenda up to and including Addenda No. 3."

Any changes in the plans or specs after the contract is awarded are handled by *change orders*. These will reflect changes needed to correct architectural mistakes, to comply with building codes, or to add a feature requested by the owner. Or they may be added to simplify the work.

All change orders will include the following:

- A description of the change
- Who will do the work
- Who will pay for the change
- Who is authorizing the change

Typical procedure is for the architect or designer to send the proposed change, with the owner's approval, to the contractor who'll be responsible for the work. They'll ask for a quote, and sign a change order if the price is satisfactory. It becomes, in effect, a mini-contract for the new work. Figure 11-7 shows a typical change order.

After the change order (and a sketch, if required) covering the work is sent to each party affected, it becomes part of the contract documents. Sometimes the plan set must be changed. On the job, it's common practice to sketch in the change on the working plan set as soon as it's received. It's the responsibility of the architect or designer to make sure each contractor gets a revised set of plans as soon as practical.

Most contractors welcome change orders. It gives them a chance to add a little extra profit to the job. Since the men and materials are already on the job site, they can charge a price that includes a high profit margin. Even so, it's likely to be lower than the price quoted by a competitor if he had to come in and do the work. Change orders can be especially advantageous in remodeling work. Existing building conditions may require field verification, and then much of the work will be handled through change orders.

Materials Substitution

Job specifications are often quite detailed when it comes to the kinds of equipment and materials suitable for a job. General materials are acceptable if they meet certain standards, such as Underwriter's Laboratories or trade and testing group standards common to the industry. The materials come to the job site plainly labeled or stamped that they meet the appropriate code or standard.

Pricing Sheet Nos.	Section of the Job	Code	Standard Materials Cost	Special Materials ①	Labor Man Hours Non-Factored	Job Factor +or-	Job Factor MH or % ②	Non-Prod. Labor Factor +or-	Non-Prod. Labor Factor MH or % ②	Productivity Factor +or-	Productivity Factor MH or % ②	Net Labor Hour Adjustments +or-	Labor Man Hours Factored	Estimating Comments Relative to Construction Phases
to														
to														
to														
to														
to														
to														
to														
to														
to														
to														
Miscellaneous Material														
	Totals		$		%									

CALCULATE TO THE NEAREST DOLLAR AND WHOLE MAN HOUR.

DIRECT JOB EXPENSES: (DOLLARS)

- Insurance Add'l For This Job
- Insurance—Other
- Insurance—Other
- Permit and Inspection Fees
- General Supervision ②
- Job Site Engineering and Drafting
- Storage & Job Shed
- Job Time Keeper and/or Stockman
- On Job Truck Expense
- Job Telephone
- Temporary Facilities ②
- On Job Tool & Equipment Costs
- Travel Expense
- Subsistence
- Sub Contract
- Labor Adder Dollars
- Other DJE
- Other DJE
- Other DJE
- Other DJE
- Total Direct Job Expenses $

LABOR ADDER % ANALYSIS ②

- Social Security
- Unemployment Insurance
- Workmen's Comp. Ins.
- Local Health & Welfare
- Other
- Other
- Total Labor Adder %

Labor Adder Calculation:

- Total Labor Dollar Cost $
- Labor Adder % x
- Labor Adder Dollars $
- (Insert as a Direct Job Expense)

Estimator's Evaluation of Job, Non-Prod. & Prod. Factors

Job Factor Defined	Non-Prod. Labor Factor Defined
1. Type of Building	1. Craft Coordination
2. Working Conditions	2. Study Time
3. General Contractor	3. Material Handling
4. Electrical Contractor	4. Lost Time

Notes by Estimator:

ESTIMATE SUMMARY (DOLLARS)

- Standard Materials at Cost
- Special Materials at Cost
- Labor Man Hours Factored _____
- Average Rate Per man/hr. $ _____
- Total Dollar Cost—Direct Labor (Lab. MH Factored x Ave. Rate)
- Total Direct Job Expenses

PRIME COST

- Overhead (Based on Prime Cost) @ _____% X $ _____ Prime Cost
  OR
- O.H. (Std. Mat., Labor and DJE) @ _____% X $ _____
  PLUS
- O.H. (Spec. Mat) _____% X $ _____

TOTAL JOB COST

- Net Profit Based on _____% of Total Job Cost

SUB-TOTAL OR SELLING PRICE $

- *Sales and/or Excise Taxes
- **Payment & Performance Bond
- **Interest on Financing

TOTAL SELLING PRICE $

*Can be Included in Material Cost
**Can be Included in Direct Job Expenses

Courtesy: National Electrical Contractors' Association

**Estimate summary sheet
Figure 11-5**

INSTRUCTIONS TO BIDDERS

CONTENTS

1. Definitions	9. Performance Bond and Labor and Material Bond
2. General	
3. Documents	10. Contract and Contract Documents
4. Inadequacies and Omissions	11. Withdrawal of Bids
5. Bids	12. Reservation
6. Alternatives	13. Submission of Post Bid Information
7. Substitution - Approval	14. Examination of Site
8. Bid Guarantees	

1. DEFINITIONS

A. All definitions set forth in the General Conditions of the Contract for Construction, AIA Document A201, are applicable to these Instructions to Bidders.

B. Time when bids are due shall be prevailing time in force at Milwaukee, Wisconsin, on the date set forth in the Bid Form.

C. Bidding Documents include Instructions to Bidders, the Bid Form and the proposed Contract Documents including any Addenda issued prior to receipt of Bids.

D. Addenda are written or graphic instruments issued prior to the execution of the Contract which modify or interpret the bidding documents, including drawings and specifications, by additions, deletions, clarifications or corrections. Addenda will become part of the Contract Documents when the Construction Contract is executed.

E. Bid: A complete and properly signed Bid Form to do the Work or designated portion thereof for the sums stipulated therein, supported by data called for by the bidding requirements.

F. Base Bid: Amount of money stated in the Bid Form as the sum for which the bidder offers to perform the work, not including that work for which Alternatives are also submitted.

G. Alternative: Amount of money stated in the Bid Form to be added to or deducted from the amount of the Base Bid if the corresponding change in project scope or alternative materials and/or methods of construction is accepted.

H. Separate Bid: Bid for each established category of work, the combination of which make up the total project. A separate contract will be let for each category.

I. Substitute Bid: Bid in which bidder submits alternatives or substitutes for materials or equipment specified. Substitute Bids are included with Bid, along with net difference in cost, but are not used to determine contracts, or change contract requirements. They are considered informational and can be added to contract requirements either before the agreement is signed, or added by change order after the agreement is signed.

Bid requirements index
Figure 11-6

B-2

2. GENERAL

A. Each bidder shall read Instructions to Bidders, Conditions of Contract, General Requirements and Specifications, all of which are part of the Project Manual and contain provisions applicable to successful bidder. He shall also examine all drawings, as successful bidder will be required to do all work belonging to this contract which is either shown on drawings, mentioned in his specifications, or reasonably implied as necessary to complete his contract, including removal of existing work and preparing old work to receive new.

B. Each bidder shall visit site to become acquainted with: adjacent areas; means of approach to the site; conditions of actual job site; and facilities for delivering, storing, placing and handling of materials and equipment. Bidders shall compare specifications and drawings with any work in place, and inform themselves of all conditions affecting execution of work, including other work, if any, being performed.

C. Unless otherwise specifically mentioned in various Sections of Specifications, base bids upon assumption that work will be performed during regular working hours. Owner requests firm lump sum Base Bid for work contemplated and as covered by these specifications and accompanying drawings.

3. DOCUMENTS

A. Drawings, project manuals and other documents which constitute "set" may be obtained only from Engineering Manager - Building Division, Wisconsin Telephone Co., 125 North Executive Drive, Brookfield, WI 53005.

4. INADEQUACIES AND OMISSIONS

A. No oral explanation in regard to meaning of drawings and specifications will be made and no oral instructions will be given before awarding of contract. Bidders shall bring discrepancies, omissions, conflictions or doubt as to true meaning of any part of contract documents to Owner or Owner's Representative's attention at least 10 days before due date for bids. Prompt clarification will immediately be supplied to bidders by Addendum and each Addendum shall be acknowledged on the Bid Form. All Addenda so issued shall become part of Contract Documents. Addenda issued after receipt of bids will be mailed or delivered only to selected bidder.

B. Failure to request clarification or interpretation of Contract Documents will not relieve Contractor of responsibility. Signing of contract will be considered as implication that Contractor has thorough comprehension of full intent and scope of Contract Documents.

C. Neither the Owner nor his representative will be responsible for oral instructions. Only a written interpretation or correction by Addendum shall be binding. No Bidder shall rely upon any interpretation or correction given by any other method.

5. BIDS

A. Bidders are invited to submit proposals on Base Bids and Alternatives as noted on Bid Form. See Table of Contents for Section numbers involved for each bid and Contract.

B. Bids will be received as follows:

Bid requirements index
Figure 11-6 (continued)

CHANGE ORDER

No. . . . .

Dated . . . .

OWNER's Project No. . . . .   ENGINEER's Project No. . . . .

Project . . . .

CONTRACTOR . . . .

Contract For . . . .   Contract Date . . . .

To: . . . .
                                    CONTRACTOR
You are directed to make the changes noted below in the subject Contract:

. . . .
                                    OWNER

By . . . .

Dated . . . . , 19. . . . .

Nature of the Changes

Enclosures:

These changes result in the following adjustment of Contract Price and Contract Time:

Contract Price Prior to This Change Order $ . . . .

Net (Increase) (Decrease) Resulting from this Change Order $ . . . .

Current Contract Price Including This Change Order $ . . . .

NSPE-ACEC 1910-8-B (1978 Edition)

Typical change order
Figure 11-7

Materials that lack this identification can be rejected by the architect or his inspector. If you've already installed them, it's just not your day. You're going to have to replace them, at your expense.

For equipment, the owner or architect will go even further, specifying the model, its capacities and even the brand name of the piece of equipment to be installed.

Knowledgeable HVAC contractors can sometimes get a better price on materials and equipment other than the ones designated in the specs. But you can't substitute materials unless you submit the specifications of the alternate materials to the architect for his approval before you submit your bid. Generally, the manufacturer's catalog pages are sufficient for this purpose. Indicate on your request what piece of equipment the proposed choice will replace. If the architect approves the substitution, he indicates it by stamping an approval on the sheets and returning them. Denials are generally returned with no comment but often a phone call to the architect can determine the reason for, and sometimes overcome, the rejection.

Material substitution can be a valuable tool in arriving at a low estimate. However, keep in mind that any substitutions you propose should be equal to the original specification in quality and design. Also, be sure you have an approval in writing before basing your bid on any substitutions.

Contracts

The contract is the final legal document that binds you to do the proposed work. Typically, it states that you will supply the labor and materials at a certain price to do the job as described in the contract documents. The contract will also cover items such as indemnity and insurance obligations, compliance clauses for nondiscrimination and minorities, owner's options if the work isn't performed, and progress payments.

For a small residential job, the contract may simply be your bid sheet with a few lines at the bottom indicating the date of the work, the cost, how payment will be made, and space for both parties' signatures. Larger jobs may require several separate contracts, provided by the owner as part of the contract documents.

Contractors who have worked with one another over a period of years may often do work based on an oral contract between them. An oral contract is just as binding as a written one. The problem with it is that the parties may later disagree about what the terms really were. Here's our recommendation: Don't do it. Always get it in writing.

No matter what type of contract you're dealing with, don't sign on the dotted line or agree to anything until you understand your obligations completely. Be sure the contract spells out work requirements, payment schedules and any other items that will affect your ability to do the work. If necessary, get legal help to clear up any doubts before signing. Once you sign, you're committed to the contract. Not knowing what the contract meant has never been accepted in court as a defense. It probably never will be.

No matter how friendly your relationship with a general contractor or owner, disagreements will come up. The whole purpose of having a good contract is to prevent these disputes from ending up in court. No one really wins in a lawsuit except the attorneys. Still, there's no contract in the world that can cover every possible mishap on the job. So don't count on it as being the catch-all for job problems. Above all, avoid contracting with either another contractor or an owner who has a poor business reputation. You'll probably end up in court trying to collect. And then you'll only get part of what you have coming.

Sometimes a *letter of intent* is used to get a job started before all the contract documents are ready. This is a letter of authorization by the owner for a limited amount of work. It's as legally binding as a contract for the amount of work it covers.

Subcontracting

On many of your HVAC jobs, you'll be acting as a subcontractor. You send your bid to a general contractor who adds 10 or 15% for project supervision and troubleshooting and then submits it to the owner. The general contractor will assume your bid is accurate and correct according to the plans and specs. If he gets the job, he'll sign separate contracts with his subcontractors.

These contracts usually have an "umbrella" clause which specifies that all the contract documents used for the job apply to the subs as well as the general. See Article 1 in Figure 11-8. The subcontract arrangement doesn't give you any leeway with mistakes or errors. You're still liable for the work as defined in the contract documents. You have to abide by any applicable requirements

Sub-Contract Agreement

THIS AGREEMENT, made this_____day of_____A.D. 19___,
by and between_____ hereinafter called the
Contractor, and_____ hereinafter called the
Sub-contractor.

For the consideration hereinafter named, the Sub-contractor agrees with the Contractor, as follows:

ARTICLE 1. WORK: The Sub-contractor agrees to furnish all material and perform all work necessary to complete____

_____

_____

_____

_____

At: _____

     ADDRESS   CITY   COUNTY   STATE

For: _____

     OWNER OR OWNERS

according to the general conditions of the contract, as per the drawings and specifications, and amendments and/or changes
to either (details thereof to be supplied as needed) prepared and identified by _____,
Architect, and to the full satisfaction of said Architect.

ARTICLE 2. TIME: The Sub-contractor agrees to promptly begin work as soon as notified by the Contractor, and to
complete the work as follows: _____

_____

_____

_____

_____

ARTICLE 3. EXTRAS: No deviations from the work specified in the contract will be permitted or paid for unless a
written extra work or change order is first agreed upon and signed as required.

ARTICLE 4. ASSIGNMENT: No assignment of this sub-contract agreement is permitted without prior written
permission from the Contractor.

ARTICLE 5. INSURANCE: The Sub-contractor agrees to obtain and pay for the following insurance coverages:
Workmen's Compensation, Public Liability, Property Damage, and any other insurance coverage which may be necessary as
required by the Owner, Contractor, or State Law.

ARTICLE 6. TAXES: The Sub-contractor agrees to pay any and all Federal, State, or Local Taxes which are, or may
be, assessed upon the material and labor which he furnishes under this contract.

IN CONSIDERATION WHEREOF, the Contractor agrees that he will pay the Sub-contractor, in _____
_____payments, the sum of_____
_____Dollars ($_____)
for materials and work, said amount to be paid as follows: _____ per cent (_____%) of all labor
and material which has been fixed in place by the Sub-contractor, to be paid on or about the _____
of the following month, except the final payment, which the Contractor shall pay to the Sub-contractor within
_____ days after the Sub-contractor shall have completed his work to the full satisfaction of the
Architect or Owner.

The Contractor and the Sub-contractor for themselves, their successors, executors, administrators and assigns, hereby
agree to the full performance of the covenants herein contained.

IN WITNESS WHEREOF, they have executed this agreement the day and year first above written.

WITNESS _____   SUB-CONTRACTOR _____

                                                       BY _____

                                                       STATE LICENSE NO. _____

WITNESS _____   CONTRACTOR _____

                                                       BY _____

                                                       STATE LICENSE NO. _____

Tops Form No. 3461—Revised
Litho in U.S.A.

Sub-Contract Agreement
Figure 11-8

or stipulations specified in the documents.

The contract between you and the general contractor establishes your legal relationship and job responsibilities. But it doesn't bind the owner or architect. Still, the owner or architect usually reserves the right to reject a proposed subcontractor. In fact, they usually reserve the right to reject any bids outright, at their own discretion. The owner will also have a clause protecting himself in the event a dispute arises between the sub and the general.

Payment Schedules

Payment schedules are the part of the contract that spell out when you get paid. For small jobs, it's usually a lump sum paid when the work is done. A larger job will have payments made as the work progresses, up to 95% of the total value. This gives the contractors some income to pay for materials and labor as the work goes on, instead of having to borrow or supply from their assets the entire amount needed for the job.

The typical procedure for progress payments starts on the job. You or your foreman will make out a daily or weekly report of the work that's done. Once a month you'll summarize these reports and turn the summary over to the owner. He in turn gives them to the architect or designer for his approval. If they're approved, you get your check.

The final 5% is held back up to 30 days after the work is completed. This allows time for the architect to make a final inspection. If there aren't any problems, the remainder is paid. The exact timetable and rate of payment is specified in the contract. If it isn't, don't sign the contract until it does include some pay arrangement that satisfies you. Try to arrange payments that work to your advantage. This means working as much with the owner's money as possible.

Also have the payment schedule tied to *your* work progress, not anyone else's progress or the percent of job completion. This is important when you're a subcontractor on a job. As a sub, you won't have any control over other contractors or the overall work progress.

You'll have to supply *lien waivers* if the owner requests them. Figure 11-9 shows a typical waiver of lien. These are legal documents that verify that you've paid your bills for the job involved. This protects the general contractor and the owner from lawsuits for unpaid bills. Filing a false waiver is evidence of fraud and a serious offense.

Finally, if you're a sub on a job and payment to the general is held up by the owner because of something the general did or didn't do, you're entitled to payment on demand from the general.

Bidding

Your bid is a formal offer to do the job at the price stated. Bids are sometimes called proposals or quotes. Figure 11-10 shows a typical proposal or bid form. No matter what you call it, however, the submitted bid is legally binding once the bid closing date is reached. Prior to that date, you can usually withdraw your bid without any problem. Most owners or architects who request bids will include withdrawal instructions and stipulations in their bid information.

Besides withdrawal terms, the contract documents or specs will have other rules regarding bids. A bid found in violation is usually rejected, even if the contractor isn't to blame. The owner may reserve the right to reject bids outright and start all over. This is often done if the bids all come in above the job cost estimate drawn up by the architect.

Sometimes the contract documents will require bids on work that can be done in two or more different designs. The plans and specs will have all the potential designs. The owner or architect will decide which way to go when he knows what each one will cost. For the contractor, this means preparing alternate estimates. You'll prepare a *base bid* for the complete job, based on one of the designs in question. Then you'll draw up *alternate bids* covering the alternate designs. These alternate bids are submitted along with the base bid.

In the Plans and Specifications section earlier in this chapter, we talked about material substitution. For example, the architect or designer will specify a certain brand of equipment for a job. In most cases, the contractor has the option of proposing a different brand of equipment if it costs less and will do the same job. These substitutions must be approved by the architect or designer. If you want to substitute, submit a base bid including the specified equipment and attach a sheet showing the bid based on the approved equipment substitution. These are called *substitute bids*. Don't confuse them with alternate bids.

WAIVER OF LIEN

For value received,_____ hereby waive_____rights and claims for lien on land
and on buildings about to be erected, being erected, erected, altered or repaired and to the appurtenances thereunto,

for _____ owner___,
by _____ contractor___,
for _____
same being situated in _____County, State of _____, described as _____
_____
_____

for all labor performed and for all material furnished for the erection, construction, alteration or repair of said
building and appurtenances. _____
_____

FORM 1362 WAIVER OF LIEN   WISCONSIN LEGAL BLANK CO. MILWAUKEE 26402

Waiver of Lien
Figure 11-9

The bid instructions may call for other items besides the job cost estimate. For example, some owners will ask for work schedules, start and finish times, proof of financial stability, or certificates of insurance. Don't get so wrapped up in figuring the job cost that you overlook other requirements. And don't overlook the bid deadline. If it's permissible to mail your bid, send it by registered mail. If you hand-deliver it, allow yourself plenty of time. One contractor we know of lost a $500,000 job when he missed a bid deadline by five minutes due to car trouble.

Bid Record

Keep a bid record if you do more than a few bids a month. A 3" x 5" card file works fine. Record on the card the job name, amount of bid, who did the estimate, when and where submitted, the amount, and its outcome. File the bid records by job name.

These bid records help you to keep track of your competition. If one company gets most of the jobs, you should ask yourself why. Do they use the same

suppliers time after time? Are their crews more efficient? Do you know any of their job superintendents or key estimators? Remember, imitation is the sincerest form of flattery. There's nothing to prevent you from duplicating a competitor's methods if they'll work for your company. Or from hiring their key personnel if you can offer something they aren't getting from their present employer, such as a promotion or a larger salary.

Bid Security and Performance Bonds

The owner may require a *bid security,* sometimes called a bid guarantee. That's a dollar amount you submit with your bid to insure that you'll enter into the contract at the quoted price. It's usually in the form of a check or a bid bond, and it's returned once the contract is signed. The amount is generally a percentage of the job cost. The bidding instructions will include the amount and conditions of the bid security. Consult with your attorney or another knowledgeable contractor if you're not clear about these instructions.

240

	Proposal	Proposal No.
FROM		**Sheet No.**
		Date

Proposal Submitted To	**Work To Be Performed At**
Name_____	Street_____
Street_____	City_____State_____
City_____	Date of Plans_____
State_____	Architect_____
Telephone Number_____	

We hereby propose to furnish all the materials and perform all the labor necessary for the completion of

_____
_____
_____
_____
_____
_____
_____
_____
_____
_____
_____

All material is guaranteed to be as specified, and the above work to be performed in accordance with the drawings and specifications submitted for above work and completed in a substantial workmanlike manner for the sum of

Dollars ($).

with payments to be made as follows:

_____
_____

Any alteration or deviation from above specifications involving extra costs, will be executed only upon written orders, and will become an extra charge over and above the estimate. All agreements contingent upon strikes, accidents or delays beyond our control. Owner to carry fire, tornado and other necessary insurance upon above work. Workmen's Compensation and Public Liability Insurance on above work to be taken out by_____

Respectfully submitted_____

Per _____

Note — This proposal may be withdrawn by us if not accepted within   days

ACCEPTANCE OF PROPOSAL

The above prices, specifications and conditions are satisfactory and are hereby accepted. You are authorized to do the work as specified. Payment will be made as outlined above.

Accepted_____ Signature_____

Date_____ Signature_____

TOPS FORM 3450   LITHO IN U. S. A.

Contract proposal
Figure 11-10

A *performance bond* is money you post to guarantee that you'll complete the job. If you don't, the money is forfeited and the bonding company may hire another contractor to finish the work. Generally these bonds are negotiated between the contractor and a bonding company after his bid is accepted, but before the contract is signed. The owner reserves the right to require bonding as a condition of suitability of the contractor. The owner won't dictate the name of the bonding company you select but will reserve the right to approve or disapprove your choice. The cost of the bond is paid by the contractor and usually amounts to one or two percent of the bid amount.

Whether you post your own certified check or use a bonding company, try to get the return date for your check tied to completion of your work as described in the contract. Avoid, if possible, a return date based on other contractors' work. Sometimes this may be difficult, because the owner will prefer to keep the bonds in force for up to one year after the final certification of the project. You'll have to take each job as it comes, but keep in mind that if you can shave off a few months on your bond, you've saved a few dollars in job cost.

Bid Invitations and Solicitations
Formal bid solicitations are found in the classified ad sections of your local newspapers. They're also found in construction magazines such as *Western Builder* and *The Dodge Report*. After you've been in business for a while and built up a good reputation, you may get calls asking you to submit a bid on jobs in your area.

Basically, a bid invitation will describe the job and tell you where to go for more information. On request, you'll get a bid package that includes bid instructions, specs and plans. On some jobs, rather than sending out many sets of specs and plans, the owners will make a set available for inspection at a given location. After this preliminary inspection, if you decide you'd like to bid, you can pay a fee to obtain a complete set of contract documents.

Bid Openings and Lettings
It's important to note the time and place when bids are opened, or let. Some contracts require that an officer of your company be present at the opening. Being late, or not showing up at all, can disqualify your bid even if it's the lowest.

The opening time is also important, since it acts as the cut-off date for bid withdrawals, submission of proof of financial stability and the like.

If you're not familiar with the bidding process, it's a good idea to attend a letting for a job like the ones you hope to get. You'll become familiar with the bidding process and may pick up hints on how to better prepare your bids.

Preparing the Estimate and Bid
The best tool for preparing solid estimates and bids is experience. Of course, you get that experience by preparing estimates and bids. It's demanding and frustrating work, especially in the beginning when you'll have lots of questions that don't seem to have good answers.

Not everyone can be a good estimator. It takes an ability to handle paperwork, calculations, minute detail, and pressure. You have to combine this with a solid knowledge of crew productivity, and labor and material costs. Since the labor rates on a job are often fixed, either by unions or government regulation, you may have to bargain fiercely with suppliers to get an edge on your competitors. Finally, you have to take all of this and put it down in black and white on the forms provided. Hopefully, it all works out as the low bid.

In the next section we'll give you some guidelines on the various steps involved. Eventually many of these will become second nature to you. But there's one key task that only you can do. It's up to you to keep accurate job records. Do it from the beginning for all jobs, and follow through to the very end. Sometimes that end will be at the bidding table, where you lost out. That's O.K. Bid instructions will include a time and place where the accepted bids can be inspected after the contract is signed. Find out and record who got the job and the winning bid amounts. You may not see how the winning bid came out lower, because the bids don't require a breakdown of dollar amounts. But perhaps you can pick up a hint from a materials supplier you know. Or you might know someone who's working for the general contractor or a sub on the job. A few conversations with these people, along with some personal observations as the job moves along, can tell you a lot.

On the other hand, don't become obsessed with a lost bid. Your competitor may have shaved his profit margin lower than you want to go. He may have had some materials suitable for the job on hand, purchased at a time when the price was low.

Learn what you can, record it and use it the next time you bid. There'll always be a next time, as long as you're smart enough to be around for it.

We've already talked about the importance of keeping accurate records on the jobs you do get. But we'll say it again. Good job records are one of the best tools you have for preparing future bids and estimates. Keep track of hours, who runs the crews, materials cost and job problems for each job you do. Write it down and file it. When you get another job like it, review the one on file and you'll find all kinds of places where you can tighten up your estimate.

Avoiding Estimate Pitfalls
Every estimate starts with the plans and the specs. Here are some basic rules to follow:

• Be sure plan sets, bid instructions and specs are complete before you begin work. Make sure all addenda are included. Visit the job site and follow any other preliminary instructions. It's a good idea to visit the job site even if it isn't required in the bid instructions.

• Read the HVAC specs completely. Note and underline information that may change job costs or cause potential problems on the site. Most often this will happen when a spec requires an installation out of the ordinary. Your crew will often assume the installation should be done the way they've been doing it. But if the specs say otherwise, it'll be wrong.

Figure 11-11 contains some fairly straight forward specs for a typical job. Look at Section 15600, *Chimneys:* It describes how the chimneys should be installed, but it doesn't tell you how many chimneys there are. You'll have to look it up on the plan. It says they must extend five feet above the roof. Let's assume the chimneys you've installed in the past only had to extend three feet above the roof. Is this five foot requirement a new code regulation, owner preference, or a misprint? It's up to you to know and bid accordingly. Suppose you assume it's a misprint and estimate only three feet of chimney instead of five. Later, after you get the job, you find out it is five feet. You can easily eat the cost of two feet of chimney if there are only a couple of them on the job. But what would you do if it's a condo project with 24 chimneys, zero clearance type, and your crew has installed them at three feet? And the carpenters have followed your crew and framed them in at the three-foot height? If this sounds impossible, it's not. It actually happened on a job.

Also read through all the other sections in the specs. Information in these divisions can affect your costs. Not all the HVAC work will be in the HVAC section. Yet its cost will be included in your bid, whether you plan for it or not.

• Examine the HVAC plan completely. Also look at the site plan and building architectural plan and elevations. Pay close attention to any notes and details. Make sure all chases, soffits and openings are where they should be for your ductwork, equipment and piping.

• Don't assume or correct any information. Check errors, omissions, and doubts with the architect for clarification. Get the correction in writing.

• Start your take-off at the upper left of the top plan sheet. Work from left to right and top to bottom. Work carefully and logically. Make sure you're using the correct scale when taking off duct runs and like items. Also make sure that the scale shown on the plan is correct. You can do this by comparing the measured scale distances from the HVAC plans with the architectural plan scales.

• As you transfer each item from the plan to your take-off sheet, check it off the plan with a colored pencil. During the recheck, preferably done by someone else, each item should be checked off again with another color. Don't simply check off the old checks. Check off the item.

• Beware of interruptions. They'll be impossible to avoid, but note carefully where you stopped so you can pick up at the same place.

• Transfer any items from the specs that affect your estimate to the take-off sheet. Again, work carefully. Don't skip any item. Check them off as you record them.

Pricing Materials
When your take-off is complete, you need to price the materials. On small estimates, this can be done right on the take-off sheet. But it's often easier to use a pricing sheet. This allows you to group together like items, figure the amount needed, and apply a per-unit cost.

15101 HANGERS, ANCHORS AND SUPPORTS

1. Furnish and install supports and hangers for all equipment installed under this Contract. Strap material is not approved. The Contractor: furnish all materials necessary to an approved installation of all supports. Space hangers 8' maximum for all lines. Provide suitable inserts or fasteners.

15600 CHIMNEYS

1. Furnish and install "UL" Class B Metabestos "RV", or equal, chimneys with Belmont top, roof flashing, fire stop spacers, tee and support collars. Extend chimney five feet above the roof.

15601 BREECHING

1. Furnish and install 16 gage galvanized breeching for the heating equipment, water heaters and laundry dryers.

15730 INSULATION

1. Ductwork: Insulate supply and return ductwork as indicated on the plans Insulate all outside air and combustion air ductwork. Insulation equal to Owens Corning Fiberglass FRK 25 ED-100, 2" thick for combustion air and outside air, 1" thick for other ductwork; installation to be as recommended by manufacturer for air conditioning ductwork. For round supply duct, see ductwork specification. Line return ducts with 1" acoustic liner as indicated on the plans.

Breeching: Insulate horizontal and vertical breeching with Owens Corning Type I thermal insulating wool, 1" thick, hold in place with wire tiles and wire mesh jacket.

15731 DUCTWORK

1. Furnish and install sheet metal ductwork, etc., as shown on the drawings and as specified herein, all in conformity as to materials and workmanship with the Code and Sheet Metal Contractors' Association.

2. Ducts unless noted: Galvanized and braced to prevent buckling and breathing and in general, ducts are routed above ceilings and or built in walls and chases unless otherwise noted on the plan.

3. Use sweep type elbows where possible having an inside throat radius equal to duct width in turning plane. Where space limitation does not permit, use turning vanes. Round supply duct to be Thermaflex M-KA insulated, UL-181.

4. Ductwork buried below grade equal to P.V.S. steel with non combustible plastic coating as manufactured by Foremost Manufacturing Co., Southfield, Michigan. Duct to be "UL" listed and FHA listed.

Typical job specs
Figure 11-11

15741 FURNACES

1. Furnaces equal to Trane or Carrier with two-stage gas valve, intermittent pilot ignition, centrifugal fan, direct 3-speed drive or adjustable V belt drive, insulated casing, disposable filters, magnetic starter.

15742 DUCT HEATERS

1. Heaters to be equal to Trane or Modine AGA certified, two-stage gas valve, high temperature limit control, intermittent pilot ignition, 24 volt control system, adjustable V belt drive, centrifugal blower, insulated casing, disposable filters, magnetic starter.

15743 COOLING COILS

1. Coils equal to Trane or Carrier with insulated casing, aluminum fins, copper tubes thermal expansion valve, tested at 225 psig.

15744 REFRIGERANT PIPING

1. Furnish and install copper piping. Insulate suction lines with 3/4" thick Rubatex insulation with ASTM E-84 flame spread rating not over 25 and smoke development not over 50. Run piping concealed in the building construction.

2. Evacuate and charge systems with refrigerant.

15745 AIR HANDLING UNITS

1. Fan coil equal to Trane complete with two circuit refrigerant coil, thermal expansion valves, insulated casing, centrifugal fans with adjustable V belt drive, magnetic starter, vibration isolators, cooling relay, disposable filters, electric heating coils with air flow switches and limit control.

15746 CONDENSING UNITS

1. Units equal to Trane or Carrier with filter dryer, crank case heater, internal motor thermal overloads, magnetic starters, compressors, high and low pressure limit controls, relief valve, time delay relays for dual compressor units, low ambient operation to 0 F. All cooling systems to meet EER standards of the State Code.

15820 EXHAUST FANS AND SUPPLY

1. Exhaust Fans: ILG, Carnes, Greenheck, AirMaster, Broan or approved equal complete with backdraft damper. Fans to have disconnect means inside housing. EF-1 to have BI blades.

Typical job specs
Figure 11-11 (continued)

Work carefully. Make sure you transfer the correct items and amounts to the pricing sheet. Check and recheck all your figuring to avoid mistakes in costs. Even an honest, small mistake in a number or a decimal point can disqualify your bid.

Buying Materials

Material costs are very important to your total bid. Obviously, the lower the price you pay for acceptable materials, the lower you can bid and still hold a reasonable profit margin. Some HVAC contractors consider only price when buying materials. They'll check with many suppliers, even very large suppliers in distant cities. Others feel more comfortable dealing only with one local supplier, going to another source only when absolutely necessary.

The best method is to apply good judgment to either situation. Certainly you want the lowest cost. But price alone isn't the only consideration. Supplier cooperation, good delivery and consistent service are also important. If you channel enough of your business through a local supplier, he knows you and you've gained another source of information about local trends and job opportunities. And you'll certainly have job emergencies that can be eased considerably with a friendly supplier's help. You'll also have an easier time dealing with delivery foul-ups, damaged materials and other problems if you buy locally.

On the other hand, local dealers sometimes can't meet the prices you can get by placing a large order for standard materials like ductwork directly with the manufacturer or a large wholesaler. If the price is really critical, all other things being equal, you almost have to take the best buy. But don't abandon your local supplier. Make sure he gets a large enough share of your business to keep you a valued customer. It will work in your favor in the long run.

Any supplier who does business with you can help you out by giving you advance notice for price increases for commonly-used materials. He may be willing to sell you what he has on hand at the old price. This gives you a chance to save some money, stockpiling the material (assuming you have the place to do it) until you need it. But beware of buying *anything* from suppliers who are:

- unfamiliar,
- doing business by phone from another state,
- passing themselves off as a ''name'' supplier or jobber,

- trying to collect the bill COD.

In recent years, more and more businesses have been ripped off by bogus companies dealing by phone from distant states. These ''WATS line hustlers'' sell materials by offering prizes or claiming to be licensees of big companies. So far, they're usually selling supposedly low-cost advertising novelties and office supplies, but who knows what they might try to get into next. They're virtually impossible to prosecute by local authorities because of the distances involved. Follow this rule of thumb: *Know your suppliers and don't pay for anything with cash.* Using monthly billing gives you leverage over the account if problems develop. Also make sure you have a solid check-in procedure for materials, both on the job and at your warehouse.

Get in writing any materials quote that you'll use in a bid. Make sure it's guaranteed far enough past the bid opening date so you can place your order at the guaranteed price. Without this protection, the cost of the material could go up after you've bid at the old price. You'd have to absorb the extra cost out of your profit.

Some suppliers handle many different types of HVAC materials and equipment. On large jobs they may ask you simply to send them your take-off sheet or plan set and they'll return a package or job lot quotation. For example, one supplier may give you a quote for all the ductwork, registers, grilles, louvers and dampers for a job. Another may give you a quote for all the furnaces, fans and temperature controls. We have no objection to this method as long as the supplier itemizes his quote. Be sure to check the size and specifications of the materials and equipment against your take-off sheet to avoid mistakes. Stay away from package or job lot pricing because you won't know where you could save dollars on any one item. You'll be at a disadvantage when competing with contractors who get itemized prices.

Figuring the Bid

After you've established the materials cost for the job, you'll be able to complete your estimate and bid. One of the oldest and most reliable methods for bidding in construction is the *lump sum*. We discussed it briefly early in this chapter. This method has been used for years on jobs of all sizes. Although many refinements have been added to fine-tune the procedure, it always relies on the same basic formula: materials cost will equal a certain percent of the total cost.

Assume that your pricing sheet shows the materials for a certain job will cost $18,500. Add your labor cost to the materials cost to find your job cost. How do you find the labor cost? In years past, most contractors assumed that materials cost would equal 60% of the base cost and labor would equal 40%. In recent years, however, those figures have reversed. Many contractors now believe that materials will cost 40%, with labor costing 60%.

The exact ratio will depend on several factors, including the going wage rate for your area and current material prices. But the productivity of your crews is the most important factor. How do you determine their productivity? By going to your job records and finding the answers. Is the job you're bidding on similar to any you've done before? If not, how is it different? What can you do to the labor/materials ratio to protect yourself but still remain competitive? If the job is similar to ones you've done before, ask yourself these questions:

1) What were the total manhours on that job?

2) What were the labor costs?

3) What were the material costs?

4) How long ago was the job completed? Have the wage and materials rates changed since then?

5) Will the same crew be available if you get the new work?

Analyze all these factors, then come up with a suitable materials-to-labor ratio. For example, we're going to assume we've got a good crew for our bid. This job is like several they've done before. We think our crew's labor rate will be 52%. That means our material percent is 48. That's 8% higher than the accepted rate of 40% for materials. But keep in mind that although it's a higher percent, it's *not* a larger amount. Our example will explain this.

To find the job cost, divide the materials cost by the percent of materials cost:

Other crews
$$\frac{\$18,500}{.40} = \$46,250$$

Our crew
$$\frac{\$18,500}{.48} = \$38,541$$

To find the base labor cost, multiply the job cost by the percent of labor cost:

Other crews
$$\$46,250 \times .60 = \$27,750$$

Our crew
$$\$38,541 \times .52 = \$20,041$$

Note that in each case, the materials cost has been the same. But by improving the labor cost through increased productivity, the total job cost is lowered. If we could lower our materials cost below that of our competitors, the job cost would be lower still.

A quick way of checking the labor cost is to divide it by the gross hourly wage rate for the area. This will give us the total manhours. By comparing this figure with the manhours for similar jobs in our job records, we can see if our labor cost is correct.

Although a lump sum bid that's well written will get you work, many contractors prefer to use the *unit method* of estimating. In this method, you use cost data books to build the estimate. These books break down all the many HVAC items used on a typical job and assign an average material and labor cost to them. These costs are averaged from various parts of the country to get a broad coverage. Then percentages for overhead and profit are figured in to get the total cost per unit.

For example, you would look in the Heating section of the cost data book for the entry for make-up air units. It might look like Figure 11-12.

Make-up air vent — indoor suspension, natural/LP gas, direct fired, standard control

| | Bare Costs | | | Total, inc. |
	Mat.	Inst.	Total	Overhead and Profit
2,000 cfm, 168 MBH	2,700	160	2,860	3,200
6,000 cfm, 502 MBH	3,875	315	4,190	4,725
For filters, add	5%			
Discharge louver, add	15%			

Cost data book entry
Figure 11-12

The cost data books will also give listings of average manhours needed to install the item. To complete a bid, locate the items on your take-off sheet in the cost data book and transfer the correct dollar amounts to the estimate. Remember that the cost data book amounts are averages used to reflect costs across the country. You will probably have to modify their costs to reflect all the items we spoke of earlier: local economic conditions, local labor costs, your company overhead, and other variables. However, these books do give you a good idea of average costs and installations. They also tell how they've calculated the unit costs so you can tailor them to job needs in your area. Here are three publishers of cost data books for HVAC:

Robert Snow Means Co.
100 Construction Plaza
Kingston, MA 02364

Richardson Engineering Services
P.O. Box 1055
San Marcos, CA 92069

Craftsman Book Company
6058 Corte del Cedro
P.O. Box 6500
Carlsbad, CA 92008

After figuring the job cost, we'll have to add on the overhead and profit. There are standard percentages, typically 22% for overhead and 10% for profit. But we'll adjust these figures based on our experience. Let's assume the job overhead on this project won't amount to anything. Our fixed overhead, since we're still a small company, will be about 17%.

Next comes profit. Assume we're pretty confident that our labor cost is lower than our competition's. Let's also assume that HVAC work has been fairly strong in our area. We don't expect any

of the really big outfits to bid on this job since they're going full bore on other projects. This is important since big, established contractors can often afford a slimmer profit margin than newer companies saddled with a business mortgage. Also, a check of our job records shows that we've won jobs of this type with bids that were substantially lower than the next bidder. In other words, we left a lot of money "lying on the table." So we're going to take a chance and set our profit margin at 14%. Our total bid then will be as follows:

Materials	$18,500
Labor	+ 20,041
Job cost	$38,541
Overhead (17% x $38,541)	6,552
Profit (14% x $38,541)	+ 5,396
Total bid	$50,489

Contingencies

On a larger job or any job that contains some estimating uncertainties, you'll want to add a few percentage points to the total bid for contingencies. This will allow for problems or costs that come up unexpectedly. Construction work seldom goes as planned and most of the surprises are unpleasant ones. They'll add to job cost and reduce your profit. Typical are delays caused by bad weather or delivery foul-ups, resulting in overtime for your crews. Although a contingency fund won't cover any major disasters, it often helps to pick up smaller items that may have been missed in the estimate.

If the work is a bid job with a big possibility of surprises, such as a remodeling project, you may want to go quite a bit higher than one or two points. A contingency fund as high as 20% may not be out of line. Again, it's a judgment that will depend on the job. In our sample bid just shown, we didn't include it at all, since we've had lots of experience in that kind of job. And we used a high profit percent which could also act as a contingency fund if absolutely necessary.

The Quotation or Final Bid

The final bid, sometimes called the quotation, is your final offer to provide all the labor, materials and know-how to do the work described in the contract documents. The quote can be simple or detailed, depending on the bid instructions. Although not all the items will apply to every bid, the following checklist will be helpful. The bid should include these items:

• The base bid and addenda covered by that bid

• Any alternative bids requested

• If needed, brief statements covering:

1) Work schedules
2) Start up and completion dates
3) Material substitutions and their approvals
4) Any exceptions to items in the contract documents and the reasons for them. (Check this with the architect first.)

• Bid security or other financial certification if required

• Payment schedule

• Signature of the contractor

Check your final bid several times for accuracy and completeness. Check the requirements in your bid instructions. Use the bid forms provided unless you're sure you are free to use your own. Look for the date, time and place specified for the receipt of the sealed bid. Follow these instructions carefully. In some cases, bid extensions are given for good cause. But don't count on it.

Bid Errors

It takes a lot of patience and hard work to turn in an error-free bid. But it's worth it. Bids with errors are disqualified. Sometimes the whole bidding process must then be repeated. This is bad for all bidders, since everyone now knows what the competi-tion's costs are. The next letting means lower quotes by everyone, cutting possible profits.

Some of the most common errors found in a bid are ordinary arithmetic mistakes, inaccuracies in copying figures from one sheet to another, misplacing a decimal point, or simply forgetting a line or column. Avoid these mistakes by working systematically and carefully. Have someone check the arithmetic, transfers and the completeness of your work. Some contractors draw up bid checklists to make sure they haven't skipped any items. The problem with relying on a bid checklist is that no two bids are exactly the same. So if you use only your checklist to locate errors, you may still miss a key item.

Errors in judgment are the most difficult to avoid. Failure to anticipate rising labor and material costs, failure to recognize job site conditions and problems, and overestimating crew productivity are all errors in judgment. Again, good job records are your best defense. Judgment is also needed for the intangibles that relate to bid work. The overall state of the economy in your area affects all the contractors you bid against as well as yourself. The amount of work available also determines how many bids will be submitted for any one job. As a rule, competition is intense in the spring of the year and lessens as the year goes on. Provided, of course, that there's enough work so that some contractors have all they can handle.

Use any legitimate method to increase your likelihood of getting the job. Use all of them to keep your competitive edge.

Cost Management and Labor Supervision

So far in this chapter we've spent a lot of time on how to get work. Now let's assume you've gotten some. What happens next? Is it time to relax and celebrate? Perhaps. But don't overdo it. You still have to bring the job in at your bid cost and on time. The actual labor and material costs will have to reflect the estimated ones. Of course, if you can do it for less, you'll increase profitability even more. If you do it for more, well, there goes some or all of your profit. In time there goes your company.

Cost management is nothing but a term for sticking to your estimates. It begins with what we've been stressing, an accurate estimate. If you've bid a job too low, you'll need cost management more than ever. But it won't recoup losses on a job that was just plain bid too low.

But let's assume your estimate was well done. That means your materials were quoted to you in writing at a good price. What cost-management steps are necessary to bring that job in at a profit?

1) Check all invoices and bills to make sure your billed price is the same as the quoted one.

2) Insure and lock up materials stored on the job site or in your warehouse. Protect yourself against loss due to theft or weather by having a fenced-in area or building. Owners will usually cooperate if there's an existing building on the site. Be careful if you're sharing this with other contractors or companies since this means that many more people have access to the premises.

3) Keep equipment and materials away from areas of heavy traffic where they might get easily damaged. Even if it's not your fault, you'll still have the hassle of dealing with the delay and replacement.

4) Make sure that materials and equipment are delivered according to schedule. Start calling several weeks in advance. Don't wait until it doesn't arrive to begin finding out why.

5) Establish an inventory system and keep it up to date. At any point, for any job, you should know what equipment's been installed, what's been delivered and awaits installation, and what's been ordered but not delivered. Keep watch on items that come out of your own shop, such as ductwork, power tools, hanger straps, nuts and bolts, and other common items. Avoid waste on the job. Although these items may not cost much compared to the overall job cost, they do add up.

Labor Management

While materials cost management is important, the biggest problem on many jobs is labor. Materials cost is fairly easy to peg down and control. Not so with labor. In fact, on almost all jobs that lose money, the greatest loss went to labor that exceeded the estimate. Whether your original estimate is low or not, you'll want to keep a tight lid on labor costs. If you can avoid overtime and get your crews off the job earlier than planned, you'll greatly increase your profit. For a job that was bid too low or where problems develop, it's absolutely essential

to make sure you're getting all you can out of your crew.

The key to labor cost control is to maximize crew productivity. Exactly how you do this has been the subject of thousands of hours of discussion and hundreds of books offering all kinds of advice and suggestions. Many contractors we've spoken with recommend firm, continuous supervision of crews to maintain productivity. This doesn't mean that you or your foreman have to watch them every minute. Any crew that feels they're viewed with suspicion will quickly become hostile. It does mean the crew must be aware of the amount of work expected out of them and that someone will be on hand to make sure they keep up the pace.

You may be able to get greater productivity out of your workers by providing job security. In the past, construction companies would often get by with a skeleton crew of key people. They'd bring in tradespeople from the union hall as they needed them. It was a system that worked and still continues to be satisfactory in many parts of the country. But many companies saw both productivity and quality fall off over a period of time. Many construction outfits have found it beneficial to keep the same workers on their payroll for as long as possible. Provided, of course, that those workers are doing the job.

More and more experts now believe that productivity and work quality increase if workers are treated more as individuals. Under the old management methods, workers complained that they were kept in the dark. They were told what to do but not given reasons why. Most workers want to know what's going on. They also want a chance to voice their feelings and input ideas into the companies where they work.

More and more companies involve their workers through profit-sharing plans. Your accountant can give you advice on this if you feel it would benefit your company.

Some companies hold weekly meetings, where everyone is encouraged to discuss job and corporate problems and how they might be solved. A word of warning, though. Keep a firm hand on how these meetings progress. Don't let them become battlegrounds for personal arguments or nitpicking on individuals. Stress that the purpose is to bring out positive ideas on how people can work together in better ways. Also, be sure you follow through on decisions that are made. Failure to do so will quickly destroy any confidence your workers have in your sincerity.

The final decision on how to handle your crews is, of course, up to you. The two most important rules we can offer are:

- Be fair but firm

- Communicate with your workers

Try to enlist their help rather than make demands on them. If they see you as a tough, solitary, money-grubbing boss, they'll respond in kind. They'll think all you care about is making money. But if you work for them and support their efforts, you'll win their loyalty. Everyone who does a good job likes to be told so. A word of thanks, a night out on the boss for a job well done, a Christmas bonus, all of these things can go a long way towards building a hard-working crew. As business becomes more and more competitive, labor productivity become extremely important. You've spent a lot of time learning your trade and learning the HVAC business. You should spend at least as much time to develop your people into happy, productive employees.

Putting It All Together
We've covered a lot of material about both the technical and management sides of HVAC contracting. We're confident that the data and suggestions we've presented are sound. They should help make you a better contractor. No one will ever be able to sit down and write a book with the last word in HVAC. No one will ever be able to tell you exactly what to do and what not to do to be a successful contractor. Each job is different, and each one will present you with new challenges and new ways to succeed. But the steps you use, the lessons you learn, will remain the same time after time.

How well you succeed will depend in large measure on your attitude. Although each day brings us news of economic problems such as growing federal deficits, increasing imports, labor problems and dozens more, there's still room in this country for those who are willing to work hard to get ahead. There's no question that the risks are great. But so are the rewards, both in money and in self-satisfaction.

One final word: when you have a problem, put yourself on the other side of it. Better yet, set it off by itself and walk around it. Look at it from top, bottom and all sides. Often you'll find the solution, simply by seeing it from a different point of view. Experience is nothing more than building up a file of problem-solving. And that's what you have to do when you're the boss.

Good luck.

Index

Other Practical References

Plumbers Handbook Revised

This new edition shows what will and what will not pass inspection in drainage, vent, and waste piping, septic tanks, water supply, fire protection, and gas piping systems. All tables, standards, and specifications are completely up-to-date with recent changes in the plumbing code. Covers common layouts for residential work, how to size piping, selecting and hanging fixtures, practical recommendations and trade tips. This book is the approved reference for the plumbing contractors exam in many states. **240 pages, 8½ x 11, $18.00**

Estimating Plumbing Costs

Offers a basic procedure for estimating materials, labor, and direct and indirect costs for residential and commercial plumbing jobs. Explains how to interpret and understand plot plans, design drainage, waste, and vent systems, meet code requirements, and make an accurate take-off for materials and labor. Includes sample cost sheets, manhour production tables, complete illustrations, and all the practical information you need to accurately estimate plumbing costs. **224 pages, 8½ x 11, $17.25**

Plumber's Exam Preparation Guide

Lists questions like those asked on most plumber's exams. Gives the correct answer to each question, under both the Uniform Plumbing Code and the Standard Plumbing Code — and explains why that answer is correct. Includes questions on system design and layout where a plan drawing is required. Covers plumbing systems (both standard and specialized), gas systems, plumbing isometrics, piping diagrams, and as much plumber's math as the examination requires. Suggests the best ways to prepare for the exam, how and what to study and describes what you can expect on exam day. At the end of the book is a complete sample exam that can predict how you'll do on the real tests. **320 pages, 8½ x 11, $21.00**

Planning and Designing Plumbing Systems

Explains in clear language, with detailed illustrations, basic drafting principles for plumbing construction needs. Covers basic drafting fundamentals: isometric pipe drawing, sectional drawings and details, how to use a plot plan, and how to convert it into a working drawing. Gives instructions and examples for water supply systems, drainage and venting, pipe, valves and fixtures, and has a special section covering heating systems, refrigeration, gas, oil, and compressed air piping, storm, roof and building drains, fire hydrants, and more. **224 pages, 8½ x 11, $13.00**

Residential Wiring

Shows how to install rough and finish wiring in both new construction and alterations and additions. Complete instructions are included on troubleshooting and repairs. Every subject is referenced to the 1987 National Electrical Code, and over 24 pages of the most needed NEC tables are included to help you avoid errors so your wiring passes inspection — the first time. **352 pages, 5½ x 8½, $18.25**

Builder's Office Manual Revised

Explains how to create routine ways of doing all the things that must be done in every construction office — in the minimum time, at the lowest cost, and with the least supervision possible: Organizing the office space, establishing effective procedures and forms, setting priorities and goals, finding and keeping an effective staff, getting the most from your record-keeping system (whether manual or computerized). Loaded with practical tips, charts and sample forms for your use. **192 pages, 8½ x 11, $15.50**

Manual of Electrical Contracting

From the tools you need for installing electrical work in new construction and remodeling work to developing the finances you need to run your business. Shows how to draw up an electrical plan and design the correct lighting within the budget you have to work with. How to calculate service and feeder loads, service entrance capacity, demand factors, and install wiring in residential, commercial, and agricultural buildings. Covers how to make sure your business will succeed before you start it, and how to keep it running profitably. **224 pages, 8½ x 11, $20.25**

Estimating Electrical Construction

A practical approach to estimating materials and labor for residential and commercial electrical construction. Written by the A.S.P.E. National Estimator of the Year, it explains how to use labor units, the plan take-off and the bid summary to establish an accurate estimate. Covers dealing with suppliers, pricing sheets, and how to modify labor units. Provides extensive labor unit tables, and blank forms for use in estimating your next electrical job. **272 pages, 8½ x 11, $19.00**

Handbook of Modern Electrical Wiring

The journeyman electrician's guide to planning the job and doing professional quality work on any residential or light commercial project. Explains how to use the code, how to calculate loads and size conductors and conduit, the right way to lay out the job, how to wire branch and feeder circuits, selecting the right service equipment, and much more. **204 pages, 5½ x 8½, $14.75**

Residential Electrical Design

Explains what every builder needs to know about designing electrical systems for residential construction. Shows how to draw up an electrical plan from the blueprints, including the service entrance, grounding, lighting requirements for kitchen, bedroom and bath and how to lay them out. Explains how to plan electrical heating systems and what equipment you'll need, how to plan outdoor lighting, and much more. If you are a builder who ever has to plan an electrical system, you should have this book. **194 pages, 8½ x 11, $11.50**

Electrical Blueprint Reading

Shows how to read and interpret electrical drawings, wiring diagrams and specifications for construction of electrical systems in buildings. Shows how a typical lighting plan and power layout would appear on the plans and explains what the contractor would do to execute this plan. Describes how to use a panelboard or heating schedule and includes typical electrical specifications. **128 pages, 8½ x 11, $13.75**

Carpentry in Commercial Construction

Covers forming, framing, exteriors, interior finish and cabinet installation in commercial buildings: designing and building concrete forms, selecting lumber dimensions, grades and species for the design load, what you should know when installing materials selected for their fire rating or sound transmission characteristics, and how to plan and organize the job to improve production. Loaded with illustrations, tables, charts and diagrams. **272 pages, 5½ x 8½, $19.00**

Contractor's Guide to the Building Code

Explains in plain English exactly what the Uniform Building Code requires and shows how to design and construct residential and light commercial buildings that will pass inspection the first time. Suggests how to work with the inspector to minimize construction costs, what common building short cuts are likely to be cited, and where exceptions are granted. **312 pages, 5½ x 8½, $16.25**

BUSINESS REPLY MAIL
FIRST CLASS   PERMIT NO. 271   CARLSBAD, CA

POSTAGE WILL BE PAID BY ADDRESSEE

Craftsman Book Company
6058 Corte Del Cedro
P. O. Box 6500
Carlsbad, CA 92008-0992

NO POSTAGE
NECESSARY
IF MAILED
IN THE
UNITED STATES

BUSINESS REPLY MAIL
FIRST CLASS   PERMIT NO. 271   CARLSBAD, CA

POSTAGE WILL BE PAID BY ADDRESSEE

Craftsman Book Company
6058 Corte Del Cedro
P. O. Box 6500
Carlsbad, CA 92008-0992

NO POSTAGE
NECESSARY
IF MAILED
IN THE
UNITED STATES

BUSINESS REPLY MAIL
FIRST CLASS   PERMIT NO. 271   CARLSBAD, CA

POSTAGE WILL BE PAID BY ADDRESSEE

Craftsman Book Company
6058 Corte Del Cedro
P. O. Box 6500
Carlsbad, CA 92008-0992

NO POSTAGE
NECESSARY
IF MAILED
IN THE
UNITED STATES